MUSIC AND THE STARS

Music and the stars

Mathematics in medieval Ireland

Mary Kelly & Charles Doherty
EDITORS

FOUR COURTS PRESS
for the Royal Society of Antiquaries of Ireland

This book was set in 10.5 on 12.5 point Ehrhardt by
Mark Heslington, Scarborough, North Yorkshire for
FOUR COURTS PRESS
7 Malpas Street, Dublin 8, Ireland
www.fourcourtspress.ie
and in North America for
FOUR COURTS PRESS
c/o ISBS, 920 N.E. 58th Avenue, Suite 300, Portland, OR 97213.

A catalogue record for this title
is available from the British Library.

ISBN 978-1-84682-392-3

SPECIAL ACKNOWLEDGMENT
This publication has received support from
the Department of Arts, Heritage and the Gaeltacht

Printed in Great Britain
by Antony Rowe Ltd, Chippenham, Wilts.

Contents

Illustrations

TABLES

APPENDICES

FIGURES

PLATES (BETWEEN P. 172 AND P. 173)

Abbreviations

Auct. Ant.	*Auctores antiquissimi*
AD	Anno Domini
AM	Anno Mundi
Annals of Inisfallen	*The annals of Inisfallen*, ed. Seán Mac Airt (Dublin, 1951)
Annals of Ulster	*The annals of Ulster (to AD 1131), part i: text and translation*, ed. S. MacAirt & G. Mac Niocaill (Dublin, 1983)
Bede's *Chronica maiora*	C.W. Jones (ed.), *Beda Venerabilis. Opera didascalica. 2. De temporum ratione* (Turnhout, 1977)
Bede, *Hist. Eccl.*	*Historia ecclesiastica gentis Anglorum*, Charles Plummer (ed.), *Baedae opera historica* (Oxford, 1896; repr. 1969)
bk	book
Bruno Krusch, *Studien* I	Bruno Krusch, *Studien zur christlich-mittelalterlichen Chronologie: Der 84jährige Ostercyclus und seine Quellen* (Leipzig, 1880)
Bruno Krusch, *Studien* II	Bruno Krusch, *Studien zur christlich-mittelalterlichen Chronologie: Die Entstehung unserer heutigen Zeitrechnung* (Berlin, 1938)
cap.	capitula
CCM	Corpus Christianorum Continuatio Mediaevalis
CCSL	Corpus Christianorum Series Latina
Clm	Codices latini monacenses
CMCS	*Cambridge Medieval Celtic Studies* – 1993 > *Cambrian Medieval Celtic Studies*
CSEL	Corpus Scriptorum Ecclesiasticorum Latinorum
DIA	Boethius, *De institutione arithmetica*: H. Oosthout and J. Schilling, *Anicii Manlii Severini Boethii De arithmetica* (Turnhout, 1999)
DIAS	Dublin Institute for Advanced Studies
DMLCS	*Dictionary of Medieval Latin from Celtic sources* (Royal Irish Academy)
DOC	*Liber de ordine creaturarum*: M. Díaz y Díaz (ed.), *Liber de ordine creaturarum: un anónimo irlandés del siglo VII* (Santiago de Compostela, 1972); M. Smyth, 'The seventh-century Hiberno-Latin treatise *Liber de ordine creaturarum*: a translation', *Journal of Medieval Latin*, 21 (2011), 107–22.
DRC	*De ratione conputandi*: Dáibhí Ó Cróinín in Maura Walsh &

	Dáibhí Ó Cróinín (eds), *Cummian's letter 'De controversia paschali' and the 'De ratione conputandi'* (Toronto, 1988)
EIHC	Dáibhí Ó Cróinín, *Early Irish history and chronology* (Dublin, 2003)
ESOF	Euro Science Open Forum
Félire Óengusso	Whitley Stokes, *The martyrology of Oengus the Culdee* (London, 1905)
Gennadius, *De dogm. eccl.*	Gennadius of Marseille, *De ecclesiasticis dogmatibus liber sive diffinitio ecclesiasticorum dogmatum* (*Patrologia Latina*, vol. 42, cols 1211–21)
IA	Irish Augustine *De mirabilibus sacrae scripturae*: F.P. [Gerard] MacGinty, 'The treatise *De mirabilibus sacrae scripturae*' (PhD, UCD, 1971)
ISOS	Irish Script on Screen: www.isos.dias.ie
JRSAI	*Journal of the Royal Society of Antiquaries of Ireland*
Lib.	*Liber*
MGH	*Monumenta Germaniae Historica*
n.s.	new series
NLI	National Library of Ireland
NUI	National University of Ireland
OPW	Office of Public Works, Ireland
OUP	Oxford University Press
PL	Jacques-Paul Migne (ed.), *Patrologia Latina* (Paris, 1844–55)
pr	Boethius' *De consolatione philosophiae* is written in prosimetric style, i.e., partly in prose, partly in verse: 'pr' refers to the prose sections; the metrical sections are indicated by 'm'
PRIA	*Proceedings of the Royal Irish Academy*
Pseudo-John	E.T. Silk (ed.), *Saeculi noni auctoris in Boetii Consolationem philosophiae commentarius*, Papers and Monographs of the American Academy in Rome, vol. 9 (Rome, 1935), pp 305–43
'Remigius'	A commentary associated with (but probably had nothing to do with) Remigius of Auxerre. See E.T. Silk, *Saeculi noni auctoris in Boetii Consolationem philosophiae commentarius* (Rome, 1935)
repr.	reprint
RIA	Royal Irish Academy
RSAI	Royal Society of Antiquaries of Ireland
UCD	University College Dublin
UCG	University College Galway

Contributors

CHARLES BURNETT, Professor of the History of Arabic/Islamic Influences in Europe at the Warburg Institute, University of London, has written several books and over 150 articles on the transmission of Arabic Learning to Europe in the Middle Ages and the early modern period. Among these influences are music and musical instruments, and astronomy and astrology. He was educated at Cambridge University (1968–70) and has held visiting professorships at the University of California, Berkeley, the Ludwig-Maximilian University of Munich and the École Pratique des Hautes Études in Paris.

HOWARD B. CLARKE, formerly of the School of History (now the School of History and Archives), UCD, continues to work on the history of medieval Dublin, on the Irish Historic Towns Atlas, on Evesham Abbey and its cartularies, and on the Bayeux Tapestry.

CHARLES DOHERTY has recently retired from UCD, where he lectured in the School of History. He was president of the Royal Society of Antiquaries of Ireland (2009–13). He has published on hagiography, early economics, settlement, the impact of the Vikings upon early Ireland, and the kingship of Tara. He edited 'Latin writing in Ireland (*c.*400–*c.*1200)' in vol. i of *The Field Day anthology of Irish writing* edited by Seamus Deane. He is currently a Fellow in the Centre for Advanced Study in the Norwegian Academy of Science and Letters, Oslo.

SIOBHÁN FITZPATRICK, a graduate of UCD, has been librarian of the Royal Irish Academy since 1997. She has worked since 1980 in Dublin libraries, including the Department of Early Printed Books, Trinity College, Dublin, and the Library and Information Service of the Electricity Supply Board. Siobhán served as Hon. Gen. Secretary of the RSAI (1999–2001) and is a past president of the Library Association of Ireland. Publications include *Treasures of the Royal Irish Academy library* (2009), with Bernadette Cunningham; *Pathfinders to the past: the antiquarian road to Irish historical writing, 1640–1960* (2012), with Próinséas Ní Chatháin and Howard Clarke; and *Science at the Royal Irish Academy: 'uniting whatever is pleasing with whatever is useful'* (2012).

ANTHONY HARVEY is editor of the Royal Irish Academy's *Dictionary of Medieval Latin from Celtic sources* (*DMLCS*). He holds a PhD in early medieval Irish and British linguistics and literacy from Cambridge University's Department of Anglo-Saxon, Norse and Celtic. Major publications have been: the first volume of the *DMLCS*, jointly compiled with Jane Power and published in 2005 as *The non-classical lexicon of Celtic Latinity, Letters A to H*; and the *St Patrick's Confessio hypertext stack*, jointly edited with Franz Fischer, which went live on the Internet in 2011 as www.confessio.ie, a comprehensive digital resource for investigating the saint's own writings.

DAVID HOWLETT, a graduate of the University of Montana (1966), studied on a Rhodes Scholarship at Corpus Christi College, Oxford (BA, MA, DPhil), in Medieval Studies. From 1975, he was assistant editor of the *Supplement to the Oxford English dictionary* at OUP and from 1979 editor of the British Academy's *Dictionary of Medieval Latin from British sources*, consultant to the Royal Irish Academy's *DMLCS*, a member of the *Comité Du Cange* responsible for the *Novum Glossarium Mediae Latinitatis* and the *Archivum Latinitatis Medii Aevi*, and a research fellow of the Faculty of Classics in the University of Oxford. He has written extensively about the influence of Latin, Greek and Hebrew traditions on the literatures and music of the British Isles.

MARY KELLY holds an MA from the School of History and Archives, UCD. She co-edited *Glendalough: city of God* (2011) with Charles Doherty and Linda Doran. She is a member of the RSAI and organized the conference 'Music and the stars' for that society in 2012. She works for the OPW at Glendalough Visitor Centre.

RACHEL MOSS lectures in the Department of History of Art and Architecture, TCD. She is president of the Royal Society of Antiquaries of Ireland (2013–17). She has published on various aspects of medieval art and architecture and is editor of and principal contributor to *The art and architecture of Ireland: medieval period,* c.*400–1600* (forthcoming).

DÁIBHÍ Ó CRÓINÍN studied at UCD, the Ludwig-Maximilians-Universität, München, and the School of Celtic Studies in the Dublin Institute for Advanced Studies, and has been lecturing at the National University of Ireland Galway (formerly UCG) since 1980. He edited the Royal Irish Academy's *A new history of Ireland*, 1: *Prehistoric and early Ireland* (2005), and among his other publications are *Cummian's letter 'De controversia paschali' and the 'De ratione conputandi'* (1988) with Maura Walsh; *Early Irish history and chronology* (2003), and *Whitley Stokes (1830–1909): the lost Celtic notebooks rediscovered* (2011). Since 1982, he has co-edited with Donnchadh Ó Corráin the *Journal of the Medieval Academy of Ireland*. He is a Fellow of the Society of Antiquaries in London and a member of the Royal Irish Academy and of the Irish Manuscripts Commission.

PÁDRAIG P. Ó NÉILL is James Gordon Hanes Distinguished Professor of the Humanities in the Department of English and Comparative Literature at the University of North Carolina at Chapel Hill. His research focuses on medieval Ireland's intellectual contacts with Britain and the Continent.

MARINA SMYTH is medieval studies librarian at the University of Notre Dame in Notre Dame, Indiana. On obtaining her BSc from UCG, she worked several years in the editorial department of the American Mathematical Society in Providence, Rhode Island, before becoming a freelance translator of scholarly books. Her doctoral thesis from the Medieval Institute at the University of Notre Dame is entitled *Understanding the universe in seventh-century Ireland* and appeared later with the same title in the series Studies in Celtic History (1996). A Fellow of the Medieval Institute, she publishes in the area of early Hiberno-Latin culture and its influence into the later Middle Ages.

ROBERT D. STEVICK, Professor of English, Emeritus, University of Washington, is still rewarded with discoveries in the mathematical matrix of early Irish art; he recently e-published *Insular art forms: their essence and construction* to help others grasp this element in these early arts and extend findings in this dazzling and unarcane tradition; and he deeply hopes that new understanding of these ancient arts may generate revival of the formal skills of that splendid past.

MAURA WALSH completed her undergraduate degree in the Catholic University of America, Washington DC, and the MPhil in medieval studies in UCD. Her thesis, a critical edition with translation and commentary on Cummian's Paschal Letter, was subsequently published with co-author Dáibhí Ó Cróinín as *Cummian's letter 'De controversia paschali' and the 'De ratione conputandi'* (1988). She became interested in computers and has taught computer skills for many years. She now lives in Galway, where she is programme director of the Galway Early Music Festival.

IMMO WARNTJES lectures in medieval history at the University of Greifswald (Germany). He received his PhD from NUI Galway in 2007, and his thesis *The Munich computus, text & translation: Irish computistics between Isidore of Seville and the Venerable Bede*, was published in 2010. In 2011, he was research fellow at the German Historical Institute (London), and in 2011–12 at the Monumenta Germaniae Historica (Munich). His research interests include early medieval computistics, the use of languages in the early Middle Ages, succession to high offices; his teaching covers Frankish, German, English and European medieval history, Latin palaeography and Medieval Latin.

Foreword

When the Royal Society of Antiquaries of Ireland was founded in 1849 (then named the Kilkenny Archaeological Society), its main interest was in the material remains of the past. The majority of papers published in its journal related to monuments and artefacts; practical activities included the excavation of endangered sites and consolidation of unstable ruins; and its headquarters soon became home to a growing museum of objects gathered by members and deposited both for safekeeping and for the enjoyment and education of others.

This was a time when the 'proper business' of the antiquary (in the words of William Borlase) was still 'to collect what is dispersed, more fully to unfold what is already discovered, to examine controverted points, to settle what is doubtful, and by the authority of monuments and histories, to throw light upon the manners, arts, languages, policy and religion of past ages'. The examination of 'monuments and histories' typically engaged a formalist approach; description, establishment of typologies and the tracing of chronological development. Contemplation of the philosophical and intellectual contexts that prompted particular innovations or characteristics in material culture was seldom considered, nor the technological advances that permitted ideas to be realized. As Anthony Harvey points out in the afterword to this volume, during the Renaissance, scientist and antiquary were often one and the same, but by the nineteenth century they had diverged into largely separate disciplines.

Our approaches to investigating the past have, happily, developed since that time. No longer are we content to simply note the recurring pattern of proportionate systems in medieval buildings, or to describe recurrent number symbolism in the pages of illuminated gospels. Now we must question why. Was the engagement of systems of geometric proportions, as described by Robert Stevick in this volume, simply a means of achieving a particular aesthetic, or was it imbued with a deeper significance? How were numbers and number symbolism read and understood by early patrons and audiences? The essays in this book explore not only the material remains of the early medieval period so familiar to the Irish antiquary, but also the intellectual context that informed them.

Advances in modern scientific techniques, from materials analysis to remote sensing, lay far beyond the dreams of those innovators examined in Howard Clarke's essay on advances in science and technology up to 1500. The application of new technologies to the discipline of archaeology is of inestimable value, widening our understanding of the past, so that increasingly 'antiquarian' and scientist are once more merging into a single entity. Although many of the new techniques available to the archaeologist are beyond the resources of a voluntary, privately-funded organization such as the Royal Society of Antiquaries of Ireland, highlighting the long and proud history of science in Ireland is not.

The inception of this book lies in a conference held in 2012, organized by Mary Kelly and former president of the society, Charles Doherty. Their ability to coax such a distinguished group of scholars together to deliver papers is a real credit to both. So too is their achievement in delivering such a lively and erudite collection of essays before the conference participants / authors have had a chance to draw breath, ensuring that the research and ideas presented in the book are still new and exciting.

The sizeable attendance at the conference, generously supported by the Department of Arts, Heritage and the Gaeltacht and the Heritage Council, is testament to the growing interest in the significant scientific innovations of Irish medieval scholars. The Royal Society of Antiquaries of Ireland gratefully acknowledges the continuing support of the Department of Arts, Heritage and the Gaeltacht in seeing the proceedings through to publication. It is hoped that these essays will not only shed light on hitherto obscured corners of Irish learning, but will encourage further enquiry into this rich seam of history.

Rachel Moss
President
Royal Society of Antiquaries of Ireland

Preface

The Royal Society of Antiquaries of Ireland has, since its inception in 1849 as the Kilkenny Archaeological Society, played a pioneering role in the struggle to safeguard the national monuments we take for granted today and in the preservation of less tangible aspects of our heritage. This pioneering spirit was much in evidence in 2012 when the society participated in the largest scientific meeting in Europe – Euro Science Open Forum (ESOF) – which was held in the National Convention Centre, Dublin. For the following week, the society had organized 'Music and the stars: mathematics in medieval Ireland', the first international conference on the history of science in medieval Ireland, hosted by the National Library of Ireland on 17–18 July. This conference was designated Partner Conference to ESOF 2012 and the papers presented at that conference, and now greatly expanded, are the subject of this volume.

The society's session at ESOF 2012, 'Is the future of mathematics medieval?', was expertly chaired by Anthony Harvey and the speakers were Immo Warntjes, David Howlett, Lorraine Hanlon and Emer O'Boyle.[1] Patrick Cunningham, the then chief scientific adviser to the government and principal organiser of ESOF 2012, observed that the society's proposal for participation in ESOF contained 'an interesting mix of science and history'.[2] The purpose of the session was threefold: to highlight the contribution of scholars in medieval Ireland to the history of science; to demonstrate, with reference to the work of medieval and modern scholars, that the arts and sciences have not always rested in opposing camps; and that integration is of benefit to both.

Immo Warntjes drew attention to scholars in early medieval Ireland who, by their advances in the science of computistics, laid the groundwork for the development of a systematic approach to scientific enquiry. David Howlett demonstrated that Irish scholars, at ease in both literary and mathematical spheres, had the skill to merge computistical data with texts of literary merit: texts that bear witness to a world far removed from any notion of mutual exclusivity. As if to transpose that medieval intellectually inclusive world to a modern context, Lorraine Hanlon and Emer O'Boyle confirmed through their groundbreaking project, 'Tunnelling art and physics', that collaboration between artist and scientist results in challenges and insights that are unforeseen, beneficial to both and can bridge the chasm in understanding that often lies between the scientist and the wider public.[3] Pursuant to this, and as the title of the session

1 Anthony Harvey, David Howlett and Immo Warntjes are contributors to this volume. 2 Pers. comm., 26 Jan. 2011. 3 Lorraine Hanlon is Associate Professor of Astronomy at the School of Physics, UCD. Emer O'Boyle is artist in residence at the School of Physics, UCD, and she devel-

suggested, a final question was raised as to how the humanities and sciences may be more closely integrated in modern education, thereby changing the way mathematics is perceived by the general public.

On completion of higher level education, the medieval student could read numbers much as we read words. Each number has its own value, as a letter has a sound, and, as a combination of letters, produces a word that is a sound representing an action, object or idea and thus enters the realm of representation, so combinations of numbers express the interwoven complexities that, according to Boethian number theory, describe the fabric of the universe. For example, a 12 represents that many units, but when it is considered in relation to an 8 it is *sesquialter* to it, that is, the 12 is made up of the 8 + half of the 8. The 12 now expresses a concept separate to the mere number of units it represents: numbers are read like words. As the contributors to this volume show, numbers and letters, number relations and words are inextricably linked for the medieval scholar. In a system where letters represent numbers – roman numerals, Greek alpha-numeric symbols – a visual link is well-established. Gematria, which imbues mystical meaning to the alpha-numeric value of words, offered another interpretation of number.

Grammar, logic, dialectic, exegesis provide the platform for medieval intellectual enquiry – the intellectual wallpaper of the medieval mind; but that paper is patterned with number. Somehow this fluency with number has been lost to us. These then are the themes, elaborated and extended to the realms of artistic expression, music, astronomy and the reckoning of time, that form the basis for exploration in *Music and the stars: mathematics in medieval Ireland*.

While individual works of great scholarship throw much light on particular aspects of Ireland's scientific history, a diligent exploration of broader academic discussions on the history of science of the medieval period suggests that Ireland had no role to play – this stands in stark contrast to the evidence provided by Ireland's extant literary and artistic achievements. Ireland's scientific history marks a continuity of achievement that stretches over one-and-a-half thousand years and bodes well for our continued contribution to science. This volume acknowledges that history.

Mary Kelly

oped and runs 'Tunnelling art and physics', the undergraduate collaborative module between the School of Physics, UCD, and the National College of Art and Design, Dublin. See www.ucd.ie/innovation/newsevents/tedxucd/speakers/emeroboyle: accessed 10 June 2013.

Acknowledgments

The Royal Society of Antiquaries of Ireland is an independent, voluntary society[1] and it would not have been possible for the society to fund participation in ESOF 2012, the conference 'Music and the stars' or this volume from its own resources. The society acknowledges the support of Jimmy Deenihan TD, Minister for Arts, Heritage and the Gaeltacht, and the staff at his department, in particular Chris Flynn and Keith Ashmore. The society is grateful to the Heritage Council of Ireland for the funding that it provided under the Heritage Education, Community and Outreach Grant Scheme 2012. The Seminar Room of the National Library proved a magnificent venue for 'Music and the stars' and the society wishes to thank Fiona Ross, director of the National Library of Ireland, and the staff at the NLI, especially Orla Sweeney and Bríd O'Sullivan, for their care and professionalism, which ensured that the conference ran smoothly. To all of the speakers and chairpersons of the ESOF 2012 session and the conference, the society extends sincere thanks for the great generosity of spirit with which they have given of their time and hard-won knowledge – without them, none of this could have happened. This is a generosity shared by Tim O'Neill, who we thank for his early Irish numbers. Thanks are due to the conference attendees, whose attention and interest ensured lively discussion. The editors wish to thank Michael Potterton, Four Courts Press, for his patience and the immense care he has expended on this book. The editors would also like to thank Colette Ellison, former director of the RSAI, for her encouragement and sound advice during the initial stages of this project. Colette was succeeded in the role of director by Niamh McCabe, and the editors wish to thank Niamh for her tireless energy, utter courtesy, efficiency and great good humour throughout the remainder of this long project.

1 The society employs a director and a librarian on a part-time basis.

Introduction

CHARLES DOHERTY

During the first century AD, as the Roman conquest of Britain was under way, mass-produced objects from Roman factories began to appear in Ireland. These were objects such as brooches, pins, barrel padlocks and coins. Latin words borrowed into the Irish language (pronounced with a Welsh accent) over the next few centuries give clear evidence of the extent of technological innovation that was introduced to Ireland as a result of Irish contact with the Roman Empire and its aftermath.

The most important borrowing was the Christian religion. At first, the Irish became aware of Christianity on a casual basis – through trade, as mercenaries in the armies of Rome, through intermarriage, through contact with slaves – and some became converted. Pope Celestine sent an official mission from Rome in 431 (led by Palladius) to those incipient Christian communities in Ireland. Palladius and his Gaulish followers would seem to have been active in the east of Ireland and in the midlands. Of his mission, only a shadow survives, subsumed within the cult of St Patrick. Patrick had been captured in an Irish raid on Roman Britain and spent six years in slavery in Ireland. He later returned and led a mission across the northern districts of Ireland. We have no certain dates for Patrick, but the notice of his death in 492/3 may be correct. These fifth-century missions brought official Christianity and the language of the church to the country. Some of the Irish would have been aware of writing (on coins and on inscriptions in Britain) from the first or second century AD, and such early contacts may have inspired the invention of the ogham 'alphabet'. It was the coming of Christianity, however, that brought literacy, not merely to all those who were directly involved with the church, but also to some of the laity such as those sons of kings who accompanied Patrick on his missionary journeys. In the documents written by Patrick we get our first references to letters that clearly went back and forth between Ireland and Britain – directed not only to clergymen but also on one famous occasion to the soldiers of a leader called Coroticus. Literacy in Latin allowed the Irish to become members of the Christian community of the then known world and to participate in its debates even if they lived at the remotest edge of that world – a conceit that was used by Columbanus. It introduced them to the predictive sciences and philosophy.

The rectangular church building (with its proportions and orientation) was a visible sign in the landscape of a new order. The church's liturgy of foundation linked heaven and earth in a new way. Trade in wine was now essential for the celebration of the Mass; and the importation of books and ecclesiastical objects provided a model and stimulus for home production. Since the Irish had never been conquered by Rome, they had to learn Latin as a foreign language. The strongest manuscript evidence for the Late Latin grammars of Rome survives in Irish contexts. As the writers began to move outside the realm of liturgy and produce teaching aids, it became necessary to consider, from a very early period, how best to reproduce the sounds of the Irish language using the Latin alphabet – and thus began a fascination with language that was characteristic of early Irish scholars.

The needs of the church demanded the adoption of new technology. The manufacture of good parchment is not trivial; and the creation of two registers of script equivalent to our upper and lower case was the result of the judicious borrowing and manipulation of various elements of the scripts of the Roman world – and are found in our very earliest manuscripts with perhaps the formal upper case being slightly earlier.

Very little written material survives before the early seventh century and the evolution of church and society before that time may be seen, if at all, as through a mist. But it is during the seventh century that historians feel that they are on reasonably solid ground and can speak about who is doing what and to whom in the political sphere. It is clear that a very ancient order is in retreat. If it is appropriate to use the term 'tribal', then those communities so described were giving way to dynasties – and indeed the most successful dynasties of the period may have been those that had sided with the church. Some ancient churches of the missionary period were in decline in parallel with the communities they represented, while others supported by up-and-coming dynasties were in a position to advance their own political agenda. By the late seventh century it was clear that Armagh was the chief church in Ireland. Her rival, Kildare, was in a shrinking, but still powerful, province of Leinster; and in a compromise was given headship of the female foundations throughout the island.

Both churches were heavily involved with the major dynastic families of their respective provinces. The Uí Dúnlainge, who had recently come to power in Leinster, provided a king of the province whose brother was bishop and abbot of Kildare. The Fotharta, the tribe from which St Brigit came, fought in the vanguard of the Leinstermen and supplied abbesses in Kildare. The bishop of Armagh, Tómméne, who died in 661, was the son of the king of the Airther (Uí Bressail branch) – one of the Airgialla tribes that lived in the vicinity of Armagh. This family was entrenched in the monastic city. Ecclesiastical power and secular power went hand in hand – and this was a pattern that was repeated in a variety of ways throughout the island. It was by way of such relationships that wealth was directed towards the church and allowed for the development of schools and the importation of materials to support and stimulate them. Such was the level of support that churches could take students from Anglo-Saxon England and provide them with board and lodgings and books while they studied. They flocked to study under teachers who were renowned and whose textbooks were the best available.[1] There was a growing network of Irish monastic houses on the Continent and it was through them that the Merovingian prince, Dagobert, was sent to Ireland to avoid assassination. Ireland was not isolated then, nor was she at any period during the pre-Norman centuries. Such contacts allowed for a continuing flow of objects and ideas into the country.

Early agricultural implements such as sickles and ploughshares are similar to those found in Roman Britain – their point of origin. The grain-drying kiln was also a borrowing. The recent discovery of a crop-processing complex (on the N18 Gort to Crusheen road scheme) at Curtaun in Co. Galway provides much information. This example was close to a large ringfort and was stone-lined. It had multiple flues and was rebuilt on a number of occasions. It was clear that the farmer had experimented to achieve optimal results. This was not simply passive borrowing.[2]

The watermill was introduced and spread rapidly. Beekeeping too was an innovation. Special laws were created to resolve the problems that arose in relation to the operation of these new technologies. Water had to be conveyed across land in order to supply the mill and this involved agreements with landowners. Who had the right to grind grain in the mill? How was joint ownership to be organized? Since beekeeping is mentioned in hagiography, in stories in which the saint, returning from Britain, carries the bees he had been attending with him to Ireland, it is very likely that the church was responsible for this innovation. The need for wax for candles would have been of prime importance as well as the production of honey. The law concerning bees regarded them as grazing animals. If a swarm landed in a neighbour's property, who had the right to the swarm? These were the kinds of practical issues that were involved. The adjustments that such innovations demanded of native law (including that of the position of the new clergy in society) required ingenuity and imagination – and indeed may have stimulated the need to record the law on parchment.

Grain was normally ground on a hand quern – usually by slaves. The introduction of the watermill allowed this task to be carried out mechanically. But this was a massive investment that only the very wealthy, or an institution such as the church, could afford. More mills have been discovered in Munster for this early period than in the rest of Europe; and excavations allow us to judge their sophistication. At High Island in Co. Galway, a reservoir was created in a natural hollow, a stream led this water to a millpond and from the pond a channel fed

1 Jean-Michel Picard, 'Bede and Irish scholarship: scientific treatises and grammars', *Ériu*, 54 (2004), 139–47. 2 www.nra.ie/Archaeology/BestofNRADiscoveries: accessed 5 June 2013.

water to the mill. When the water had turned the paddles in the wheelhouse, it exited by a tailrace over the cliff. Excavation on the N6 Ballinasloe to Athlone road scheme revealed a mill at Kilbegly in Co. Roscommon. This lay a short distance from the ancient church. The timbers of the undercroft of the mill survived and also the millpond and sluices, all preserved in the peat bog. This monastic mill belonged to the major church of Clonmacnoise and was probably on one of the church's estates.

Even more remarkable are the tidal mills that have been discovered. The earliest excavated tidal mill in the world was discovered at the monastery of Nendrum in Co. Down. It was built between AD619 and 621. A second mill was built in the late seventh or early eighth century and a third has been dated to AD787/8. Literary references suggest that this church was a dependency of Armagh and may have supplied grain to the major monastic city as well as for its own use. The technical sophistication of this mill complex is astonishing. The first mill was fed by seawater collected in a pond with a surface area of 6,500 square metres. The pond was retained behind an embankment that cut off part of the foreshore of a sheltered bay to the northeast of the monastery. The base of the surviving embankment was 8m wide and 110m long. The excavators described it as 'a massive structure of complex construction'.[3] The three successive mills and their ponds at Nendrum demonstrate an increasing sophistication in milling technology over the period of their use. The waterproofing techniques, using various clays and mosses in the embankments, together with the use of timber, wattles and stone in stabilizing the structures shows an acute awareness of the properties of materials. The same may be seen in the evolution of wheel and paddle design. The massive sandstone flume of Mill 3 measures 3.41m long by 98cm at its maximum width and is 'presently unparalleled among other known horizontal mills anywhere in the world'.[4] It has been estimated that theoretically Mill 3 would have produced energy of between 1,750 and 2,250 kilowatt-hours per annum.[5] The tide mill at Little Island east of Cork city had a twin-flume horizontal-wheeled mill dating to *c.*AD630 and it had a vertical undershot waterwheel alongside it.[6] Excavations on the N25 Waterford City Bypass revealed an early seventh-century vertical watermill at Killoteran – the earliest to be identified in Ireland to date.[7]

The organization of labour for such projects would have been formidable. Cogitosus, who wrote a 'life' of St Brigit about AD650, describes a miracle of a

3 Thomas C. McErlean, Caroline Earwood, Dermot Moore and Eileen Murphy, 'The sequence of Early Christian period horizontal tide mills at Nendrum monastery: an interim statement', *Historical Archaeology*, 41:3 (2007), 63–75 at 66. 4 Ibid., 70–1. 5 Ibid., 74. 6 Colin Rynne, 'Waterpower in medieval Ireland' in Paolo Squatriti (ed.), *Working with water in medieval Europe* (Leiden, 2000), pp 1–50. 7 I. Russell, C. Moore, A. O'Sullivan and V. Ginn, 'Killoteran 9: early medieval vertical watermill' in James Eogan and Elizabeth Shee Twohig (eds), *Cois tSiúire: nine thousand years of human activity in the Lower Suir Valley* (Dublin, 2011).

river diverted from its own course to another (§30). The incidental information in this story is of the greatest interest. The king of the country (*patria*) where Brigit lived made an edict for the building of a solid wide road (*via firma*). He was a king of various peoples and provinces. Lots were drawn to decide which people would build a section of the road. The workforce was divided into kinship groups and households (*per cognationes et familias*). Brigit's people (presumably the Fotharta and or the tenants of Kildare) were browbeaten by a stronger group to take on the very difficult stretch that had been allotted them; and the miracle hangs on this. The road was to have foundations of branches, rocks and earth in the softer sections. It was to be sufficiently strong to bear the weight of charioteers (*quadrigas*), horsemen (*equites*), chariots (*currus*) and wagon-wheels (*plaustrorum rotae*). Although it is not stated, it is likely that a bridge was built across the river. Given the close relationship between the king of Leinster and the monastery, the road may very well have been built in the interest of Kildare as well as the king. It is clear from the 'life' that Kildare was a major centre of pilgrimage at this time and such a road would have been of great benefit to the church.

A massive road and wooden bridge were built downstream of the monastery of Clonmacnoise *c.*804. It can be demonstrated that the Southern Uí Néill were heavily involved in the affairs of the monastery at this time and in this instance too the interests of church and state went hand in hand. The bridge was at least 120m long and 5m wide. The road that led to the bridge from the west was 3–5m wide and could be traced for approximately 2.4km. The bridge rested on two rows of about twenty-five pairs of large oak posts (30–5cm in diameter) that were spaced at intervals of 5–6m running across the river. These posts were driven into the soft mud of the river to a depth of 3.5m. They were 'stabilized on the riverbed by means of ingenious carpentry': a beam and horizontal planks were mortised to each post to act as a base-plate at 3.5m from the blunt tip of the post to prevent it from sinking any further into the soft clay. The above-water section of the bridge no longer survives, but it is thought that it may have stood 4–5m above the surface of the water.[8] This bridge could certainly have taken the traffic listed in Cogitosus' 'life' of Brigit quoted above. It was comparable to the longer Ravning Enge Bridge built in 979–80 in Vejle river valley, 10km south of Jelling in Denmark. This bridge was built by Harald Bluetooth as one of his many building projects in his attempt to secure control of Scandinavia. It was 760m long and 5m wide and it has been calculated that it could bear a stress of five tons and that the designers 'must have made geotechnical studies prior to construction'.[9] The designers of the bridge at Clonmacnoise must have made

8 This account has been taken from Aidan O'Sullivan and Donal Boland, 'Underwater archaeological excavations of an early medieval bridge in the River Shannon': www.science.ulster.ac.uk/crg/nua/research/shannon/shannon.htm: accessed 6 June 2013. 9 http://natmus.dk/en-/historisk-viden/danmark/moeder-med-danmarks-oldtid/the-viking-age/kongeligt-byggeri/b

similar studies prior to its construction. The earliest excavated timber bridge in England is the eleventh-century bridge at Hemington, near Castle Donnington, Leicestershire.[10]

Other major works were the construction of fish traps or weirs, both of wattle and stone that have been discovered as a result of recent coastal surveys in Europe. One wooden trap at Chapel Island in Strangford Lough in the north of Ireland gave a radio-carbon date of AD711–889; and it has been suggested that it may have been linked to the monastery of Nendrum, discussed above. In the same lough, other weirs were dated to the eleventh, twelfth and thirteenth centuries – those in stone may have belonged to the Cistercian house of Grey Abbey founded in 1193. Others have been surveyed in the estuary of the Shannon. These were relatively small, with fences 20–30m in length. In Strangford Lough, they could be as long as 300m. They were generally V-shaped (although there were variations) with a trap of wattle, a pound, a net or other device at the end to catch the fish. They trapped fish mostly on the ebb tide, but some also operated on the flood tide. Knowledge of tides, channels, inflowing rivers and seasonal movements of fish as well as a capacity to handle the materials was necessary.[11] Given the large volume of wattle used, access to managed woodland was essential. While the servile class probably did the actual work, it is likely that lord, king or church managed the organization of the activity. While fish traps are known from the Mesolithic, *c.*5000BC, it is possible that the coming of Christianity with its fast days was a stimulus to the building of such structures.

The most substantial and sophisticated buildings were churches. The earliest description is that by Cogitosus in the 'life' of Brigit mentioned above. In this, he describes a large rectangular building that had been recently renovated. At the eastern end, it had a screen running from wall to wall – an iconostasis. It was painted with images and had wall hangings. There were doors at either end giving access to the chancel in which the bodies of Brigit and the first bishop, Conlaeth, reposed in sarcophagi (decorated with gold, silver, gems and precious stones) on either side of the altar. Chandeliers of gold and silver were suspended above. The archbishop, priests and monastic chapter entered through the door on the right of the chancel screen. The abbess and her nuns and faithful widows entered through the door on the left to take communion. The church was divided down the middle by another partition running from the west wall to the chancel screen. This arrangement created three chapels – the chancel and two in the nave of the church. In the south chapel, there was a highly ornamented door

roen-over-ravning-enge: accessed 6 June 2013. **10** Susan Ripper and Lynden P. Cooper, *The Hemington bridges: the excavation of three medieval bridges at Hemington Quarry, near Castle Donnington, Leicestershire* (Leicester, 2009). **11** Aidan O'Sullivan, 'Place, memory and identity among estuarine fishing communities: interpreting the archaeology of early medieval fish weirs', *World Archaeology*, 35:3 (2003), 449–68 at 456–61.

in the side (probably close to the west wall of the church) through which the priests and male population entered. In the north chapel, there was a similarly placed door through which the nuns and female population entered. Cogitosus called this building a basilica in which 'a large congregation of people of varying status, rank, sex and local origin, with partitions placed between them, prays to the omnipotent Master, differing in status, but one in spirit'. The great church at Armagh was also called a basilica. There were also 'different images presenting a variety of carvings and colours', which suggests that the church may have been decorated with painted wooden statues as well.[12] This large wooden church with its many windows reached a great height and the nearest analogy may be the stave churches of Norway. The earliest stone churches (the first is mentioned in the seventh century) were small. Larger mortared churches were built from the tenth century onwards and were skeuomorphs of the wooden churches. They were comparable in size to many of the churches in Anglo-Saxon England. As Tómás Ó Carragáin has argued, and as is confirmed by the numbers killed or burned in church buildings during Viking attacks, the small stone church is not the norm in early Ireland.[13] Wood was the preferred building material and when mortared stone churches became more common from the tenth century onwards the builders felt compelled to reproduce in stone the characteristic features of wooden construction.

In the same 'life' of Brigit, a miracle is related concerning a millstone that was sent rolling downhill from a high inaccessible place after it had been cut from the living rock and completely fashioned on site. It survived its journey and finally came to rest on an even place (after having negotiated deep water) from where it could be drawn by oxen to the mill. From this story, it is clear that Kildare had a watermill that burned down. One millstone survived – the one that careered safely downhill and it was subsequently placed near the ornate door of the cashel that surrounded the church as witness to the miracle. In Redhills near Kildare, there is evidence of ancient quarrying among the rocks and it was probably there that the millstone was manufactured. This story shows that Kildare in the mid-seventh century had specialist stonecutters and millwrights who could assemble a watermill. There was a miller and his assistants who operated it daily. It was at some distance from the monastery, beside a river. It would seem that those who lived in the vicinity could have their grain ground there (presumably for a price). The reference to the heathen man (a druid) whose grain could not be ground at the mill because he was a pagan may refer to those who were not numbered among the monastic tenants.

12 Sean Connolly and J.-M. Picard, 'Cogitosus's "Life of Brigit": content and value', *JRSAI*, 117 (1987), 5–27 at 25–7, §32. 13 See the very important work of Tomás Ó Carragáin, 'The architectural setting of the Mass in early medieval Ireland', *Medieval Archaeology*, 53 (2009), 119–54 at 120–1; idem, *Churches in early medieval Ireland: architecture, ritual and memory* (New Haven, CT, 2010).

It is clear from the laws and evidence discussed above that there was much specialization in craftwork in early Ireland. Stonecutters were necessary in the manufacture of millstones. It is clear that sculpturing in stone was practised during the Iron Age – one can point to the Turoe stone, Co. Galway, the standing stone from Mullaghmast in Co. Kildare and the Castlestrange stone from Co. Roscommon to cite some of the most famous, and also to the stone heads and other figures. It is likely that most sculpture was in wood and that stonework was reserved for the most important objects. Free-standing wooden crosses are mentioned by Adomnán and Bede and were probably common by the early seventh century. By comparison, the free-standing stone cross took time to develop and, in its earliest manifestation, was a skeuomorph based on wooden originals.[14] It is possible that greater resources were needed to commission crosses in stone and that it was not until the seventh century that such resources were available.

As Dorothy Kelly has demonstrated, the crosses were constructed on mathematical principles:

> While it stops short of a consideration of perspective, beginning with the third dimension, depth, the analysis undertaken here suggests that our high crosses were designed in accordance with ancient principles of proportionality. These principles were put into effect using mathematical formulae which themselves were well-known in classical antiquity, including irrational ratios, especially $1:\sqrt{2}$. The familiarity of the architects of the crosses with these concepts, and the use of the Roman foot, says much about Ireland's contacts with early medieval Europe and through these with the world of classical learning. The methods of design-construction also have ancient roots, as old as the early Iron Age (Henry 1965, 218; Richardson 1984, 28–9); the methods which developed out of these roots, modified through contacts with the Anglo-Saxon world, were widely used in Ireland and Britain in the early Christian period. The methods may well be described as 'rigidly geometrical' but as Henry observed with typical acuity (1963, 106) it is important to distinguish between the well-perfected tool and the mind which uses it. The extraordinary variety of crosses designed using 'rigid' rules is testament to the unfettered creativity of the minds of their architects. Truly, they may be said to have followed, not only the principles, but the spirit of Vitruvius (*The ten books on architecture*, ch. II.I) […][15]

14 Dorothy Kelly, 'The heart of the matter: models for Irish High Crosses', *JRSAI*, 121 (1991), 105–45. 15 Dorothy Kelly, 'A sense of proportion: the metrical and design characteristics of some Columban High Crosses', *JRSAI*, 126 (1996), 108–46 at 145.

She suggested that such principles are to be found in book production (referring to other scholars) and also in metalwork. This had been revealed earlier by Robert Stevick in his own work on book production and metalwork and is further demonstrated in his essay in this book.[16]

That other iconic monument in the Irish landscape – the round tower (*cloigtheach*) – also demonstrates a mastery of building technique. While there were possibly bell towers of wood, our first reference to stone towers is during the first half of the tenth century. These are conical – a shape that was ideal in avoiding wind damage. They range between 18m and 40m in height and from 12m to 18m in circumference. While they stand on remarkably shallow foundations, approximately eighteen survive in an almost complete state. The entrance door was about 3m above ground level. This may have been placed to avoid weakening the base. But it also made access to a building that housed the relics of the church more difficult for those intent on theft. The towers tapered from base to top. Some of the finest towers were built of ashlar blocks that fitted precisely and their curved shape is consistent as the diameter of the tower decreased with height. A conical stone cap was placed on top. Ratios must have been used in their planning and construction. Roger Stalley has suggested that they had a ratio of height to circumference of 2:1.[17] This is an area that requires further study. It seems that a system similar to French curves was needed to provide for consistency in the profile of these buildings. By the time that they make their appearance in the tenth century in the company of stone churches and developed high crosses, they are constituent parts of a slowly emerging artefact – the town. These monuments and the settlements in which they are located are capable of exegesis and, as such, they may be read against the exegesis of the manuscripts. These were 'Cities of God' – a counterpart of the Heavenly Jerusalem. Some outgrew their initial frugal monastic or pastoral function, acquired lands, animals, tenants and properties to become complex sites in which the purely religious ruling elite was a relatively small element within a largely secular environment. Kings had houses within them and exploited their vast resources. Smaller ecclesiastical settlements could probably be described as villages; and the same claim has been made for secular settlements as a result of recent archaeological research.[18]

16 See also Robert Stevick, *The earliest Irish and English bookarts: visual and poetic forms before AD1000* (Philadelphia, 1994); idem, 'The Echternach Gospels' evangelist-symbol pages: forms from "the two true measures of geometry"', *Peritia*, 5 (1986), 284–308. 17 Roger Stalley, *Irish round towers* (Dublin, 2000); idem, 'Sex, symbol and myth: some observations on the Irish round towers' in C. Hourihane (ed.), *From Ireland coming: Irish art from the early Christian to the late Gothic period and its European context* (Princeton, NJ, 2001), pp 27–47. See also Brian Lalor, *The Irish round tower: origins and architecture explored* (Dublin, 1999). 18 Finbar McCormick, Thom Kerr, Meriel McClatchie and Aidan O'Sullivan, *The archaeology of livestock and cereal production in early medieval Ireland, AD400–1100*, EMAP Report 5.1 (Dec. 2011), pp 101, 109. www.emap.ie/documents/EMAP_Report_5_Archaeology_of_Livestock_and_Cereal_Producti

When the Vikings arrived at the end of the eighth and beginning of the ninth century, Ireland was a rich island. That they failed to overrun large parts of the country during their raids and campaigns of the ninth century is witness to the power of organization of the greater kings and the church. As was the case during the Roman period, the Irish were avid borrowers of technology from the Norse. New terms for ships and fishing were borrowed into the Irish language as contact between the two peoples increased. Accommodation with the Norse came from the earliest period of contact through the taking of slaves, intermarriage, trade and the adoption of Christianity. By the eleventh century, the Urnes style (Scandinavian animal art form) was in use among Irish artists and indeed some of these artists had Norse names. Some of the Viking coastal settlements had developed into international trading towns, such as Dublin, Waterford, Limerick and Cork. By the beginning of the twelfth century, they were increasingly under the control of native kings.

Ship building was an occupation in eleventh-century Dublin. The evidence for this comes from one of five ships deliberately scuttled in the 1070s in a channel in Roskilde Fjord in Denmark to provide a barrier against possible attack on the town. The keelson of the ship was fashioned from an Irish oak felled in May–June 1042. This was the year in which Diarmait mac Máel-na-mBó came to power in Leinster. He dominated Leinster until he was killed in battle in 1072. He installed his son Murchad as king of Dublin in 1052, a position he held until his death and burial there in 1070. Diarmait supplied the Anglo-Saxon exiles at his court, Harold and Leofwine Godwinson, with a fleet to aid the recovery of their territories. Murchad led a fleet to the Isle of Man in 1061. So, during this period, Skuldelev 2 (the *Sea Stallion from Glendalough* is the replica of this ship) saw many campaigns in the Irish Sea. She was repaired with oak from the British Isles in the 1060s. This warship is one of the biggest yet found. She had sixty rowers/warriors and with other sailors would have had a complement of eighty men. She may have taken part in the attacks by Harold Godwinson's sons on southern England in 1068 and it has been suggested that her final resting place may have followed the old ship's voyage to Roskilde carrying Harold's sons and daughter on a diplomatic mission in 1068–9 in an attempt to restore Danish or Anglo-Saxon rule in England.[19] By the eleventh and twelfth centuries, ships had come to play an important role in warfare and trade in the Irish Sea controlled by kings aspiring to be kings of all Ireland. The words *long-both* 'shipyard' or 'boat-shed', *long-thech* 'boat-house' may be the equivalent of the Norse *hrof*, and *naust* found in Icelandic literature for structures in which boats were housed, built or repaired.

From the early eighth century onwards, the Uí Néill dynasty was normally

on_WEB.pdf: accessed 10 June 2013. **19** www.vikingeskibsmuseet.dk/en/research/ship-archaeology/the-skuldelev-ships-i: accessed 10 June 2013.

the strongest in Ireland. They dominated the north and the midlands. They subdued Leinster, partitioning the province on numerous occasions to keep it weak. The strongest Uí Néill kings also attempted to control Munster. A few were sufficiently successful to give themselves the title *Rex Hiberniae*, or *Rí Érenn*. Máel Sechnaill mac Máele Ruanaid, who died in 862, and his son Flann Sinna (d. 916) were patrons of high crosses, using them as propaganda documents on which to display this title. During the pre-Norman period, only very exceptional kings could make such a claim. Flann Sinna also built the church that is now the cathedral at Clonmacnoise.

The impact of the Norse upon Ireland brought about an acceleration in the changes that were already taking place in society. The Norse raids and campaigns allowed great kings to become ever stronger in periods of emergency, and power gained was not easily relinquished. By the eleventh century, the Uí Néill were no longer as powerful as they had been and new dynastic families rose to dominance in each of the provinces. The political structure in Ireland now resembled much more closely that of England or continental Europe; allowing for its own peculiar social norms. It was during this period that a monarchy of Ireland under one king might have emerged, but this process was interrupted by the Normans. The apparent ease with which the Normans conquered areas in the mid-twelfth century may be explained partly by the fact that they faced a society whose structures were not very unlike their own. The Gregorian Reform of the church, when it made an impact upon Ireland in the twelfth century, was a major factor in bringing about a change in the nature of the monastic schools. It may have been an attempt to realign with the new ideas from Europe that the clergy of Ireland in a synod in 1162 ruled that 'no one should be lector in a church in Ireland, except an alumnus of Ard Macha'. The head of that school was Flann Ua Gormáin, who died in 1174. His death notice describes him as

> arch-lector of Ard-Macha and of all Ireland, a man learned, observant in divine and human wisdom, after having been a year and twenty learning amongst the Franks and Saxons and twenty years directing the schools of Ireland, died peacefully on the 13th of the Kalends of April [March 20], the Wednesday before Easter, in the 70th year of his age.

No doubt Flann's curriculum contained the most up-to-date scholarship from Europe.

Despite the image of Ireland as existing in conservative tribal isolation that appears with monotonous regularity in some scholarly works, Ireland was never out of contact with the continent of Europe. At the very height of Norse activity, Irish scholars went to and from Irish foundations in Europe and some of those foundations may have been governed from Ireland. For example, the Annals of Ulster record that in 825 'Colmán son of Ailill, abbot of Slane and of other monasteries in France and Ireland, died'. Of course, Colmán may not have been abbot of these monasteries concurrently (multiple office-holding was common at the time), but, whether he was or not, there is an implication that there must have been close contacts within his monastic federation. The letter from Alcuin to (most likely) Colcu Ua Duinechad, a learned scholar who died in Clonmacnoise in 796 and author of the *Scúap Chrábaid*, 'Broom of Devotion', gives some indication of the kind of information that was shared among clergymen at the time.[20] In such circumstances, scholarship would have flowed in both directions. As is clear from the essays in this volume, the work of Irish scholars on the Continent cannot be divorced from what was happening in Ireland; and, in the absence of material surviving from Ireland itself, cannot be an indication that Irish schools had suffered a decline during the Viking period. The fact that Irish scholars were so well prepared before going to the Continent allowed them to exploit the new sources that they discovered there. Johannes Ériugena is the greatest example of this. That works of scholarship created by the Irish on the Continent came back into Ireland may be seen in those manuscript fragments that have survived from Ireland in the eleventh and twelfth centuries. For example, a fragment of the *Ars Grammatica* attributed to Clemens Scottus was part of a schoolbook from Glendalough (MS Egerton 3323, fo. 16).[21] Oxford, Bodleian Library, MS Auct. F.3.15 is in four parts – part I of which is in Irish hands of the late eleventh or early twelfth century.[22] Part I contains the Latin translation of Plato's *Timaeus* and may also derive from Glendalough. Parts III–IV have an epitome and long extract from Ériugena's *Periphyseon* and may come from north-east Ireland, possibly the Cistercian abbey of Newry.

Irish contacts with the Continent were renewed in the tenth century under the Ottonian emperors at Waulsort and Metz and other places. In 975, the Irish monastery of St Martin's at Cologne was assigned to Irish monks. Cologne became a major centre of Irish activity during the eleventh century – as did Würzburg. It is not surprising that advanced mathematical tables are to be found on folio 18 (MS Egerton 3323), also from Glendalough and a sister folio to folio 16 mentioned above. These are derived from the *De Abaco* of Gerbert of Aurillac. Notes from the bottom of the page recording the death of Tuathal on the night of Monday May 14 or early on Tuesday May 15 1106 show that it was in use in the monastery at that time. As pointed out above, mathematics – its ratios, its symbolism – was woven into the fabric of Irish technology in buildings, books, artefacts and exegesis. As Ériugena demonstrated in his philosophical system, it was the language of the universe.

20 Tomás O'Sullivan, 'Texts and transmissions of the *Scúap Chrábaid*: an Old-Irish litany in its manuscript context', *Studia Celtica Fennica*, 7 (2010), 26–47 at 26. 21 Ludwig Bieler and Bernhard Bischoff, 'Fragmente zweier frühmittelalterlicher Schulbücher aus Glendalough', *Celtica*, 3 (1956), 211–20. 22 Francis John Byrne, *A thousand years of Irish script: an exhibition of Irish manuscripts in Oxford libraries arranged by Francis John Byrne* (Oxford, 1979), pp 14–15.

Saints, scholars and science in early medieval Ireland

DÁIBHÍ Ó CRÓINÍN

The coming of Christianity to Ireland in the fifth century brought about many changes in Irish society, one of the most profound being the introduction of the Latin language.[1] Ireland had never been a part of the Roman Empire, and therefore had never acquired the apparatus of Roman government, which included Latin as the everyday *lingua franca* not only of administrators but also of the population at large. The Rome that Irishmen revered was, in the words of the great sixth-century Irish missionary saint, Columbanus, not the Rome of the Caesars but the Rome of SS Peter and Paul.[2] Within a century of receiving formal Christianity, however, Irish scholars had acquired a remarkable mastery of Latin, but it was the Latin of the Bible and the Church Fathers rather than of Virgil. The image of Ireland as a haven of Classical Latin literature (and even Greek) in the decades following the fall of the Roman Empire has been greatly exaggerated,[3] but the reality, while more modest, is no less impressive in its own way.

Whether the initial impetus owed anything to the activities of the first continental missionaries, led by Bishop Palladius, is impossible to say. That Gallican mission was believed – until recently – to have left no traces, either in surviving manuscripts or in any texts associated with Palladius and his followers. In fact, however, that is not the case. A corpus of six documents (two letters and four substantial treatises) has come down to us, which I believe were the work of Palladius.[4] Whether those texts exercised any influence on the earliest Irish writers in Latin – whether, indeed, they were even known to them – has not yet been established. The only other fifth-century writings to survive, Patrick's confession and letter, left no discernable mark on later Irish writings in Latin, except insofar as Patrick's works display a mastery of what has been called Biblical Style by its discoverer (David Howlett), and that Biblical Style was to become a distinguishing feature of later Hiberno-Latin prose compositions.

1 For a general discussion, see Dáibhí Ó Cróinín, 'Hiberno-Latin literature to 1169' in idem (ed.), *A new history of Ireland*, 1: *Prehistoric and early Ireland* (Oxford, 2005), pp 371–404. 2 G.S.M. Walker (ed. and trans.), *Sancti Columbani opera* (Dublin, 1957; repr. 1971), p. 48. 3 Most recently – and with spectacular popular success – by Thomas Cahill, *How the Irish saved civilization* (New York, 1995). 4 See Dáibhí Ó Cróinín, 'Who was Palladius, "First Bishop of the Irish"?', *Peritia*, 12 (2000), 205–37. I argued previously – erroneously, as I now believe – that an Easter table referred to in seventh-century texts (but ascribed to one 'Patricius') might, in fact, be a precious memory of Palladius: see Ó Cróinín, 'New light on Palladius?', *Peritia*, 4 (1986), 276–83, repr. in idem, *Early Irish history and chronology*, hereafter *EIHC* (Dublin, 2003), pp 28–34.

Any rough chronological survey of the earliest Hiberno-Latin and vernacular Old Irish literary compositions would demonstrate the extraordinary range and variety of subjects treated by the earliest Irish writers, but it is possible to narrow the focus of such a survey somewhat by pointing out that – already by *c.*600 at the latest – Irish monastic schools based their teachings on a threefold division of the curriculum: bible-studies (exegesis), Latin grammar and computus (the science and study of Easter-reckoning). The thirty texts listed in Appendix 1.1 (chosen at random, but intended to illustrate the best as well as the best-known early writings) are only the tip of the iceberg; it would be equally possible to select an entirely different group of thirty texts to illustrate our point. As one speaker at the Dublin conference pointed out, the list comprises only texts of ecclesiastical provenance; a list of equally early vernacular tracts from the corpus of surviving brehon laws and from early Irish literature would illustrate a separate but parallel native tradition, for the most part untouched by Latin literacy and culture.[5] For the purposes of this survey, however, we shall confine ourselves to writings of ecclesiastical provenance.

It is reasonable to suppose that there was continuity of Latin literacy in Ireland from the fifth century on, but it is surely significant that the two earliest demonstrably Irish compositions on our list (nos 2 & 3), the computistical note of Mosinu maccu Min, and the letters and tracts of Columbanus (both of Bangor, Co. Down), dateable respectively to the periods *c.*575 and *c.*585–600, should reflect the intense engagement with Easter calculations and the history of the Paschal controversy in the early church that was to be a hallmark of Irish scholarship in the sixth and seventh centuries.[6]

By contrast, the hymn in praise of Patrick traditionally attributed to his disciple Secundinus (*Audite omnes amantes*, no. 4) is hagiographical in purpose, and demonstrates how that genre of ecclesiastical composition had already become established in the Irish church. That said, the hymn is now believed to be either late sixth century or early seventh century, the work of Colmán Alo (of Lynally, Co. Offaly, d. 610), rather than of the fifth-century Secundinus.[7] Even so, it already attests to a very respectable grasp of Latin language and metrics at that early date; it is a precursor of the Latin poems preserved in the late seventh-century Antiphonary of Bangor (no. 27). A possible rival in terms of dating is the remarkable poem *Altus prosator* ('Ancient Creator', no. 29), a sort of 'Paradise Lost' ascribed to Columba (Colm Cille), founder of the monastery of Iona (d. 597). However, that work is believed by modern scholars to be of seventh-century date (where I have placed it in our list).

5 See Francis John Byrne, 'Seventh-century documents', *Irish Ecclesiastical Record*, 108 (1967), 164–82. 6 On the computistical note, see Dáibhí Ó Cróinín, 'Mo-Sinu maccu Min and the computus at Bangor', *Peritia*, 1 (1982), 281–95, repr. idem, *EIHC*, pp 35–47. On Columbanus, see Ó Cróinín, 'Hiberno-Latin literature', pp 374–5. 7 See Ludwig Bieler, 'The hymn of St Secundinus', *PRIA*, 55C (1952/3), 117–27. The case for attributing the composition to Colmán of Lynally was made by James Carney, *The problem of Saint Patrick* (Dublin, 1961), pp 40–6.

The note concerning Mosinu maccu Min's Bangor computistical studies is an occasional piece, a once-off glimpse of an Irish monastic teacher at work. It is only with the figure of Columbanus, originally of Bangor (Co. Down), but later founder of monasteries in Gaul and Italy (d. 615), that the first real evidence emerges of substantial writings in Latin by Irishmen. Six of his letters survive, along with a number of poems, sermons and two monastic rules. He mentions two mentors by name, the British writer Gildas (mid-sixth-century?) and Finnian (Uinnianus), possibly Irish, but more likely British in origin. Gildas' *De excidio Britanniae* has left definite traces on Columbanus' prose style.[8] Indeed, the evidence of surviving manuscripts makes clear the debt to British teachers in general in the formative stages of the Irish church, but Columbanus' complete mastery of Latin, in a variety of different prose styles, as well as his command of both quantitative and stressed metres,[9] demonstrates for the first time the full range of native Hiberno-Latin skills.

Sometime in the mid-seventh century the arrival in Ireland of Isidore of Seville's writings spurred a massive production of Hiberno-Latin compositions on every imaginable subject, across the full spectrum of the monastic curriculum. Newly acquired grammatical texts from Late Antiquity led to a surge of renewed interest in that field also, and Irish writers perfected a new type of instructional handbook, the elementary grammar, for use with beginners in Latin, which led in turn to more advanced study using 'exegetical' grammars. By combining the methods of biblical exegetes and Latin grammarians in one text, Irish teachers perfected an instructional technique that was clearly very successful. This range finds expression in prose and verse compositions throughout the seventh century: saints' lives and instructional literature (nos 5, 11, 20, 21, 22, 24, 26, 28 – hagiography, and nos 1, 8, 13, 14, 15, 16 – instructional), biblical commentaries and Latin grammars (nos 9, 18, 19, 23, 30 – exegesis, and no. 29 – grammar),[10] canon law and handbooks of penance,[11] beside a rich variety of poems devoted to biblical learning and computistics (nos 8, 14, 15).

8 See Michael Winterbottom, 'Columbanus and Gildas', *Vigiliae Christianae*, 30 (1976), 310–17. 9 The case for Columbanus' authorship of these poems (and hence that for a mastery of quantitative verse in Latin) has been vindicated by David Howlett, 'Insular Latin writers' rhythms', *Peritia*, 11 (1997), 53–116. 10 The field of Latin grammar has generated an entire corpus of writings too extensive to discuss here; see, for example, Louis Holtz, 'Irish grammarians and the Continent in the seventh century' in Howard Clarke and Mary Brennan (eds), *Columbanus and Merovingian monasticism* (Oxford, 1984), pp 135–52; Vivien Law, *The Insular Latin grammarians* (Woodbridge, 1982) and the critical reviews of that book in *Studia Hibernica*, 22/3 (1982/3), 149–56; *CMCS*, 6 (winter 1983), 100–1; *Peritia*, 2 (1983), 312–16, and *Celtica*, 16 (1984), 182–6. 11 Not listed in the Appendix because too extensive, but see Hermann Wasserschleben (ed.), *Die irische Kanonensammlung* (Giessen, 1874; Leipzig, 1885; repr. Aalen, 1966) and Kathleen Hughes, *Early Christian Ireland: introduction to the sources* (London, 1972), pp 65–95; for penitentials, see J.T. MacNeill and Helena M. Gamer, *Medieval handbooks of penance* (New York, 1938; repr. 1990).

One of the earliest of these compositions, Cummian's letter on the Paschal question (632/3), is remarkable for its rich patristic sources (some of them unique) and the collection of ten different Easter tables on which its author was able to draw. It is treated of at greater length elsewhere in this collection.[12] Cummian is very likely also the author of a handbook of computus, *De ratione conputandi* ('On the reckoning of computus', no. 10), which is regarded as one of the most sophisticated and comprehensive manuals on the subject from the entire early Middle Ages.[13] The DRC and a number of other contemporary Irish computistical textbooks set the pattern for all subsequent study in the field, including the Venerable Bede's two great works On Time.[14] Nowhere is this more clearly to be seen than in the fact that, of the texts listed in Appendix 1.1, almost a third are computistical in subject-matter (nos 2, 3, 7, 8, 10, 17, 25), and at least three of the others, while dealing with biblical exegesis in the main, clearly evince a profound knowledge of computistics in their discussions and refer specifically to the science (nos 18, 23, 30).[15]

The field of study that appears to have generated the most literature in the Irish schools, however, is that of biblical commentary or exegesis. Notwithstanding the silly statement of one recent author that 'the evidence for exegetical activity in Ireland is practically non-existant' [sic],[16] there is an enormous body of writings (not all of it scintillating in quality, it must be admitted!) that shows that biblical study was a mainstay of the Irish monastic curriculum.[17] Not the least striking aspect of these writings is the very wide range of patristic and post-patristic sources to which the Irish authors make reference.[18] Equally impressive is the remarkably uniform language of these

12 See Walsh, this volume. 13 Dáibhí Ó Cróinín in Maura Walsh and idem (eds), *Cummian's letter 'De controversia paschali' and the 'De ratione conputandi'* (Toronto, 1988), pp 113–213, and Ó Cróinín, 'A seventh-century Irish computus from the circle of Cummianus', *PRIA*, 82C (1982), 405–30; Bonnie Blackburn and Leofranc Holford-Strevens, *The Oxford companion to the year* (Oxford, 1999), p. 813 ff. 14 See now Immo Warntjes, *The Munich Computus: Irish computistics between Isidore of Seville and the Venerable Bede and its reception in Carolingian times* (Stuttgart, 2010), and the same author's review of Calvin Kendall and Faith Wallis (trans.), *Bede: 'On the nature of things' and 'On times'* (Liverpool, 2010), *Medieval Review*, 12 August 2001, see https://scholarworks.iu.edu/dspace/bitstream/handle/2022/14623/12.08.01.html?sequence=1. 15 It may be pointed out in this case as well that the numbers of surviving computistical texts from the Irish church are so huge that the subject cannot be adequately treated here. For a comprehensive review of the material, see now Warntjes, *The Munich Computus*, introduction. 16 Michael Gorman, 'The myth of Hiberno-Latin exegesis', *Revue Bénédictine*, 110 (2000), 44–85 at 55. See my response in 'Bischoff's "Wendepunkte" fifty years on', *Revue Bénédictine*, 110 (2000), 204–37, and the other critical literature against Gorman's view cited there. 17 See esp. Bernard Bischoff's seminal essay, 'Wendepunkte in der Geschichte der lateinischen Exegese im Frühmittelalter', *Sacris Erudiri*, 6 (1954), 189–279; rev. ed. in Bischoff, *Mittelalterliche Studien: Ausgewälte Aufsätze zur Schriftkunde und Literaturgeschichte* (3 vols, Stuttgart, 1966, 1967, 1981), i, pp 205–73. 18 There is a useful brief summary in Martin McNamara, 'Patristic background to medieval Irish ecclesiastical sources' in Thomas Finan and Vincent Twomey (eds), *Scriptural interpretation in the Fathers: letter and spirit* (Dublin, 2008), pp 253–81.

different authors. Though scholars have happily used the term 'Hiberno-Latin' to describe the language of Irish texts from this period, in fact Irish Latin was indistinguishable in grammar and syntax from its Continental counterpart – a testament to the efficacy of teaching in Irish schools.

What most impresses, however, is the sheer number of Hiberno-Latin writings in the sixth and seventh centuries, and the range of their subject-matter. There is far more surviving from Ireland in this period than from any other European country, including Spain. These commentaries are mundane, on the whole, with their authors generally happy to list excerpts from their sources, adding occasional linking comments but without any pretence at originality; a good example would be the Genesis commentary cited as the last item on our list.[19] What marks out our Hiberno-Latin authors, however, is that they drew on a huge variety of Late Latin, biblical and patristic sources, in addition to unorthodox writings like the commentaries of the heresiarch Pelagius, and a remarkable number of biblical apocrypha nowhere else available.[20]

In the works of one or two authors there is a genuine streak of originality,[21] and that is the case with no. 16 (Irish Augustine, *De mirabilibus sacrae scripturae*, 'On the miraculous things in Scripture', dated internally to 654/5) and no. 23 (Irish Isidore, *Liber de ordine creaturarum*, 'The book on the ordering of creation', later than the *De mirabilibus*, but close to it in date).[22] These two commentaries represent the high point of Irish scholarly achievement in the seventh century, and their unique contribution is discussed elsewhere in this collection (by Marina Smyth). That achievement, however, represents in turn the product of a collaborative movement that involved scholars in different parts of Ireland exchanging their ideas and commenting on them in a most remarkable way.[23] It is not going too far to say that some of these texts represent 'schools' of biblical exegesis, where biblical study was raised to a level of sophistication that attracted not only native students but also individuals from abroad.[24]

19 See Dáibhí Ó Cróinín, 'A new seventh-century Irish commentary on Genesis', *Sacris Erudiri*, 40 (2001), 231–65. 20 See Martin McNamara, *The Apocrypha in the Irish Church* (Dublin, 1975). 21 There is a very useful set of excerpts (with commentary) from the texts discussed here in John Carey, *King of mysteries: early Irish religious writings* (Dublin, 2000). On the originality of the thinking in the *De mirabilibus* especially, see idem, *A single ray of the sun: religious speculation in early Ireland* (Aberystwyth, 1999), pp 47–53, 58–63, 71. 22 Manlio Simonetti, '*De mirabilibus sacrae scripturae*. Un trattato irlandese sui miracoli della Sacra Scrittura', *Romanobarbarica*, 4 (1979), 225–51; Damian Bracken, 'Rationalism and the Bible in seventh-century Ireland' in A.M. Luiselli Fadda and Éamon Ó Carragáin (eds), *Le isole britanniche e Roma in età romanobarbarica* (Rome, 1998), pp 129–70; Marina Smyth, 'The dove and the star: enduring ideas from seventh-century Ireland', *Peritia*, 20 (2008), 98–134, and the literature there cited; eadem, 'The date and origin of *Liber de ordine creaturarum*', *Peritia*, 17–18 (2003–4), 1–39, and eadem, 'The seventh-century Hiberno-Latin treatise *Liber de ordine creaturarum*: a translation', *Journal of Medieval Latin*, 21 (2011), 107–22. 23 See Paul Grosjean, 'Sur quelques exégètes irlandais du viie siècle', *Sacris Erudiri*, 7 (1955), 67–98. 24 For example, see the Venerable Bede's famous reference to the Frank, Agilberct, later bishop of Paris, who spent several years in

The Irish also began to gloss their Latin texts in the vernacular, very quickly passing to full texts in Old Irish. The most remarkable example of this phenomenon is the Cambrai Homily (no. 13, probably mid-seventh century), a bilingual Latin-Irish text combining excerpts from the gospels, Pauline Epistles and Gregory the Great's gospel homilies with a parallel text in Old Irish whose language is extraordinarily archaic.

The oldest known manuscript with original bilingual Latin-Irish glosses dates to *c*.700,[25] but the most famous is the Würzburg codex of *c*.800 containing Pauline epistles with a huge number of glosses in both Latin and Old Irish. This probably belonged to Clemens Scottus, master of the palace school at Aachen in Charlemagne's time, who ended his days at the shrine of the Irish saint Kilian in Würzburg.

Alongside these 'ordinary' expressions of Latin culture in Ireland, however, there was also the extraordinary: the so-called *Hisperica Famina* ('Western sayings') make their appearance in the mid-seventh century.[26] A 'culture fungus of decay' (as one scholar described them), these bizarre colloquies are a pastiche of the pedantic hypercorrectness of some of the grammarians, mocking the high-falutin' language and vocabulary of the schools. They are probably not to be taken too seriously (and may not be Irish at all in origin). Another, more exotic, flowering of Irish Latin culture, however, definitely is a native concoction: the bizarre writings of Virgilius Maro 'the Grammarian' (no. 18) are an extraordinary rodomontade of mock learning that pokes glorious fun at the pomposities of the self-same schoolmen.[27] The fun was probably well intended, but the Irish humour may have been lost in transmission, with the result that Virgil's more exuberant pronouncements sometimes made their way into the works of seriously minded computists and biblical commentators, so that Continental men of learning, when they came face to face with such oddities in the eighth century, tended to look askance at Irish learning.

Hiberno-Latin scholars nevertheless enjoyed a very good reputation when they travelled across Europe, following in the steps of Columbanus. The eighth

Ireland *scripturarum legendarum gratia* ('for the purpose of studying Scripture'); Bede, *Hist. Eccl.*, Lib. iii, cap. 7, Charles Plummer (ed.), *Baedae opera historica* (Oxford, 1896; repr. 1969), p. 140. **25** See Dáibhí Ó Cróinín, 'Early Echternach manuscript fragments with Old Irish glosses' in Georges Kiesel and Jean Schroeder (eds), *Willibrord. Apostel der Niederlande, Gründer der Abtei Echternach. Gedenkgabe zum 1250. Todestag des angelsächsischen Missionars* (Luxembourg, 1989), pp 135–43, repr. in idem, *EIHC*, pp 133–44. Note that the fragments concerned represent a computistical collection. **26** Michael Herren (ed.), *Hisperica Famina* (2 vols, Toronto, 1974 & 1987). See also John Carey, 'The obscurantists and the sea-monster: reflexions on the *Hisperica Famina*', *Peritia*, 17–18 (2003–4), 40–60. **27** See the literature cited in Ó Cróinín, 'Hiberno-Latin literature', pp 388–91. The most recent treatment is in Rory Naismaith, 'Antiquity, authority and religion in the *Epitomae* and *Epistolae* of Virgilius Maro Grammaticus', *Peritia*, 20 (2008), 59–85, and Damian Bracken, 'Virgil the Grammarian and Bede: a preliminary study', *Anglo-Saxon England*, 35 (2006), 7–21.

century saw the appearance on the Continent of men like Dicuil (author of a remarkable cosmographical work, *Liber de mensura orbis terrae*, as well as computistical and grammatical texts) and Dungal of Pavia, Muredach Scottus 'most learned of all men' (in his own estimation, at any rate), and Joseph Scottus, friend of Alcuin. Even more remarkable, however, was the generation of scholars that followed them in the ninth century, especially Sedulius Scottus of Liège and his circle of friends, and the most famous of them all, Iohannes Eriugena ('Irish-born'). These men were the superiors of their Continental contemporaries not only in terms of Latin learning but also in their knowledge of Greek. Eriugena in particular was, by common consent, the finest intellect of his generation.

In their Latin poetry (and Greek poetry in Eriugena's case) Sedulius and Eriugena demonstrate a complete mastery of the language. Sedulius too, with his 'Handbook for princes' (*De rectoribus Christianis*), also established a genre that was to have lasting influence in the area of political philosophy. Irish scholars have traced the origins of such 'Mirrors' back to the mid-seventh-century *De duodecim abusiuis (saeculi)* (no. 15). In John's case, his philosophical works (especially the *Periphyseon*, 'On the nature of things') reveal a mind that had no equal in Europe in his time, and a unique grasp of Greek philosophy. But their achievements were predicated on the work of their sixth- and seventh-century predecessors, the men whose writings made of Ireland – in the words of the great Eoin Mac Neill – 'A pioneer of nations'.

APPENDIX 1.1 CHRONOLOGY OF EARLY IRISH WRITINGS, *c.*AD550–*c.*AD700

1.	Uinnianus/Finnian (of Clonard?), Letter to Gildas	*c.*545?
2.	**Mosinu moccu Min (Bangor) 'Greek' computus**	***c.*575?**
3.	**Columbanus (Bangor), Epist., *Tres tomi, Abortiua luna***	***c.*585–600?**
4.	Hymn of Secundinus (*Audite omnes amantes*)	*c.*600?
5.	Dallán Forgaill, *Amrae Choluim Chille*	*c.*597?
6.	**Cummian (Clonfert?), Letter on the paschal controversy**	**632/3**
7.	**Nonae Aprilis (computistical verses)**	***c.*640**
8.	**Anonymous, Poem on the Six Ages (*Deus a quo facta fuit*)**	**645**
9.	Cummian, Commentary on Gospel of Mark	*c.*650?
10.	**Cummian (of Clonfert), *De ratione conputandi***	***c.*650?**
11.	Ultán moccu Conchobair (Ardbraccan), *Vita Brigitae*	*c.*650?
12.	Anonymous, Cambrai Homily (Latin + Old Irish)	*c.*655?
13.	Ailerán (of Clonard), *Interpretatio mystica … Christi*	*c.*655?
14.	Ailerán, Verses on Eusebian canons (*Quam in primo*)	*c.*655?
15.	Pseudo-Cyprian, *De duodecim abusiuis*	*c.*655?
16.	**Pseudo-Augustine, *De mirabilibus sacrae scripturae***	**654/5**
17.	**Anonymous, 'Irish Computus'**	***c.*658?**
18.	Virgilius Maro Grammaticus, *Epistolae & Epitomae*	*c.*660?

19.	Anonymous, Commentary on the Catholic Epistles	*c*.660?
20.	Cogitosus (Kildare), *Vita Brigitae*	*c*.660?
21.	Cummeneus Albus (Iona), *Liber de uirtutibus s. Columbae*	*c*.660?
22.	Béccán mac Luigdech (Iona), Verses on Colm Cille (Irish)	*c*.670?
23.	**Pseudo-Isidore, *Liber de ordine creaturarum***	***c*.680?**
24.	Muirchú maccu Machtheni (Armagh), *Vita Patricii*	*c*.680?
25.	**Anonymous, Munich Computus**	**689**
26.	Tírechán (Armagh), *Itinerarium Sti Patricii*	*c*.690?
27.	Anonymous (Bangor), Antiphonary of Bangor	*c*.690?
28.	Adomnán (Iona), *Vita Sti Columbae*	*c*.700?
29.	*Altus Prosator* ('Paradise Lost')	*c*.670?
30.	Anonymous, *Pauca in Genesim*	*c*.700?

Biblical commentary: nos 9, 13, 14, 16, 19, 23, 30
Hagiography: nos 4, 5, 11, 20, 21, 22, 24, 26, 28
Computistical: nos 2, 3, 6, 7, 8, 10, 17, 23, 25
Devotional: nos 12, 27
General: nos 1, 15

Boethius in early Ireland: five centuries of study in the sciences

PÁDRAIG P. Ó NÉILL

Anicius Manlius Torquatus Severinus Boethius (*c.*480–526), one of the last great scholars of late Antiquity and perhaps the most influential in the centuries that followed, is mainly known for his 'Consolation of Philosophy'. He also composed a series of non-philosophical works, many of them inspired by Greek originals. The present essay will survey the evidence for knowledge of Boethius' works in early medieval Ireland from the seventh to the twelfth century while attempting to measure the influence they had on Irish scholarship, especially in the areas of computistics, mathematics, music and natural philosophy. Since it is to be hoped that further evidence awaits discovery (perhaps in the large body of Hiberno-Latin biblical exegesis), the present essay can only claim to be a preliminary survey at best. But first an overview of Boethius' works, which can be divided into four main categories:[1]

1. MATHEMATICAL

De institutione arithmetica, *De institutione musica*, and *De institutione geometria* (fragments of which have survived). Apparently, he also composed a work on astronomy, as implied by Isidore of Seville in his account of these four subjects in his *Etymologies*, which seems to be based on Boethius.[2] The four works belong to an early stage in Boethius' career, *c.*500–6. Their subject matter was envisaged by him as constituting the *quadrivium*, the four subjects to be undertaken in a liberal arts education, following the triad of grammar, logic and rhetoric, which later became known as the *trivium*. This concept of the seven liberal arts as the basis of an education became the mainstay of the medieval curriculum.

2. LOGICAL

Boethius translated six Greek works on this subject: Porphyry's *Isagoge*, Aristotle's *Categories*, *On Interpretations*, *Prior Analytics*, *Topics* and *Sophistical*

1 See J. Marenbon (ed.), *The Cambridge companion to Boethius* (Cambridge, 2009), pp 303–10.
2 Gerbert of Aurillac (writing in the late tenth century) claims to have seen a manuscript of Boethius on *astrologia* at Bobbio.

Refutations. He wrote commentaries on several of them, in addition to another on Cicero's *Topics*; and also composed a series of short works on syllogisms, topical argument and *differentiae*. As suggested by their subject matter and sources, these works are not textbooks or aids for beginners in the *trivium*, but investigations of the higher stages of logic and, unlike the first category of his works, they never gained the status of official textbooks in the medieval curriculum.[3]

3. THEOLOGICAL

Five short treatises comprising works on the Trinity, the basic tenets of the catholic faith and refutations of heresy.

4. PHILOSOPHICAL

The 'Consolation of Philosophy', composed in the period 523–6, is a prosimetrum and a work *sui generis*. Some three centuries after its composition, it had acquired status and fame as perhaps the most influential work of the Middle Ages after the Bible. However, the 'Consolation' does not seem to have been studied in the Continental schools before the Carolingian period and it apparently remained unknown in Ireland and England until the late ninth century at the earliest.[4]

Judging by the surviving evidence, not all of these Boethian works seem to have been used in early medieval Ireland; indeed some of them may never have reached there. Moreover, for those works known to have been in Ireland it is difficult to detect patterns or trends of influence because the manuscript evidence is both meagre in content and uneven in chronological spread. On the negative side, no evidence has yet been identified for the use in Ireland of Boethius' theological works and precious little for the logical works until the early twelfth century. Traces of the latter category are found in two Irish manuscripts from that period: Oxford, Bodleian Library, MS Auct. F.3.15 and Florence, Biblioteca Medicea Laurenziana, MS Plut. 78.19.[5] Another logical

3 See C.J. Martin, 'The Logical textbooks and their influence' in Marenbon, *Cambridge companion*, pp 56–84 at p. 80, who observes that Boethius 'bequeathed to the Middle Ages confused and fragmentary accounts of the logic of conditional propositions and of the use of the topics in the discovery and justification of arguments'. He also notes that following Abelard's discoveries, Boethius' *De hypotheticis syllogismis* 'ceased to have any influence from the middle of the twelfth century'. 4 See A. Papahagi, 'The transmission of Boethius' *De consolatione philosophiae* in the Carolingian Age', *Medium Aevum*, 78:1 (2009), 1–15, who argues (at 5) that the propagation and study of the 'Consolation' was begun at Fleury *c*.800, and that 'there is no hard evidence that the *Consolatio* was available anywhere in Britain before the year 900'. 5 On

work, the commentary on Cicero's *Topics*, was proposed by Robin Flower as the source for the topos of person, place, time and cause, which enjoyed considerable use among medieval Irish scholars as a means of giving 'a great air of verisimilitude to the facts related'.[6] However, this well-known rhetorical device is not unique to Boethius and could have come from a number of other sources. That leaves for consideration the mathematical works and the 'Consolation'. The latter, although strictly speaking philosophical in content, frequently gave rise among the medieval commentators to discussions of the natural sciences, thus serendipitously lending itself to the theme of the present conference.

The mathematical works

From their favourite non-biblical book, Bishop Isidore of Seville's *Etymologies* (bk III.ii), the Irish would have learned (if they had not already done so directly) that Boethius had translated 'mathematical works', a term that, in context, means the study of arithmetic, geometry, music and astronomy, corresponding to the four subjects of the *quadrivium*, the stage after the *trivium* in the scheme of the seven liberal arts proposed by Boethius. Isidore seems to have known these treatises; certainly, his discussion of the arithmetical, geometrical and musical mean (*Etym.* bk III.viii and xxiii) comes from Boethius, *De institutione arithmetica* (II.43–4 and 47; hereafter referred to as *DIA*). No less significant is a short chapter in the same book of the *Etymologies* where Isidore discusses the importance of the study of numbers. 'The reckoning of numbers', he says, 'ought not to be despised, for in many passages of sacred writings it elucidates how great a mystery they hold'. He gives the examples of the numbers 6 and 40, the first of which signifies the six days of creation, the second the duration of the fasts of Moses, Elijah and Christ. These examples also illustrate that his appeal is to the usefulness of numerology for the understanding of Scripture. To this function he adds a more practical one: 'It is even our lot to depend on the discipline of numbers to some extent when we dispute about the course of the months, and when we recognize the duration of the turning year [...]. Take away the computation of time, and blind ignorance embraces all things; those who are ignorant of the method of calculation cannot be differentiated from the other animals' (*Etym.* bk III.iv).[7]

Seventh-century Irish scholars evidently took Isidore's exhortation to heart. Significantly, the two earliest witnesses to the use of *DIA* in Ireland, both from

the first, see P.P. Ó Néill, 'An Irishman at Chartres in the twelfth century: the evidence of Oxford, Bodleian Library, MS Auct. F. III.15', *Ériu*, 48 (1997), 1–35 at 19, 22; on the second, see below, p. 32. 6 R. Flower, 'Quidam Scotigena .i. discipulus Boëthii', *Ériu*, 8 (1916), 150–4, which identified the topos in the Middle Irish work (probably late eleventh century) known as *Aislinge Meic Conglinne*. The same article also argued for the influence of the 'Consolation' (bk I, pr 1) on an eleventh- or twelfth-century commentary on the *Félire Óengusso*. 7 Translated by S.A. Barney et al., *The etymologies of Isidore of Seville* (Cambridge, 2006), p. 90.

that century, draw on the work within the context of computistical studies. The first of these is the so-called 'Sirmond dossier', a medley of letters, excerpts, tracts, *argumenta*, Easter tables, and lunar-and-solar-letter tables, all dealing with computistical subjects and ultimately designed to determine accurately the date of Easter. C.W. Jones, who re-discovered the 'Sirmond dossier' in Oxford, Bodleian Library, MS 309 (*s.*XI), demonstrated that its contents represent a type of computus favoured by the Irish.[8] Subsequent work by Dáibhí Ó Cróinín refined Jones' inventory of the contents of the 'Sirmond dossier' manuscript[9] and, on the basis of a dating-clause in one item, argued that the collection originated in south-east Ireland (Co. Waterford) in 658.[10] Folios 149r–64r of this manuscript contain a substantial portion of *DIA*, identified as *De Boetio*.[11] Judging by its acephalous opening '*<expli>cabit*', as well as its ending with an incomplete word '*sub<iecta>*', the Sirmond text originally covered even more of *DIA*.

The second witness to the influence of *DIA* is another Irish computistical work, *De ratione conputandi*, a treatise on computistics, probably also dating from the seventh century. As pointed out by its recent editors, Maura Walsh and Dáibhí Ó Cróinín, this work cites *DIA* at least three times, borrowing from bks 1.1, 1.2 and 2.4.[12] The fact that these passages are cited verbatim suggests direct access to the original; that none of them corresponds to the chunk of the same work in the Sirmond dossier makes it likely that a full text of *DIA* was available in seventh-century Ireland.[13] How *DIA* reached Ireland remains a matter of conjecture, since the early history of its transmission is obscure; virtually all of the surviving manuscripts belong to the ninth century or later. The one exception is a bifolium from a Bobbio manuscript, dating to the sixth or seventh century, now housed at Turin, Biblioteca Nazionale Universitaria, MS F.IV.1

8 C.W. Jones, 'The "lost" "Sirmond dossier" manuscript of Bede's computus', *English Historical Review*, 52 (1937), 204–13 at 209. See also his *Bedae opera de temporibus* (Cambridge, MA, 1943), pp 105–10. 9 D. Ó Cróinín, 'Bede's Irish computus' in idem, *Early Irish history and chronology* (Dublin, 2003), pp 201–12 at pp 202–3 (hereafter abbreviated *EIHC*). 10 D. Ó Cróinín, 'The Irish provenance of Bede's computus', *Peritia*, 2 (1983), 229–47 (repr. in *EIHC*, pp 173–90); idem, 'Early Irish annals from Easter tables: a case restated', *Peritia*, 2 (1983), 74–86 (repr. in *EIHC*, pp 76–86); and idem, 'The date, provenance and earliest use of the works of Virgilius Maro grammaticus' in Günter Bernt et al. (eds), *Tradition und Wertung: Festschrift für Franz Brunhölzl zum 65. Geburtstag* (Sigmaringen, 1989), pp 13–22 at pp 15–19 (repr. in *EIHC*, pp 191–200). However, the date of 658 is at best trustworthy for the single item in which it is present. 11 The excerpt covers bk I.17–bk II.26, which in the new edition by H. Oosthout and J. Schilling, *Anicii Manlii Severini Boethii De arithmetica* (Turnhout, 1999), corresponds to pp 44.72–144.22. 12 M. Walsh and D. Ó Cróinín (eds), *Cummian's letter 'De controversia paschali' and the 'De ratione conputandi'* (Toronto, 1988), pp 122.11.6–16 and 123.12.2–3 and 5–8 (referring to page, chapter and line number). 13 The Irish grammatical treatise edited by G. Calder, *Auraicept na n-éces: the scholars' primer* (Edinburgh, 1917), pp 112–13, lines 1443–53, has a discussion of perfect and imperfect numbers which may come from *DIA* I.18, but the borrowing was likely effected via Isidore's *Etymologies* and the section in question is much later than the original eighth-century core of the *Auraicept*.

(fasc. 3),[14] which contain *DIA* II.46.26–47.5; 50.37–46; and 50.67–79. Bobbio's close links with Ireland at that very time make it possible that that monastery served as a conduit for the transmission of *DIA* to Ireland. Another possible conduit would have been Visigothic Spain, where *DIA* (and the other mathematical works of Boethius) were evidently available in the early decades of the seventh century, as suggested by the descriptions of them found in Isidore's *Etymologies*. It is now generally accepted that this and other works of Isidore reached Ireland within his lifetime in the early decades of the seventh century.[15] At the same time it has been suggested that much of the computistical contents of the Sirmond dossier also came from Spain.[16] As already noted, that collection contained a major portion of *DIA*, which raises the possibility that the latter work might have travelled from Spain to Ireland in a computistical context.

If I have laboured the point that *DIA* was well known in seventh-century Ireland, it is in part to highlight that such access was remarkable for these islands at that time. Neither *DIA* nor any other of the works of Boethius was known in Anglo-Saxon England before 900.[17] As a student under Abbot Hadrian at the much-vaunted school of Archbishop Theodore of Canterbury in the last quarter of the seventh century, the Wessex scholar Aldhelm proudly relates (in a letter to Bishop Leuthere) that he mastered fractions but concedes that he did so only through persistent study[18] – the implication seems to be that he had no textbook such as *DIA* to guide him. Nor was *DIA* known to his contemporary, the Venerable Bede, the foremost computist of the early Middle Ages and a scholar with access to the formidable monastic library of Wearmouth-Jarrow. Indeed, given Bede's heavy reliance on the original contents of the 'Sirmond dossier', it is a major puzzle why he did not use two works that formed part of that collection, Boethius' *DIA* and Victorius' *Calculus*. Both are mathematical works that, in an age when few such texts were available, might be expected to command Bede's attention. Admittedly, Victorius' work was not of an especially high calibre – it consisted mainly of multiplication and division tables – but *DIA* was the primary text on the subject during the Middle Ages and Bede's apparent failure to avail of it is therefore all the more puzzling.

Other chronological milestones suggest that *DIA* continued to be a staple of

14 On this manuscript, see E.A. Lowe, *Codices Latini Antiquiores*, IV (Oxford, 1947), no. 450. 15 See B. Bischoff, 'Die europäische Verbreitung der Werke Isidors von Sevilla' in idem, *Mittelalterliche Studien: Ausgewählte Aufsätze zur Schriftkunde und Literaturgeschichte*, i (Stuttgart, 1966), pp 171–94 at pp 180–7. 16 Jones, *Bedae opera de temporibus*, pp 75–7 and 112–13. 17 M. Lapidge, *The Anglo-Saxon library* (Oxford, 2006), an exhaustive catalogue of the Latin sources known to Anglo-Saxon writers, contains no evidence for first-hand knowledge of Boethius before *c*.900; and its list of manuscripts containing Boethius' works has nothing dated earlier than the tenth century (pp 293–4). 18 'And so finally, sustained by heavenly grace, I learned the most difficult proofs of the subject and the reckonings of calculations, which they call fractions, through persistent study'; trans. M. Lapidge and M. Herren, *Aldhelm: the prose works* (Cambridge, 2009), pp 152–3 at p. 153.

the Irish schools. From the early ninth century two pieces of evidence appear in the great manuscript collection of the library of Sankt Gallen. The first of these, Sankt Gallen Stiftsbibliothek, MS 728, is a catalogue of the library's holdings drawn up in the second half of the ninth century. It includes a list of books in Irish script (*libri scottice scripti*) preserved in that library, one of which is described as 'Arithmetica Boetii, volumen 1' (perhaps bks I and II of *DIA*).[19] Books from Ireland certainly were preserved at Sankt Gallen, but in this instance some caution is in order, since no manuscript matching the description of the entry in Sankt Gallen 728 has survived. Moreover, a manuscript of *DIA* in an Irish hand could just as easily have been written on the Continent as in Ireland; Irish copyists were particularly active there at that time. The second piece of evidence from Sankt Gallen is contained in MS 804, the so-called 'St Gall Priscian' (*c.*850).[20] As indicated by its name, it is a copy of Priscian's *Institutiones grammaticae*, in this instance one heavily glossed in Irish and Latin. Among the Old Irish glosses, which range in date from the late seventh to the early ninth century, are two that bear witness to knowledge of *DIA*.[21] The first explains *CURTENSIS* as 'belonging to a town/court or, as in Boethius, truncated' (gl. 57a34s). The explanation attributed to Boethius probably derives from *DIA* 2.24, which uses the word *curta* in the context of a truncated pyramid. Another gloss, on *ALTER* (gl. 41b34p), explains that '*alter* (the other) is a name for duality: it is the first number that varies from unity'. This gloss, according to Hofman, is 'a perfect paraphrase' of *DIA* II.28. That the terminology of Boethian mathematics should be pressed into the service of grammatical studies in the vernacular, points to both the prestige and the currency of *DIA* in the Irish schools of the ninth century.

Another, roughly contemporary, witness to *DIA* in Ireland about this time, and the earliest physical evidence for its circulation, is a fragment of the work consisting of one leaf of parchment, preserved at Leiden, Bibliotheek der Rijksuniversiteit, MS BPL 2391a, which has been dated to the second quarter of the ninth century. Written in a characteristic Insular semi-uncial script with decoration of initials reminiscent of the style and mix of colours found in Irish Psalters,[22] it also contains some vernacular (Irish) glosses. Its Latin orthography is consonant with Irish origins. While relatively standard, it has post-Classical spellings such as *collegit* (b9) for *colligit*, *deferentia* (c31, d17) for *differentia*, *motuatur* for *mutuatur* (b20).[23] More significantly, it contains other spellings that

19 See M. Richter, 'St Gallen and the Irish in the early Middle Ages' in M. Richter and J.-M. Picard (eds), *Ogma: essays in Celtic studies in honour of Próinséas Ní Chatháin* (Dublin, 2002), pp 65–75, who suggests (p. 68) that some of these manuscripts actually may have been written at Sankt Gallen. 20 See P.P. Ó Néill, 'Irish observance of the three Lents and the date of the St Gall Priscian (MS 904)', *Ériu*, 51 (2000), 159–80. 21 What follows comes from R. Hofman (ed.), *The Sankt Gall Priscian commentary, part 1* (2 vols, Münster, 1996), ii, pp 215, 276–7. 22 The first two letters of the opening word of chapter 47, '*ARmonica*' in form and combination of colours (yellow and purple) recall the decorated initials of early Irish Psalters. 23 References

suggest a specifically Hiberno-Latin milieu; notably, *puplica* (c20) for *publica*, *longuior* (c3, c5, c8, c9) for *longior*, and *sescalter* (a17, a18, b5) for *sesqualtera*. Overall, the cumulative evidence indicates that the manuscript was written by an Irish scribe, most likely working in Ireland.

Since this fragment (Z) is the earliest surviving witness to *DIA* from Ireland, its text merits closer examination. In particular, it would be desirable to know how it relates to other contemporary copies of *DIA* from the Continent. A textual collation of Z, which contains the text of *DIA* II.46–7, with the recent critical edition by H. Oosthout and I. Schilling, reveals a generally accurate text with very few aberrant variants – that is to say, its variants, almost without exception, are attested in other ninth- or tenth-century Continental manuscripts of the work. However, it is not possible to tie **Z** directly to a particular manuscript witness or even to any of the four textual families of *DIA* identified by Oosthout and Schilling.[24] Indeed, **Z**'s text most closely resembles a group of manuscripts (ninth and tenth century) characterized by variants that combine readings shared with the four major families as well as others diverging from them, so that no clear-cut relationship can be established. More narrowly, **Z** shares some significant variants with **M** and **G**, namely the readings *in eandem proportionem* (a15) (critical text, *in eadem proportione*, 192.19)[25] and *uno altrinsecus* (b27) (critical text, *unum ab altrinsecus*, 193.43), both of which variants occur in **M**; as well as *a duobus* (d5) (critical text *duobus*, 195.10), a variant in **G** and **M** only. Yet elsewhere **Z** has *procreatur* (c15), where **M** has *procreabitur* (194.54). And **Z** lacks the chapter heading (*titulus*) to bk II.47, which is found in **G** and **M**. Overall, the 'mixed' text of **Z** resists localization, recalling a similar phenomenon with Irish Gallican Psalters in their relationship to Continental Psalters from about the same period.

In her edition of the Leiden fragment, Maartje Draak skilfully reconstructed its prior textual history.[26] From a critical error in the main text, the deliberate omission of a crucial numeral for which a space was left blank – it was subsequently (and incorrectly) supplied by another, contemporary, scribe[27] – she argued that the main scribe was a mere copyist rather than a student of mathematics. And she concluded that the glosses were also copied from **Z**'s exemplar, explaining that had the scribe been an engaged and knowledgeable glossator he would surely have commented on the problem posed by the missing crucial numeral with some kind of note. In so arguing, she plausibly postulated the

to this manuscript (in round brackets) are identified by the column and line number of M. Draak's edition, 'A Leyden Boethius-fragment with Old-Irish glosses', *Mededelingen der Koninklijke Nederlandsche akademie van Wetenschappen*, n.s. 11:3 (1948), 115–27. **24** Oosthout and Schilling, *Boethii De arithmetica*, pp v–vii, who use the siglum **Z** for the Leiden fragment. **25** Page and line number references to the Oosthout and Schilling edition. **26** Draak, 'A Leyden Boethius-fragment', 4–5. **27** Who rashly filled the void with *tria* (instead of *duo*). As noted by Draak, p. 5, n. 13, the script and the ink of *tria* is different from that of the main scribe.

existence of a second copy of *DIA*, the exemplar for the Leiden fragment that included the latter's glosses.

Her edition of **Z** included its thirty-six glosses, sixteen of which are in Old Irish. Since she said nothing about the character of the Old Irish glosses, it may be worth teasing out their possible implications for the study of *DIA* in Ireland. Their language is classical Old Irish, which makes them contemporaneous with, or even earlier than, the manuscript that carries them.[28] Thus they bear witness to the serious study of *DIA* in Ireland *c.*800, if not before. That study was evidently pursued in the vernacular at a fairly sophisticated level as suggested by the presence of Old Irish words of a technical nature, designed to address the numerous mathematical terms and concepts contained in *DIA*. The simplest accommodation to this process of translating from Latin to Irish are direct lexical borrowings such as *cyb* (gl. 7) (Lat. *cybus*, 'a cube') and *dupuil* (gll. 34 and 35) (Lat. *duplus*, 'double'). Other Irish words in the glosses point to semantic borrowing, a process whereby a previously existing Irish lexeme with a 'native' meaning borrows another level of meaning from a Latin word, as in *errannas* ('a mathematical division') (gl. 1) and *du-foxla* in the sense of 'takes' ('twice the cube') (gl. 10). Another gloss, *sainredach* (gl. 2), was characterized by Draak as a 'translation gloss' (by which presumably she meant a lexical equivalent); but it is also, arguably, a 'grammatical' gloss in that it is intended to reflect the genitive plural of its lemma, *singulorum*. That, in turn, would suggest that the glossator was thinking in Irish as he glossed the Latin text of *DIA*.

The only other surviving Irish witness to *DIA* dates almost three centuries later than **Z**. It also is a fragment, now Trinity College Dublin, MS 1442, containing bk II.19–44 and written in an Irish hand of the late eleventh or early twelfth century.[29] Despite the large chronological gap that divides the Leiden and Trinity fragments, one would hardly know it except by the difference in script (semi-uncial replaced by pointed minuscule). Both have the same layout of double columns with generous margins; both are glossed in a combination of Latin and Irish; and both provide the same kind of explanatory glosses designed to clarify the text, using technical terms from the vernacular. The Trinity fragment, however, presents a somewhat more sophisticated text in two respects: it has several illustrative diagrams of geometrical shapes (entered on the margins in dry-point); and it contains glosses that offer alternative readings to those of the main text. Notwithstanding these differences, both fragments seem to reflect

28 Note, for example, the preservation of the Old Irish relative construction, *amal as n-erranas* (gl. 1) and *cyb as nessam dó* (gl. 10), as well as the verbal of necessity in *tobarthi* (gl. 19); spellings such as *cosmil* (gl. 1), *dáu* (gl. 4), *immurgu* (gl. 4); and the rare superlative of *ír* ('long') in *hírem* (gl. 10). 29 See W. O'Sullivan, 'Manuscripts and palaeography' in D. Ó Cróinín (ed.), *A new history of Ireland*, 1: *Prehistoric and early Ireland* (Oxford, 2005), pp 511–48 at p. 544 (and pl. 55 – not pl. 57 as given); and P.P. Ó Néill, 'A Middle-Irish note on Boethius's *De institutione arithmetica*', *Éigse*, 35 (2005), 1–8.

the same kind of pedagogy, suggesting continuity in methods of teaching *DIA* in Ireland from the ninth to the twelfth century.

A second member of Boethius' four mathematical works, *De institutione musica* (*DIM*), was known in Ireland at least by the twelfth century and perhaps even as early as the ninth. The 'St Gall Priscian' manuscript, in addition to two Irish glosses based on *DIA*[30] also has a Latin gloss (gl. 3a34uu) explaining that motion is a prerequisite to sound, and citing as its authority, the text *in .II. libro boetii de musica* (actually *DIM* bk IV). The gloss was not entered by the main glossators; nevertheless, its scribe belonged to the ninth century and was Irish, though his hand shows Continental influence.[31] The latter circumstance leaves open the possibility that he acquired his knowledge of *DIM* on the Continent. The twelfth-century evidence comes from Oxford, Bodleian Library, MS Auct. F.3.15, a composite manuscript assembled probably in the second quarter of the twelfth century, all of it written in Ireland by three different scribes. On the recto of folio 20, a folium left blank by Scribe 1 after he had copied the text of Calcidius' translation of the *Timaeus*, Scribe 3 entered (perhaps a decade or two later) a musical drawing of tones and intervals.[32] This remarkable figure consists of three lambda diagrams that outline a scale of musical notes progressing through tones and semi-tones to the number 41,472 (based on the Calcidian numbers, 2, 4 and 8), while at the same time adverting in nos 2304–9216 to the rival diatonic system propounded in Boethius' *DIM* (pl. 1). This combination in a single diagram is characterized by C. Meyer as a 'relecture, en définitive platonicienne, de Boèce'.[33] It seems to be an original construct of the Bodleian scribe, judging by his comment (in a marginal note) that he would have provided a fuller diagram had he not been constricted by the narrow dimensions of the present folium.[34] If so, one has to assume a high degree of familiarity with *DIM*, which allowed him to reconcile the theoretical model of that work with the more practical system of Calcidius that was used in the production of musical instruments. By the same token, the Bodleian scribe must have known *DIA* thoroughly, not only on the evidence of his obvious facility with five-digit numbers and his ability to define the relationship between sounds in mathematical terms, but also because that work was envisaged by Boethius as an introduction to music theory. His complementary use of *DIA* and *DIM* suggests that he had learned them as components of the Boethian *quadrivium*.[35]

30 Discussed above, p. 26. 31 Information from Hofman (ed.), *The Sankt Gall Priscian commentary*, 2, p. 13, and 1, p. 24 (on the scribe). 32 See Ó Néill, 'An Irishman at Chartres', 11. 33 C. Meyer, 'Le diagramme lambdoide du MS Oxford Bodleian Library Auct. F.III.15 (3511)', *Scriptorium*, 49 (1995), 228–37. 34 *Sed non scripsi proportionaliter quia non patitur angustia parcimenaris huius folii; et qui uult scribere, scribat*; Ó Néill, 'An Irishman at Chartres', 34, n. 144. 35 See M. Masi, *Boethian number theory: a translation of the De institutione arithmetica* (with intr. and notes) (Amsterdam, 1983), p. 23.

The 'Consolation'

This is the work for which Boethius is best known and the one that gave rise to a body of early medieval commentary second only to that on the Scriptures. Strictly speaking, its subject matter is philosophical, a discussion of the nature of human life and divine providence, subjects that at first blush are hardly germane to the present volume. However, the work frequently adverts to natural phenomena such as flora and fauna and the four elements, topics that lent themselves to scientific explanation and commentary once the 'Consolation' had achieved its pre-eminent status in the schools. This rise to fame came centuries after its composition – not until the early ninth century when, according to tradition, Alcuin of York gave the 'Consolation' a privileged place in the cathedral schools of the Carolingian Empire.[36] As a result of its new status, numerous commentaries grew up around the work. Two main traditions of commentary circulated in the early Middle Ages (that is, before 1100), one associated with the Anonymus of St Gall, the other with Remigius of Auxerre.[37] The so-called 'Remigian' commentary, which reached its extant form in the early tenth century, created the dominant tradition of Boethian commentary;[38] as we will see, it was evidently available in Ireland in the early twelfth century. More controversial is another commentary, labelled by its editor as a ninth-century work that, he argued, was composed by John Scottus Eriugena.[39] His claims both as to date and authorship are extremely dubious; indeed the work has since been dated to the early twelfth century by N.M. Häring, who points to the presence in it of terms from logic of the type being developed by Abelard at that time.[40] The primary aim of these three commentaries was to clarify the meaning of the text by focusing on individual words and grammatical constructions and by providing background information about allusions in the 'Consolation' to classical history, politics, mythology and natural philosophy. For the complicated

36 For what follows, see L. Nauta, 'The consolation: the Latin commentary tradition, 800–1700' in Marenbon, *Cambridge companion*, pp 255–78 at pp 257–8. See also n. 4, above. 37 See Pierre Courcelle, *La Consolation de philosophie dans la tradition littéraire: antécédents et postérité de Boèce* (Paris, 1967), pp 275–90 and 405–6. However, see J. Wittig, 'The "Remigian" glosses on Boethius's Consolation of Philosophy in context' in C. Wright, F. Biggs and T. Hall (eds), *Sources of wisdom: studies in Old English and medieval Latin in honor of Thomas D. Hill* (Toronto, 2007), pp 168–200 and idem, 'The Old English Boethius, the Latin commentaries and Bede' in G.H. Brown and L.E. Voigts (eds), *The study of medieval manuscripts of England: Festschrift in honor of Richard W. Pfaff* (Tempe, AZ, 2010), pp 225–52 at p. 226, who concludes: 'by the tenth century there existed a body of "commentary" in two distinguishable traditions, each of which shows much common content; but each seems more likely to reflect a gradual accumulation of glosses over time rather than two original "commentaries" […]'. 38 Hereafter referred to as 'Remigius' since it probably had nothing to do with Remigius of Auxerre and likely attained its extant form through a gradual process of accretion. I use the edition of E.T. Silk in the appendix to his *Saeculi noni auctoris in Boetii Consolationem philosophiae commentarius* (Rome, 1935), pp 305–43. 39 Silk, *Saeculi noni auctoris*, pp 1–304. 40 N.M. Häring, 'Four commentaries on the *De consolatione philosophiae* in MS Heiligenkreuz 130', *Mediaeval Studies*, 31 (1969), 287–316.

metres of the poetic parts of the 'Consolation', medieval students could have recourse to a treatise composed by Lupus of Ferrière *c.*850.[41]

These early commentaries were normally written in the form of glosses, interlinear and marginal. But by the late eleventh century the Continental schools were developing more systematic, continuous commentaries that could stand on their own as independent works. They were also attempting 'the wider goal of giving the text its proper place in the liberal arts curriculum by linking it to other texts, both pagan and Christian'. The belief was that 'by climbing the stairs of the liberal arts [...] men could overcome their fallen state and come to learn the structure of the cosmos and the creator'.[42] Study of the 'Consolation', and related works, notably, Plato's *Timaeus*, Martianus Capella's *De nuptiis* and Macrobius' *Somnium Scipionis*, could begin the process, aided by Christian teaching and education. A pioneer in this development was the Norman *magister*, William of Conches, one of the foremost scholars of the twelfth-century renaissance, author of commentaries on Priscian, Plato, Macrobius, natural philosophy and, critically for our purpose, the 'Consolation' (*c.*1120). Indeed, William's commentary on the 'Consolation' became the standard work during the twelfth and thirteenth centuries.

Like his predecessors, William does not comment on every phrase, passing over long sections of the 'Consolation' without comment. Still, his approach is more systematic and comprehensive than theirs. He usually starts with a lemma, placing it in the wider context of the argument before descending to the level of the individual word. A distinctive feature of William's work is his tendency to use the commentary format for developing new areas of knowledge, especially in natural philosophy, and including even medicine. For him, the cosmos becomes not just a manifestation of the divine but a matter of investigation on the physical level to be undertaken along rational lines. As a result of William's approach, the text of the 'Consolation' becomes a vehicle 'for developing new areas of knowledge, in particular in the field of natural philosophy'.[43] Thus, William discusses a wide range of scientific topics such as the anatomy of the brain, the working of an astrolabe, the nature of lunar eclipses, the seasonal variations in daylight, the various combinations of the four elements – all the stuff of the *quadrivium* and natural philosophy, though on a higher plane. For example, a reference in the opening *prosa* of the 'Consolation' to Lady Philosophy positioning herself above Boethius' head elicits from William an anatomical discussion of the three parts of the brain and their respective functions.[44]

William's works seem to have been known in twelfth-century Ireland, as evidenced by two manuscripts: (1) Bodleian MS Auct. F.3.15, which contains (in addition to the musical diagram discussed above) a copy of Calcidius' translation

41 As did the Irish glossator of the Florence Boethius, a manuscript which is discussed below, p. 32. 42 Nauta, 'The Consolation', p. 258. 43 Ibid., p. 260. 44 Discussed below, pp 40–1.

of the *Timaeus* (hereafter referred to as 'the Bodleian *Timaeus*') and (2) Florence, Biblioteca Medicea Laurenziana, MS Plut. 78.19, a glossed copy of the 'Consolation' (hereafter referred to as the 'Florence Boethius'). Besides their large pointed minuscule script (the so-called 'national hand')[45] and a common layout of double columns of text with generous margins for glosses, both manuscripts share significant similarities in methodology. One finds in both numerous alternative readings entered as glosses, suggesting systematic textual collations of their respective works, perhaps the result of receiving more recent copies from the Continent. Both have the same mixture of brief glosses focused on individual lemmata with substantial scholia addressing the larger context and argument; and both show dependence on William of Conches – the Bodleian *Timaeus* on William's commentary on that work,[46] the Florence Boethius on William's Boethian commentary. Finally, both contain discussions of the natural sciences, the Bodleian *Timaeus* in the form of an anonymous work on the subject, the Florence Boethius in its glosses. However, since only the latter work deals expressly with Boethius, it alone will be the focus in what follows. As will be argued, the glosses of the Florence manuscript, while heavily dependent on William of Conches' commentary on the 'Consolation', reveal a somewhat ambiguous attitude towards the scholarship of the dominant *magister* of the early twelfth-century French schools.

Visually, the Florence Boethius is an impressive manuscript, with its striking initials composed of 'beast and snakes decoration', its main text written in the large angular script regularly employed for works of the highest status, and careful layout of contents. It is no less impressive for its scholarship in the form of numerous marginal and interlinear glosses that testify to scholarly engagement with the text on several levels. Since both main text and glosses seem to be written by the same scribe, they likely are coeval. If so, the dependence on William's Boethian commentary evident in the glosses would give a likely *terminus post quem* for both main text and glosses of *c.*1120.[47] Such a date would harmonize with the verdict of palaeographers and art historians that the manuscript dates to the first half of the twelfth century.[48] Altogether, the manuscript contains approximately five thousand glosses and scholia. Among the latter, William of Conches' commentary on the 'Consolation' is a significant and pervasive influence.

45 For a brief account of the palaeography of both manuscripts, see O'Sullivan, 'Manuscripts and palaeography', pp 543–4 (and pl. 57). 46 See Ó Néill, 'An Irishman at Chartres', 17–18. 47 The date proposed by L. Nauta (ed.), *Guillelmi de Conchis glosae super Boetium* (Turnhout, 1997), pp xxiii–xxv (hereafter referred to as 'Nauta, *Glosae*'). 48 For a brief description of the manuscript, see F. Henry and G.L. Marsh-Micheli, 'A century of Irish illumination (1070–1170)', *PRIA*, 62C (1962), 101–64 at 140–1 (and pl. XXa–d). On the basis of its decoration, which closely resembles that of the 'Corpus Missal' (Oxford, Corpus Christi College, MS 282), they would date the Florence Boethius as 'early twelfth century' and locate its origins possibly in the west of Ireland (p. 143).

Mindful of the rich trove of sources contained in the roughly coeval Bodleian *Timaeus*, I initially assumed that the Florence Boethius would yield similar results, perhaps in the form of extracts from other twelfth-century neo-Platonic works. However, a study of the full corpus of these glosses showed this supposition to be incorrect. Although I have by no means identified all the influences at work in the glosses, the evidence so far suggests that they depend on three main sources, all commentaries on the 'Consolation', listed here in order of importance: (1) a version of a commentary edited by Silk that was probably composed in the early twelfth century – for convenience sake, I shall call it 'Pseudo-John'; (2) William's commentary on the 'Consolation', which, while harking back to traditional scholarship on that work, was opening up new areas of investigation; and (3) a version of the 'Remigius' commentary, representing a ninth- and tenth-century tradition of glossing.[49]

From a study of the glosses followed other, more troubling, findings – troubling at least in the context of a volume whose theme is the sciences in medieval Ireland. What began as an expectation that the Florence Boethius glosses would provide insights into the Irish reception of the seven liberal arts and heavy dependence on William's frequent *excursus* on matters scientific and medical, instead yielded sobering evidence of a contrary nature. As Chaucer put it so well in *The Parliament of Fowls*, after he had spent a whole day reading Macrobius' 'the Dream of Scipio' (one of the basic texts of the neo-Platonic repertoire): 'I hadde thyng which that I nolde/ And ek I ne hadde that thyng that I wolde' ('I had something that I didn't want and at the same time I lacked the thing that I wanted').[50] How both of these undesirable conditions apply to the Florence glosses will, I hope, become clear in the analysis which follows.

I begin with the hypothesis that the Irish glossator had access to full texts of the three different commentaries on the 'Consolation' identified above. This prospect is not as far-fetched as it might seem. At least one Continental manuscript from the twelfth century has survived that contains a copy of the 'Consolation' followed by texts of the same three commentaries.[51] Internal evidence from the Florence glosses also points to the same conclusion. The borrowings from the three sources evident throughout the glosses represent a mixture of long verbatim citations and brief comments, suggesting a process that was anything but mechanical. Moreover, the quotations from these sources are invariably *à propos*. In the case of dependence on William, for example, the glossator is very selective; he tends to borrow statements that succinctly explain or summarize the master's point. For example, Boethius, *MARCOQUE*

49 All quotations from (and references to) William's commentary use Nauta, *Glosae*, which is quoted by page and line number; quotations from Pseudo-John are taken from Silk's edition (as in n. 38), and are also identified by page and line number. 50 L. Benson (ed.), *The Riverside Chaucer* (3rd ed., Boston, 1987), p. 386, lines 90–1. 51 MS H in Nauta, *Glosae*, pp lxxxviii–xci.

TULLIO CUM DIVINATIONEM DISTRIBUIT VEHIMENTER,[52] is explained by William as 'the books of divinations', and further illustrated by supplying a list of the seven kinds of divination. The Irish glossator, however, condenses this into 'Cicero shows what are the constituent parts of divination' while omitting the list.[53] Such activity suggests a scholar actively engaged with his sources.

Even more telling are several instances where the glossator combined the comments of two of his sources in a manner that suggests that he had ready access to both at once and that he was actively comparing them. For example, one of the glosses on *BIBLIOTHECA* (bk I, pr 4, 3)[54] reads *Bibliotheca .i. librorum positio et est per yroniam dictum; 'biblus' .i. papirus…, 'teche' grece, mandatum latine, inde bibliotheca ubi commendantur et ponuntur*.[55] Virtually all of this gloss, notably the proposed etymology of the lemma as a compound of *biblus/papirus* and *teche/mandatum*, comes from Pseudo-John (33.14–18), but the opening words, *librorum positio* (although close to Pseudo-John, which has *librorum repositio*) represent the rather different etymology proposed by William, who analyses the lemma as **biblus* + **thesis: In biblis enim antiqui scribebant; thesis uero dicitur positio, inde bibliotheca, id est positio librorum*.[56] Elswhere, the gloss on *TU CAUSIS ANIMAS: scilicet, sapientes PARIBUS VITASQUE MINORES* (bk III, m 9, 18–19):[57] *.i. stultas* comes from 'Remigius',[58] but an alternative gloss that follows includes a verbatim quote from Pseudo-John (187.2–3), *per animas accipe spiritus angelorum, per minores uitas animas hominum*.[59]

I emphasize the evidence for the glossator's easy access to and ready engagement with these sources because it allows for a working hypothesis that for any particular lemma in Boethius for which the three (sometimes two) commentaries supplied comment, the privileging of one source over another, whether by subordinating the latter or even omitting it, is likely to have been conscious and deliberate. By the same token, glosses in the Florence manuscript that do not have identifiable sources, or even parallels, in these three commentaries are likely to have come from the Irish glossator himself.

A potentially significant omission in the Florence Boethius glosses is the almost total absence of the numerous references to the liberal arts in general and to the matter of the *quadrivium* in particular that permeate William's commen-

52 Quotations from the text of the 'Consolation' are taken from L. Bieler (ed.), *Anicii Manlii Seuerini Boethii Philosophiae Consolatio* (Turnhout, 1957). Quotations from this edition are further identified by book number, prosa or metrum number, and subsection number for the prosa, line number for the metrum. Thus, in the present instance, the quotation comes from bk V, prosa 4, subsection 1 (= 'Bieler, *Anicii Manlii Seuerini*, p. 95'). 53 All quotations from the Florence Boethius come from my own (unedited) transcript of the glosses, and are identified simply as 'Florence MS' with the appropriate foliation. 54 Bieler, *Anicii Manlii Seuerini*, p. 7. 55 Florence MS, fo. 6v. 56 Nauta, *Glosae*, p. 80.13–15. 57 Bieler, *Anicii Manlii Seuerini*, p. 52. 58 Silk, Appendix, p. 338, gl. 199. 59 Florence MS, fo. 25r.

tary and are fairly common in Pseudo-John. For example, bk I, m 2, 9–12,[60] *VISEBAT GELIDAE SIDERA LUNAE/ ET QUAECUMQUE VAGOS STELLA RECURSUS* ('he used to gaze on the constellations of the cold new moon and on each star that turns through its changing circles – all such things he mastered and bound by number and law'), clearly offered an opportunity for astronomical comment, and predictably drew lengthy responses from both William (50.126–53.96) and Pseudo-John (20.12–21.1). By contrast, the Irish glossator simply summed up the topic with a brief verbatim extract from Pseudo-John (20.10–12), *sciebat quot annis et quot diebus unusquisque planeta suum cursum peragit* ('he used to know how many years and how many days each planet took in travelling its course').[61]

On those rare occasions when he does advert to matter about the natural sciences, the glossator preferred the much briefer and more old-fashioned comments of Pseudo-John over the more seriously scientific discussions of William. For example, at bk I, m 5, 7–9,[62] *CONDAT STELLAS LUNA MINORES [...] PHOEBO PROPIOR LUMINA PERDAT* ('The shining moon [...] closer to the sun pales and loses all her light'), the Irish glossator merely repeats the old-fashioned theory given in Pseudo-John (50.89), that the closer the moon is to the sun, the more its light diminishes, *quanto fit uicinior soli tanto plus lumine suo minuitur.*[63] By contrast, William flatly rejects this theory,[64] advancing a sustained argument (86.32–87.40) on why he thinks it is impossible. None of this matter appears in the Florence Boethius. Likewise, bk 1, pr 4, 4,[65] *CUM MIHI SIDERUM VIAS RADIO DESCRIBERES* ('when with your pointer you drew for me the paths of the stars') prompts from William a detailed account of the workings of an astrolabe (80.17–81.32), whereas the Irish glossator chooses to borrow a rather jejune comment in Pseudo-John (34.5–7), *radius est philosophorum uirga qua utebantur ad demonstrandos planetarum cursus.*[66]

Although generally ignoring the *quadrivium* part of the seven liberal arts, the Irish glossator betrays considerable interest in the *trivium*, especially the subjects of grammar and rhetoric. Grammatical notes (almost certainly of his own composition) abound in the glosses; at the simplest level they serve to disambiguate case, part of speech, or verbal tense; for example, *LACRIMIS* (bk I, pr 4, 1): *.i. ablatiuus* (fo. 6r); *NOCTIBUS* (bk II, m 8, 8): *datiuus* (fo. 18r); *IN ID AETATIS* (bk II, pr 4, 7): *aduerbialiter* (fo. 13r); and *DECORATO* (bk III, pr 4, 4):[67] *propir*[68] (fo. 21r), where the gloss signals that the lemma is not a verbal form but a proper name. Others are more complex, as in *CEDAT* (bk IV, m 5, 21):[69]

60 Bieler, *Anicii Manlii Seuerini*, p. 3. 61 Florence MS, fo. 5r. 62 Bieler, *Anicii Manlii Seuerini*, p. 11. 63 Florence MS, fo. 8v. 64 *Non ergo uerum est quod dicunt*; Nauta, *Glosae*, p. 87.39–40. 65 Bieler, *Anicii Manlii Seuerini*, p. 7. 66 Florence MS, fo. 6v. 67 Bieler, *Anicii Manlii Seuerini*, pp 6, 36, 23 and 43, respectively. 68 An Irish word, meaning 'proper name', which should be added to those in my edition of the Irish glosses, 'Irish glosses in a twelfth-century copy of Boethius's *Consolatio Philosophiae*', *Ériu*, 55 (2005), 1–17. 69 Bieler, *Anicii*

si; uel imperatiuus (fo. 35v), which suggests for the sake of the meaning that either 'si' be inserted before *CEDAT* or, alternatively, that the verbal form be read as an imperative. Where necessary, the glossator could be creative in his grammatical explanations, as in *LACERAE* (bk I, m 1, 3):[70] *lacerae pro laceratae per sincopiam* (fo. 4r). Although the equating of *lacerae* with *laceratae* actually comes from Pseudo-John (8.23) or 'Remigius' (312, gl. 3), the justification for it was provided by the Irish glossator.

He also expresses his interest in matters rhetorical, sometimes drawing attention to tropes; for example, he identifies a certain use of *TACEO* (bk III, pr 3, 19) as a *color rhetoricus* (fo. 20v), he warns that *IMPATIENTIAE MANUS* (bk II, pr 4, 19) is *metaforice dicitur* (fo. 13v), or he labels *VOLUPTARIAM VITAM DEGAS?* (bk III, pr 8, 6)[71] as *hironice* (fo. 23r). He marks the passage beginning with *SED QUONIAM* (bk I, pr 5, 11)[72] as the *epilogum*, which he proceeds to define forensically as *nouissima pars orationis quae breuiter superiora conplectitur aut ad misericordiam flectens auditorem aut ad iram commouens iudicem* (fo. 9r) ('the final part of the speech which briefly covers the previous parts either inclining the listener to pity or stirring the judge to anger').

That said, the Irish glossator seems more interested in what he regards as the potential for abuse inherent in rhetoric. In bk I, pr 1, 7–11,[73] when Lady Philosophy first appears before Boethius she vents her anger that in his time of grief he has invited the Muses to inspire him: 'who allowed these theatrical tarts (*SCENICAS MERETRICULAS*) in with this sick man?' The commentators responded in kind: Pseudo-John (16.14–25) labels the Muses as deceivers who prostitute their art, comparing them to poets who seduce listeners with their secular songs; William (37.439–47) identifies them with the poets, whose only purpose in composing is to gain fame or wealth. But the Irish glossator directs his animus at a rather different object: *nam sicut merretriculae decipiunt suos sectatores, ita secularis scientia decipit suos sectatores; computant in ea aliquid boni esse* ('just as prostitutes deceive their followers so secular knowledge deceives its followers who think that there is something good in her') (fo. 4v). Although the glossator does not clarify what he means by the term *secularis scientia*, his comment on the Boethian passage immediately following suggests that he is thinking of those who pursue the liberal arts, especially rhetoric, for secular gain. In that passage, Lady Philosophy calls the Muses *SIRENES* ('Sirens'), which Pseudo-John interprets as follows: 'they attract men to love them by reading aloud songs of the poets' (16.19–20). In borrowing this idea, the Irish glossator enlarges on it to detail the techniques of seductive poetry, while pointing out the dire consequences of rhetoric gone wrong: *sic musae poeticae hominem attrahunt per melodias, per rithmos, per dilectiones, donec auertant eum a uera ratione et uia*

Manlii Seuerini, p. 78. 70 Ibid., p. 1. 71 Ibid., pp 42, 24 and 48, respectively. 72 Ibid., p. 13. 73 Ibid., pp 2–3.

ueritatis [...] ('so the Muses of poetry attract a man by means of melodies, rhythms, and words of endearment, until they turn him away from true reasoning and the way of truth') (fo. 4v).[74]

The glossator's ambiguous feelings about the liberal arts find fullest expression in his interpretation of two crucial passages of Boethius, one at the beginning, the other at the end of the 'Consolation'. In bk I, pr 1, 1–6,[75] Lady Philosophy has just appeared to the disconsolate Boethius and her clothing is described as fashioned with the most perfect design and made from indestructible material (*INDISSOLVBILI MATERIA*). For William (27.218–22) the account is not about the latest trends in feminine fashion; her clothing symbolically represents the *quadrivium* and other aspects of philosophy. The indestructible nature of Philosophy's clothing signifies that the ideas of the individual arts cannot be dissolved or contradicted, because although sometimes perversely expounded and interpreted, they cannot be rationally contradicted; indeed reason protests against and refutes those who so expound and interpret them. The Irish glossator offers a much more muted comment based on Pseudo-John (14.7–9): *quia artes per se sunt 'indissolubilis'; uel quamuis deficient qui eas captant, ipse tamen non pereunt; licet enim pereat scientia, scibile tamen semper erit* (fo. 4v) ('For the arts in themselves are indestructible; alternatively, although those who appropriate them may fall short, the arts themselves, however, shall not perish; even though knowledge should perish, nevertheless it shall always remain knowable (*scibile*)'). A ringing endorsement of the arts – or so it would appear.

Yet, the glossator's comments on the lines immediately before tell a different story. There Boethius had described Philosophy as sometimes appearing to have the stature of an ordinary person, but at other times so exalted physically that she penetrated the heavens themselves (bk I, pr 1, 2). For William this image becomes the paradigm of his own scholarly agenda: the search for an understanding of the creator. One begins that search with natural philosophy (physical things) – on the ground, so to speak – and by means of such grounded knowledge one gains an insight into the plan of the mind behind these creations and ultimately insight into the divine potential.[76] None of this appears in the Florence Boethius; our glossator has recourse to an interpretation that equates Lady Philosophy in her normal grounded stature with St Augustine (a pillar of patristic orthodoxy) discoursing on physical things, *AD COMMUNEM: quando Agustinus de uisibilibus*; and Lady Philosophy, head in the clouds, with Augustine

74 Note also the glossator's comment on *LACERAE* (bk I, m 1, 3) with reference to the Muses: *Lacerae .i. poeticae sententiae scribendi et fingendi quod lacerant cor hominis et inconstantem redunt* (fo. 4r), which is clearly aimed at secular poets ('"Wounded": that is, poetic sentiments in composing and imagining which wound the heart of man and make him restless'). That the condemnation was borrowed from William (11.45–49) scarcely matters. 75 Bieler, *Anicii Manlii Seuerini*, p. 2. 76 See Nauta, *Glosae*, p. 25.168–86.

writing of hidden mysteries (*QUAE CUM ALTIUS CAPUD EXTULISSET: .i. quando Agustinus de inuisibilibus*; fo. 4r). In the same spirit, the Irish glossator then addresses Boethius' observation that Philosophy in her exalted state eludes the sight of mankind with the following comment: *RESPICIENTIUMQUE HOMINUM FRUSTRABATUR (decipiebat) INTUITUM: .i. quia qui mel multum comedit non est bonum ei, sic qui ultra modum scrutator est diuinitatis opprimetur*[77] ('For he who eats too much honey will not feel good; in the same way, he who excessively searches out the divinity will feel oppressed'). The glossator is citing Proverbs 25:27, *sicut qui mel multum comedit non est ei bonum sic qui scrutator est maiestatis opprimitur gloria*, almost verbatim except that he qualifies its condemnation of *scrutator* with *ultra modum*. It is tempting to see his recourse to an Old Testament passage with its aphoristic wisdom (and homely simile) not only as old-fashioned (for the twelfth century) but also as a caustic comment on William's grandiose search for divine wisdom. So much for the potential of the liberal arts, it seems to say.

A second expression of the glossator's scepticism is prompted by bk V, m 5,[78] which contrasts animals, forever looking downwards at the earth, with man gazing upwards towards the heavens. Boethius begins by describing birds, 'the lightness of whose wings beats on the winds', a passage that the Irish glossator allegorizes as follows: *SUNT QUIBUS ALARUM LEUITAS VAGA VERBERETQUE VENTOS: .i. philosophi qui de creaturarum naturis perscrutati sunt pura mentis intentione* (fo. 45v); in other words, the birds of the lower sky represent '(natural) philosophers who examine with pure purpose of mind the nature of created things'. Likewise, birds who soar to the higher levels of the air he interprets thus: *SPATIA AETHERIS: .i. de cursu solis et aliarum stellarum et uolubilitatem caeli astrologi narrant* ('astrologers who tell the course of the sun and other stars and the vastness of the heavens'). Finally, earth-bound animals who delight in impressing their footsteps on the ground, are also identified allegorically, *HAEC PRESSISSE SOLO VESTIGIA GRESSIBUSQUE GAUDENT: .i. geometricii in mensura terrae* ('That is, geometricians who measure the earth'). None of these allegorical applications (as far as I know) occurs in the commentaries.

It all sounds quite complimentary to both the animals and the practitioners of the *quadrivium* (natural philosophers, astronomers and geometricians), their allegorical counterparts – that is, until we consider the second part of Boethius' *metrum*, which deals with a higher level of being, humankind. For Boethius, what distinguishes man from the animals is that he despises the earth *(DESPICITQUE TERRAS*), words that the Irish glossator allegorically takes to mean the rejection of all *terrenas scientias* (fo. 45v). Clearly the sciences that he has in mind are the physical sciences that, by virtue of their subject matter, are lowly and

77 Florence MS, fo. 4v. 78 Bieler, *Anicii Manlii Seuerini*, p. 100.

earthbound (*terrenas*); in contradistinction, presumably, to the exalted study of the Godhead ('theology' if we want to give it a name). Not surprisingly, this stark dichotomy between the lowly sciences (the *quadrivium*) and the higher search for God has no place in William's discussion. Instead, he interprets this *metrum* as a call to meditate on divine wisdom, and sees the animals as simply inferior beings incapable of ratiocination; but in no way does he connect them with natural philosophers. William had made clear earlier in his commentary that he regarded the study of the *quadrivium* not as a mundane pursuit unworthy of man's aspirations but as the first rung of the ladder leading to knowledge of the divinity.[79]

All of this raises the question: who was this Irishman who could selectively use William's commentary, sidestepping his scientific insights and denying his heuristic approach to the liberal arts? Almost certainly he was a person of religion, as suggested by instances where he introduces God, Christian martyrs and moral interpretations into his glossing of the 'Consolation', well beyond the occasional prompts found in 'Remigius' and Pseudo-John. For example, in bk II, pr 4, 23, Lady Philosophy says to Boethius 'Let me briefly show you the hinge on which the greatest happiness turns' (*OSTENDAM BREVITER TIBI SUMMAE CARDINEM FELICITATIS*).[80] Here, William has no comment, while Pseudo-John (85.3–5) simply paraphrases the words of the speaker, interpreting *CARDINEM* as *perfectionem*. But for the Irish glossator, *CARDINEM FELICITATIS* is the *fundamentum quod est Christus* (fo. 13v), probably an echo of Ephesians 2:20, where Christ is referred to as the corner-stone and the apostles as the foundation. Likewise, commenting on bk IV, m 7, 13–31,[81] which recounts the labours of Hercules, William conscientiously goes through the history of each labour and draws the conclusion from the hero's deeds that 'by bravely conquering all earthly challenges, one comes to the joys of heaven'.[82] The Irish glossator (evidently on his own) draws a different conclusion: *Hercules maximos labores passus est et hoc pro nullo premio .i. pro seculari gloria. Si isti pro sola gloria elaborauerunt, uos debetis multum laborare cum uestrum sit ineffabile premium* (fo. 39v) ('Hercules endured huge challenges and did so with no prize in mind except worldly glory. If such people toiled for mere fame, you ought to toil greatly, since yours will be an ineffable prize'). This comment with its clear Pauline ring is the voice, not of a commentator, but of a homilist exhorting his audience to virtuous living (*uos debetis*).

That he was probably a monk is hinted at by his uneasiness about the power of rhetoric and the influence of secular poetry. Indeed, other evidence suggests that he was a member of the Cistercians, one of the Continental orders newly introduced into Ireland in the twelfth century. The Cistercians had their own culture of learning, even when it came to the study of secular works and authors.

79 See above, p. 37 and n. 76. 80 Bieler, *Anicii Manlii Seuerini*, p. 24. 81 Ibid., pp 87–8. 82 Nauta, *Glosae*, pp 286.241–2 and 287.260–1.

Thus, of the two secular works most heavily studied in the French schools during the first half of the twelfth century, Plato's *Timaeus* and Boethius' 'Consolation', they were leery of the first[83] but enthusiastic about the second. A rich manuscript tradition of commentaries on Boethius from Cistercian centres of learning in the twelfth century testifies to that enthusiasm. It is that tradition that may shed some light on the origins of the Florence Boethius and its glossator. The evidence is mainly textual, beginning with Table 2.1 below, which contains three parallel versions of William's commentary on *ASTITISSE MIHI SUPRA VERTICEM VISA EST MULIER* (bk 1, pr 1, 1),[84] 'a woman seemed to stand above my head'.

The first column (see table 2.1) gives the critical text of William's commentary, the second a variant version from **R**, and the third (the Florence manuscript gloss) represents a somewhat condensed version of William. The passage is typical of William's digressive approach: a simple reference to Philosophy standing above Boethius' head prompts an excursus on natural philosophy that focuses on the mechanism of the brain, explaining how its different sections are related to corresponding mental faculties that together make a person truly wise.[85] But, in a striking departure from William's critical text, the Florence manuscript identifies four, instead of three, faculties that make for wisdom. The same variant occurs in the Troyes copy (**R**) of William's work, where its character as an interpolation is evident from its disagreement with the opening and closing statements (which mention three) as well as its awkward location suggestive of an afterthought.[86] However, these discrepancies have been smoothed out in the Irish glossator's text, which indicates that his work post-dates the state of text represented by **R**, a manuscript copied at Clairvaux, the headquarters of the Cistercians, in the late twelfth century.[87] That date, however, should not be treated as a *terminus post quem* for the Florence Boethius, since **R**'s version of William's commentary is heavily interpolated in a manner suggesting

83 See M.-D. Chenu, 'Platon à Cîteaux', *Archives d'histoire doctrinale et littéraire du moyen âge*, 21 (1954), 99–106. A striking illustration of this bias seems to be the Florence Boethius glossator's treatment of bk III, m 11, where Boethius refers to the Platonic account of the transmigration of souls, a theory inimical to Church doctrine. William (192.70–9) addressed the notion head on, defending Plato as speaking *per integumenta*. The Florence glossator totally omits any reference to William's comments (borrowing only from Pseudo-John), even though he had used William's commentary heavily for the previous section. 84 Bieler, *Anicii Manlii Seuerini*, p. 2. 85 On William's interest in this subject, see Nauta, *Glosae*, p. xlvii. The notion of a tripartite division of the brain (though not the one expounded by William) seems to have been well known in Ireland by the eighth century: see J. Corthals, 'Stimme, Atem und Dichtung: aus einem altirischen Lehrbuch für die Dichterschüler (*Uraicept na Mac Sésa*)' in H. Birkhan (ed.), *Kelten Einfälle an der Donau*, Akten des vierten Symposiums deutschsprachiger Keltologinnen und Keltolegen Linz/Donau, 17–21 Juli 2003 (Vienna, 2007), pp 127–47 esp. p. 132. I am grateful to Johan Corthals for drawing my attention to this parallel and for kindly supplying me with a copy of his article. 86 Nauta, *Glosae*, p. cxviii, suggests that it is based on a later (genuine) passage in William's commentary (282.136–41). 87 See Nauta, *Glosae*, pp xci–xcii.

William (Nauta's edition, 19.39–53)	**Troyes, MS lat. 1101 [R]**	**Florence, MS 78.19** (fo. 4r)
(1) *Tria sunt quae faciunt sapientem: ingenium, id est naturalis uis ad aliquid cito intelligendum; ratio, id est discretio intellectorum; memoria praeteritorum [...] et si ista sciat, sapiens est.*	(1) *Tria sunt quae faciunt sapientem: ingenium, id est naturalis uis ad aliquid intelligendum; ratio, id est discretio intellectorum; memoria praeteritorum [...] et si ista sciat, sapiens est.*	(1)*[...]* ***quatuor*** *sunt quae perfecte sapientem reddunt. Quorum tria in capite sedem habent, scilicet, ingenium, ratio, memoria;*
	(3) *Est enim* ***quartum*** *quod ornat, id est eloquentia uel retorica que docet ista ornare [...]*	(2) *cuius esse ingenii uero sedes est in sinciputio, rationis in capitis medio, memoriae in occiputio.*
(2) *Ista uero tria, quae perfecte faciunt sapientiam, in capite habent sedem. Est enim in prima parte capitis cellula cerebri in qua est uis intelligendi quae uocatur fantastica [...] In medio uero capitis est alia cellula in qua est uis discernendi [...] In occipite, id est posteriore parte capitis, est alia cellula in qua uis est retinendi memoriam.*[α]	(2) *Ista uero tria, quae perfecte faciunt sapientiam, in capite habent sedem. Est enim in prima parte capitis cellula cerebri in qua est uis intelligendi quae uocatur fantastica [...] In medio uero capitis est alia cellula in qua est uis discernendi [...] In occipite, id est posteriore parte capitis, est alia cellula que dicitur memoralis in qua uis est retinendi.*[β]	(3) *Quartum uero quod est eloquentia non in capite sedem habet [...]*[γ]
α) '(1) There are three things which produce a wise man: ingenuity, that is, an innate power of understanding something quickly; reasoning, the ability to discern meanings; memory, the ability to recollect past events [...] and if he knows those things, he is a wise man. (2) Those three, which comprehensively produce wisdom, are located in the head. In the front part of the head is located the cerebral compartment in which resides the power of comprehension which is called imagination; [...] in the centre of the head is a second compartment in which resides the power of discernment; [...] in the occiput, that is, the back part of the head, is another compartment in which resides the power of memory'. (Numbering of sections has been supplied for ease of comparison.)	β) Since parts (1) and (2) of **R** agree with the critical text (translated in the previous note), I give a translation only of part (3), as follows: 'there is a fourth quality which adorns, that is, eloquence or rhetorical skill which teaches how to adorn those (three qualities)'.	γ) '(1) There are four things which produce the perfectly wise man, of which three have their location in the brain, namely, ingenuity, reasoning, and memory. (2) The seat of ingenuity is in the frontal compartment; of reasoning in the middle of the head; and of memory in the occipital compartment. (3) The fourth, which is eloquence, is not located in the brain [...]'.

Table 2.1 William's commentary on *ASTITISSE MIHI SUPRA VERTICEM VISA EST MULIER* (parallel versions).

William (Nauta's edition, 39.482–8)	Heiligenkreuz MS 130 [H]	Florence, MS 78.19 (fo.4v)
Affectus animi sunt quatuor: gaudium, ira, spes, timor[...]. isti affectus reddunt hominem anxium et[...]uires animi et uirtutes corporis extrahunt.[α]	*Affectus animi sunt quatuor: gaudium,* ***dolor****, spes, timor[...]. Isti affectus reddunt hominem anxium[...]uirtutes animi et uires corpori extrahunt[...].*	*Quattuor sunt affectus hominis: gaudium, timor,* ***dolor****, spes, Quorum quodque uehimenter hominem purgit.*[β]
α) 'There are four emotions of the soul, joy, anger, hope, fear [...] These emotions make a person anxious and [...] they drain the soul's energy and the body's vigor'.		β) 'There are four human emotions, joy, fear, sorrow, hope, any of which causes human catharsis'.

Table 2.2 William's comment on bk I, pr 1, 9, *INFRUCTUOSIS AFFECTUUM SPINIS*

a process of accretion over a considerable period. Nor did the Irish glossator likely have access to a direct copy of **R** (or its exemplar); the textual history of his dependence on William's work is a bit more complicated than that. Several other glosses of the Florence Boethius contain readings that connect them to a textual tradition of William's commentary distinct from that of **R**. The most striking example (see table 2.2) occurs in William's comment on bk I, pr 1, 9,[88] *INFRUCTUOSIS AFFECTUUM SPINIS* ('with the sterile thorns of feelings').

H and the Florence manuscript agree in listing one of the four emotions as *dolor* ('sorrow') against the critical text (and **R**) which has *ira* ('anger'). Like **R**, **H** is a Cistercian manuscript and, although copied at a foundation of that order in Austria, is probably based on French exemplars.[89] Also significant about **H** is that its contents consist of a glossed copy of the 'Consolation' plus the three commentaries on that work that were used in the Florence Boethius. Evidently, the Irish glossator had direct access to a glossed copy (or copies) of the 'Consolation' that were produced at Cistercian centres. That and the religious predilections that he reveals in his glosses suggest that he himself was a Cistercian working in an Irish foundation of that order.

To sum up: the evidence for knowledge of Boethius' works in early medieval Ireland, although reliable, is totally inadequate for a survey of five centuries. It is, moreover, extremely lopsided, both in regards the works studied (only two or three, *DIA*, *DIM* and the 'Consolation') and the intellectual context in which they were studied. At one pole is the starting point of Boethian scholarship in Ireland in the seventh century and at the other its closing in the twelfth; other

88 Bieler, *Anicii Manlii Seuerini*, p. 2. 89 See Nauta, *Glosae*, pp lxxxviii–xci.

evidence scattered throughout the ninth century serves as a sort of mid-point on the chronological spectrum. This division, however, may reflect more than mere chronology: the early tradition seems to have focused on *De institutione arithmetica* within a context of computistics, perhaps hinting that the work first came to Ireland in the seventh century as part of the literature attending the Paschal controversy.

The possibility that *DIA* was studied in seventh-century Ireland as a subject of the *quadrivium* seems unlikely for two reasons. First, all the evidence for its use in Ireland at that time (and indeed until the ninth century) occurs in isolation from the other three branches of the mathematically based *quadrivium* which Boethius envisaged as a unity. Second, it appears that the curriculum of the Irish schools, in the broadest sense, was centred on the triad of grammar, computistics and scriptural studies. It is surely significant that the author of the seventh-century computistical tract *De ratione conputandi*, who cited Boethius three times, explains in an earlier passage of his work that the four subjects essential to the church are Scripture, history, computistics and grammar.[90] The insertion of history is artificial; a nod of respect to Augustine's fourfold division of learning in *De Genesi ad litteram*, which included that subject. The remaining three, the traditional triad of Irish ecclesiastical learning, match what is known of the scholarly output of the early Irish schools – but there is no mention of mathematics or any other subject of the *quadrivium*. Nor does the triad of grammar, rhetoric and logic collectively (which much later came to be called the *trivium*) figure in the extant evidence.[91]

As to when the study of *DIA* in Ireland began to be tied to that of *DIM* (as Boethius had envisaged) cannot be determined, though the use of both works by the St Gall Priscian glossators opens up that possibility for the ninth century. By the early twelfth century, *DIA* was being studied within the context of the Boethian *quadrivium*, as suggested by the Bodleian *Timaeus* musical chart and by glosses in the Florence Boethius that refer to the *artes* collectively or individually.[92] However, unlike the former work, the latter work reflects a reserved attitude towards the *trivium* and *quadrivium* (as revealed in its glosses) that is out of step with the contemporary twelfth-century optimism of scholars such as William of Conches.

90 See Walsh and Ó Cróinín, *Cummian's letter*, p. 117.8–11. 91 G. Knappe, 'On rhetoric and grammar in the *Hisperica Famina*', *Journal of Medieval Latin*, 4 (1994), 130–62, has argued from a study of the contents of the *Hisperica Famina*, a work produced in Ireland about the mid-seventh century, that while 'the faminators did not have first-hand knowledge of the rhetorical teaching of antiquity, [t]hey seem to have been familiar with some late antique rhetorical literature [and] also with the sort of handbooks that are typical of (grammatical and) rhetorical training in late antiquity'. In sum, they seem to have 'taught grammar and basic rhetorical studies' (161). Strikingly absent is any mention of logic, the third member of the triad and the one to which Boethius in his writings gave most prominence. 92 See, for example, the Florence gloss on *PARTICULAS* (bk I, pr 1, 5): *grammatica [...] et arithmetica* (fo. 4v), where the first represents the *trivium* and the second the *quadrivium*.

Seventh-century Ireland: the cradle of medieval science?

IMMO WARNTJES

INTRODUCTION

Since the early modern period, historians of science have declared the entire period termed medieval a dark age, at least as far as Western Europe was concerned. The fall of the Roman Empire resulted, in this reading, in a loss of scientific texts and knowledge, which was only recovered in the course of the fifteenth-century Renaissance. This view was fundamentally revised in the 1920s by the great Charles Homer Haskins, who drew attention to the so-called renaissance of the twelfth century.[1] Fundamental Greek and Arabic texts became available to the Latin West through the translator schools of Spain and Sicily, and this revived Western science. In fact, the process set in slightly earlier, from around the 950s, personified by Gerbert of Aurillac, the future Pope Sylvester II, who had studied Arabic science in Spain.[2] Still, even in this reading, the period *c.*500–1000 remained dark, a period of no continuity of antique science, certainly of no innovation worth the name. This was not only due to a loss of texts, but also to a Christian, therefore theological-dogmatic approach to science that lacked the rationality of antique and renaissance scholars.

Over the past few decades, however, doubts about this interpretation have been expressed, principally on the basis that the new and extremely complex texts available since *c.*1000 could only be understood and developed further by minds capable of that task.[3] In fact, if science of the period 500–1000 is analysed

1 Charles Homer Haskins, *The renaissance of the 12th century* (Cambridge, MA, 1927; repr. 1993); idem, *Studies in the history of mediaeval science* (New York, 1924). Shortly afterwards, George Sarton produced his seminal *Introduction to the history of science*, 2:1 (Baltimore, MD, 1931). 2 For the earliest Western contact with Arabic science, see especially David Juste, *Les* Alchandreana *primitifs: étude sur les plus anciens traités astrologiques latins d'origine arabe (Xe siècle)* (Leiden, 2007), as well as the collected essays in Charles Burnett, *Arabic into Latin in the Middle Ages: the translators and their intellectual and social context* (Farnham, 2009); idem, *Numerals and arithmetic in the Middle Ages* (Farnham, 2010). For Gerbert's scientific works, see the pioneering study by Nicolaus Bubnov, *Gerberti, postea Silvestri II papae, opera mathematica (972–1003)* (Berlin, 1899); illuminating are the essays in Flavio G. Nuvolone, *Gerberto d'Aurillac, da abate du Bobbio a papa dell'anno 1000* (Bobbio, 2001), for the influence of Gerbert's scientific opus, especially those by Marco Mostert and Menso Folkerts. 3 An excellent reappraisal of early medieval mathematics can be found in Wolfgang Hein, *Die Mathematik im Mittelalter: von Abakus bis Zahlenspiel* (Darmstadt, 2010), though the importance of computus is underrated in

in detail, it becomes apparent that there was continuity of antique science, there was innovation and there certainly was scientific rationality. In this process, seventh-century Ireland played a crucial part, and this essay will focus on the role it played for the continuity, innovation and scholarly analysis of science in the dark ages and how this helped in the adoption of Arabic science in the eleventh century.

CONTINUITY

Historians of early medieval mathematics and astronomy have often stressed the continuity of these sciences from late Antiquity to the Middle Ages. In geometry, for example, a very interesting pseudo-Boethian tract appears in ninth-century manuscripts, containing excerpts from the original Boethius, and through Boethius of Euclid (obviously under omission of all proofs), but also from the writings of the Roman *agrimensores*.[4] In astronomy, the ninth century also saw the collection of the relevant excerpt from late antique authors like Chalcidius, Macrobius and Martianus Capella.[5] Even originally Greek prognostic texts were transmitted, in Latin translation and variation, since the early ninth century.[6]

Still, these examples do not, strictly speaking, prove a continuation of antique science into early medieval Western Europe, even though they have often been taken as such. The surviving excerpts of Boethian (or rather Euclidian) geometry, of *astronomica*, as well as the first occurrence of Latin *prognostica*, all

this volume, reduced to a field of applied mathematics. 4 See especially the studies of Menso Folkerts: *'Boethius' Geometri II: Ein mathematisches Lehrbuch des Mittelalters* (Wiesbaden, 1970); 'The importance of the pseudo-Boethian *geometria* during the Middle Ages' in Michael Masi, *Boethius and the liberal arts: a collection of essays* (Bern, 1981), pp 187–209, repr. in Menso Folkerts, *Essays on early medieval mathematics: the Latin tradition* (Aldershot, 2003), article VII; 'Die Altercatio in der Geometrie I des Pseudo-Boethius. Ein Beitrag zur Geometrie im mittelalterlichen Quadrivium' in Gundolf Keil et al. (eds), *Fachprosa-Studien: Beiträge zur mittelalterlichen Wissenschafts- und Geistesgeschichte* (Berlin, 1982), pp 84–114, repr. in Folkerts, *Essays*, article VIII; and also Paul Tannery, 'Notes sur la Pseudo-Géométrie de Boèce', *Bibliotheca Mathematica*, 1 (1900), 39–50; L.D. Reynolds, 'Agrimensores' in idem, *Texts and transmission: a survey of Latin classics* (Oxford, 1983), pp 1–6. 5 For early medieval astronomy, see especially the studies of Bruce S. Eastwood: *Ordering the heavens: Roman astronomy and cosmology in the Carolingian Renaissance* (Leiden, 2007); idem, *The revival of planetary astronomy in Carolingian and post-Carolingian Europe* (Aldershot, 2002); and also Bruce Eastwood and Gerd Grasshoff, *Planetary diagrams for Roman astronomy in medieval Europe, ca. 800–1500* (Philadelphia, 2004); Stephen C. McCluskey, *Astronomies and cultures in early medieval Europe* (Cambridge, 1998). 6 Prognostics has recently received renewed interest among scholars of the English language: see especially Roy M. Liuzza, 'Anglo-Saxon prognostics in context: a survey and handlist of manuscripts', *Anglo-Saxon England*, 30 (2001), 181–230; idem, *Anglo-Saxon prognostics: an edition and translation of texts from London, British Library, MS Cotton Tiberius A.iii* (Cambridge, 2011); L. Sándor Chardonnens, *Anglo-Saxon prognostics, 900–1100: study and texts* (Leiden, 2007); these studies provide excellent access to older literature.

belong to the middle and late Carolingian age and are owed directly to what is called the Carolingian renaissance.[7] Thus, the phenomenon is more appropriately described as a revival rather than a continuation of antique science. This revival is not only due to the Carolingian zeal for collecting everything scientific available (especially from the rich Italian monasteries recently integrated into the empire), but also to the introduction of a rediscovered concept of dividing science into subjects. The *quadrivium* consisting of geometry, astronomy, arithmetic and music became the centre of discussion, particularly at the court of Charles the Bald and in the leading Carolingian centres of learning of the mid-ninth century like Fleury, Auxerre and Corbie. Incited by Martianus Capella's outline of the *artes liberales*, of which the *quadrivium* formed the scientific part, in his *De nuptiis Philologiae et Mercurii*, Carolingian intellectuals began to order written knowledge within these categories. Since these subjects have survived to the present day, the retrospective focus of modern scholarship has gratefully embraced the *artes liberales* and, as a consequence, overrated their importance.[8] If medieval science is considered in these categories only, then the Carolingian age saw its revival – the continuity, however, was broken for the two centuries between Cassidorus and Charlemagne.

The concept of the *artes liberales* as well as the modern division of science into strictly defined subjects has clouded the view of modern scholarship as to what really constituted early medieval science. With the fall of the Roman Empire, the secular institutions of education and learning lost their foundation. Learning shifted into the newly developing monasteries and, with this, became decidedly Christian in character. The principal objective of learning now was to receive an understanding of God's creation, which manifested itself in two ways – the Holy Scripture and the cosmos. In order to be able to study the scripture in the form of the Vulgate, first the language in which it was written had to be learned. Latin obviously was the native tongue of those monks inhabiting sixth-century Italian monasteries and it still was the learned language of the senatorial class that dominated the bishoprics in the Merovingian kingdoms. But in regions further away from the former centre of the antique world, Latin had to be learned from scratch, and therefore a study of Latin grammar was of primary importance. With this tool at hand, the Holy Scripture itself could be read and interpreted, a task that formed the second major subject of this period, exegesis. This left the devoted Christian with an analysis of the cosmos. As became

7 For the scholarship on the Carolingian renaissance, see the literature referenced in Immo Warntjes, 'Köln als naturwissenschaftliches Zentrum in der Karolingerzeit: Die frühmittelalterliche Kölner Schule und der Beginn der fränkischen Komputistik' in *Mittelalterliche Handschriften der Kölner Dombibliothek. Viertes Symposion, November 2010* (Köln, 2012), pp 41–96 at 42–6. 8 See the discussion in Immo Warntjes, 'Irische Komputistik zwischen Isidor von Sevilla und Beda Venerabilis: Ursprung, karolingische Rezeption und generelle Forschungsperspektiven', *Viator*, 42 (2011), 1–31 at 1–3, with a review of the literature.

immediately obvious from studying the first book of the Bible, Genesis, the most important of God's creations besides the earth were the sun, moon and stars on the fourth day of creation. These were not only a manifestation of God's power, but their course ordered human life and, more importantly, the liturgical year. Accordingly, an understanding of the course of the moon and of the Julian calendar was vital for establishing the date of the most important Christian feast, Easter, which in turn ruled all other festivities in the liturgical year. The science that developed around the calculation of Easter and time-reckoning more generally became known as computus. Especially from the sixth to the eighth centuries, but also beyond the Carolingian age right up to the reception of Arabic science, computus was first the only, then the principal science within Christian learning.

If, therefore, we are to search for continuity of late antique learning into the Middle Ages, this has to be done in the Christian monastic milieu and on the basis of computistics. The central documents of this science were Easter tables. The Easter reckoning that was to become universally accepted in the Middle Ages and that is followed by the Orthodox Church to the present day was introduced into the Latin West from Alexandrian sources by Dionysius Exiguus in 525.[9] His Easter table consisted of eight columns: 1) the *anni domini*, i.e. the consecutive count of years from the incarnation of Christ; 2) the indiction, i.e. the 15-year taxation cycle of the Romans; 3) the epacts, i.e. the lunar age of 22 March; 4) the *concurrentes*, i.e. the weekday of 24 March; 5) the *cyclus lunaris*, i.e. the 19-year lunar cycle of Byzantine ordering; 6) the Julian calendar dates of the Easter full moon; 7) the Julian calendar dates of Easter Sunday; 8) the lunar age on that date. The Dionysiac reckoning was based on a 532-year Easter cycle, of which Dionysius had listed only the first ninety-five years. In his prologue to the table proper, he provided a manual for calculating a subsequent period, and he also appended *argumenta* for calculating the data of each of these columns. It is highly unlikely that Dionysius constructed these *argumenta* on his own. More plausible is the suggestion that these were invented in fifth-century Alexandria, the centre of late antique scientific learning, and were then translated by Dionysius into Latin, in a version adapted to his present year, 525.[10] Dionysius

9 Dionysius' computistical writings are edited by Bruno Krusch, *Studien*, II, pp 59–86; the older edition by Wilhelm Jan, *Historia cycli Dionysiani cum argumentis paschalibus et aliis eo spectantibus* (Wittenberg, 1718), repr. in PL 67, 453–520 is preferable in many aspects. The most recent discussion of Dionysius' *computistica* is Alden A. Mosshammer, *The Easter computus and the origins of the Christian era* (Oxford, 2008), pp 57–106; for Dionysius' *Argumenta*, see especially Immo Warntjes, 'The *argumenta* of Dionysius Exiguus and their early recensions' in Immo Warntjes and Dáibhí Ó Cróinín, *Computus and its cultural context in the Latin West, AD300–1200, Proceedings of the First International Conference on the Science of Computus in Ireland and Europe, Galway, 14–16 July, 2006* (Turnhout, 2010), pp 40–111. 10 See Immo Warntjes, *The Munich Computus: text and translation. Irish computistics between Isidore of Seville and the Venerable Bede and its reception in Carolingian times* (Stuttgart, 2010), pp xxxix, n. 85, xlv, n. 103.

therefore made available to the Latin West a set of late antique calendrical formulae. These enjoyed an exceptional popularity in the Carolingian age, but they were not rediscovered at that time. Rather, the fact that these formulae explain the underlying algorithms by reference to the present year, the development of this computistical formulary can be traced through space and time from Dionysius Exiguus right into the Carolingian age. It is therefore an ideal indicator of the continuity of late antique science, rudimentary as it may be, into the early medieval period.

Dionysius' original compilation was adapted within Cassiodorus' circle at Vivarium to the year 562.[11] The algorithms received minor additions towards the end of the sixth century, which also saw the attachment to the original corpus of the most sophisticated of these formulae. In 625, further algorithms were invented and added to the extended corpus, and this process repeated itself in 663, 675, 689 and 695, before the Venerable Bede incorporated them in his minor work on time, *De temporibus* of 703, and then also in his major computistical textbook, *De temporum ratione* of 725.[12] After Bede, such computistical formulae mushroomed in Francia.

Which role, then, does Ireland play in this story of continuity? From Italy, the extended corpus of computistical formularies apparently first went to the Iberian Peninsula. The additions of 625 appear still to have been the product of Italian computists,[13] while the 663 adaptation can be found in an evidently Spanish computus.[14] With the 675 and 689 versions, we enter the Irish stage, though the texts in question were not compiled by Irish monks.[15] As far as can

11 This text is edited by Paul Lehmann, 'Cassiodorstudien' in idem, *Erforschung des Mittelalters: Ausgewählte Abhandlungen und Aufsätze*, 2 (Stuttgart, 1959), pp 38–108 at 41–55, which is a reprint of his article in *Philologus*, 71 (1912), 278–99. For this text, see also André van de Vyver, 'Cassiodore et son oeuvre', *Speculum*, 6 (1931), 244–92 at 289–91; Otto Neugebauer, 'On the Computus Paschalis of "Cassiodorus"', *Centaurus*, 25:3 (1982), 292–302; Warntjes, '*Argumenta*', pp 67–8. 12 Bede, *De temporibus*, 14, Charles W. Jones (ed.), *Bedae opera de temporibus* (Cambridge, MA, 1943), pp 301–2; Bede, *De ratione temporum*, 47, 49, 52, 54, 57, 58, Jones, *Bedae opera*, pp 266, 269, 273–4, 278. For the other computistical formularies, see the following notes: the formulary of 695 is transmitted in Munich, Bayerische Staatsbibliothek, Clm 14725, fos 12r–14v. 13 The algorithms of 625 are transmitted in the *Fragmentum Nanciacense* (binding fragment of Nancy, Bibliothèque Municipale, 317 (356)); they are transcribed in Warntjes, '*Argumenta*', pp 96–7, with facsimile and discussion ibidem, pp 111, 69–72. As Michael Gorman very kindly pointed out to me in private communication, the Nancy fragment was part of the same manuscript as Vatican, Biblioteca Apostolica, Vat. Lat. 5755, 3–6 and Milan, Biblioteca Ambrosiana, L. 22 sup., fos 146r–147v and this fragmented codex may be the only early medieval *computistica* in Irish script that has survived. 14 Paris, Bibliothèque Nationale, Lat. 609, fo. 81v (dating clause); cf. the brief discussion in Warntjes, 'Köln', pp 77–8, with a transcription of the relevant passage. Studies of this codex by Alden Mosshammer and me are in preparation. 15 The *Computus Digbaeanus* of 675 is transmitted in only one manuscript, Oxford, Bodleian Library, Digby 63, fos 72v–79r; it is edited under Dionysius' name in Krusch, *Studien*, II, pp 75–81; for a thorough discussion of this text, see Warntjes, '*Argumenta*'. The *Computus Cottonianus* of 689 also survives in only a single codex, London, British Library, Cotton Caligula A XV, fos 73r–80r; it is partially edited by Joan Goméz Pallarès, 'Hacia una

be established, the Irish used modes of calculation different from the Dionysiac *argumenta*.[16] But Ireland also had non-Irish centres of learning. Bede, in a famous and often-cited passage, relates that many Anglo-Saxons went to Ireland for the sake of learning.[17] Some of them appear to have set up their own monasteries in Ireland. The most prominent among these was Ecgberht, an Anglo-Saxon monk of noble descent who was one of the most learned men of his generation and a staunch supporter of the Dionysiac reckoning. His foundation of Rath Melsigi soon became a major centre of monastic learning, attracting many of Ecgberht's fellow countrymen.[18] Here, the formulary of 675 seems to have been composed.[19] The formulary of 689 also appears to have its origin in Ecgberht's monastery. For quite some time, Rath Melsigi was the centre of missionary activity to the Germanic *gentes* in the lower Rhine region. In 690, the last and most successful of these missionary expeditions set out from Rath Melsigi to Frisia under the leadership of Willibrord. He became friends with the Carolingian *maiores domus*, under whose protection he organized the mission from Utrecht, and who also granted him a place of retreat in Echternach in modern-day Luxembourg, where Willibrord founded a monastery. For his mission, a number of texts were essential. Among them, those concerning the Julian calendar and the calculation of Easter feature prominently. The computistical formulary of 689 was one of these texts intended to support Willibrord and his twelve companions in their missionary activity.[20] This text, then, marked the beginning of computistical studies in Francia.[21] More specifically, it introduced the genre of computistical formularies into the Frankish realm and therewith provided for the continuation of this tradition on the Continent. Thus, Ireland, as the place of composition of this central text, played a

nueva edición de los argumenta paschalia de Dionisio el Exiguo', *Hispania Sacra*, 49 (1994), 13–31, which is badly reprinted in idem, *Studia Chronologica: estudios sobre manuscritos latinos de cómputo* (Madrid, 1999), pp 93–109 at 101–9; for a comprehensive study of this text, see Immo Warntjes, 'The *Computus Cottonianus* of AD689: a computistical formulary written for Willibrord's Frisian mission' in Immo Warntjes and Dáibhí Ó Cróinín, *The Easter controversy of late antiquity and the early Middle Ages: its manuscripts, texts and tables. Proceedings of the Second International Conference on the Science of Computus in Ireland and Europe, Galway, 18–20 July, 2008* (Turnhout, 2011), pp 173–212. **16** Warntjes, *Munich Computus*, pp clx–clxi; idem, '*Argumenta*', p. 93. **17** Bede, *Historia ecclesiastica gentis Anglorum*, III, 27 (Charles Plummer (ed.), *Venerabilis Baedae opera historica* (2 vols, Oxford, 1896), i, p. 192); cf. also Aldhelm's *Epistulae ad Wihtfridum* and *ad Ehfridum* (Rudolf Ehwald (ed.), 'Aldhelmi opera omnia' in *MGH, Auct. Ant.* 15 (1919), pp 479–80, 486–94). For the Anglo-Saxons in Ireland, see Plummer, *Venerabilis Baedae opera historica*, ii, pp 196–7; Michael Richter, 'The young Willibrord' in Georges Kiesel and Jean Schroeder, *Willibrord: Apostel der Niederlande und Gründer der Abtei Echternach* (Luxembourg, 1989), pp 25–30 at 27–9; Vera Orschel, 'Mag nEó na Sacsan: an English colony in Ireland in the seventh and eighth centuries', *Peritia*, 15 (2001), 81–107; and the following note. **18** For Rath Melsigi, see esp. Dáibhí Ó Cróinín, 'Rath Melsigi, Willibrord and the earliest Echternach manuscripts', *Peritia*, 3 (1984), 17–49, repr. in idem, *EIHC*, pp 145–65. **19** Warntjes, '*Argumenta*', pp 92–5. **20** Warntjes, '*Computus Cottonianus*'. **21** Warntjes, 'Köln', pp 56–8.

prominent role both in the introduction of science and the transmission of late antique knowledge into the Carolingian world.

INNOVATION

Probably the most important Irish achievement towards the end of the seventh century was the invention of a new literary genre, the computistical textbook.[22] The material available to seventh-century Irish computists did not explain all features of the Dionysiac system for the computation of Easter, not even the writings of Dionysius himself. As outlined above, Dionysius was principally concerned with the data found in his Easter table. He did not go beyond that. In fact, he could not have provided more details even if he had wanted to, as he did not understand the system underlying his Easter table himself. Dionysius was, first and foremost, a canonist and translator from Greek into Latin, and in the later function he was commissioned by the papal curia to make the Alexandrian system of reckoning Easter, which could only be accessed through Greek sources, available to the Latin West. This is what he did in his prologue (known as *Epistola ad Petronium*), the Easter table proper, the attached *argumenta* for calculating its data, as well as his Latin translation of Proterius of Alexandria's letter dealing with the paschal controversy of 455.[23] This appears to have sufficed at first, but then Paschasinus' letter was found in the papal archive and was sent to Dionysius so that he may explain its content.[24] The Alexandrian system, which Dionysius introduced into the West (and which henceforth was known as the Dionysiac reckoning), was based on a 19-year lunar cycle. This was the theme of Paschasinus' letter, though the bishop of Lilybaeum in Sicily did not go into its technical details. Dionysius' task, therefore, was to provide exactly such a technical explanation.

The construction of a 19-year lunar cycle appears relatively simple to a modern scientist, but this does not necessarily apply to their early medieval counterparts. In early medieval thinking, a solar year consists of $365\frac{1}{4}$ days, a lunar year of $354\frac{1}{4}$ days. Accordingly, 19 solar years consist of $19\times365\frac{1}{4} = 6939\frac{3}{4}$ days, 19 lunar years, however, only count $19\times354\frac{1}{4} = 6730\frac{1}{4}$ days. This leaves a difference of 209 days. These are compensated for by the intercalation at appropriate places of 7 lunar months of 30 days (called embolisms) and the subtraction of one lunar day (called *saltus lunae*): $7\times30-1 = 209$.

22 See especially Warntjes, *Munich Computus*, pp li–lvi; for these textbooks on computus and their Irish context, cf. also Daniel Mc Carthy, 'The study and use of numbers in early Irish monasteries' in Charles Doherty et al. (eds), *Glendalough: City of God* (Dublin, 2011), pp 223–37 at 226, 234–5. 23 For the editions of Dionysius' *computistica*, see n. 9; the *Epistola Proterii* is edited in Bruno Krusch, *Studien*, I, pp 266–78. 24 The *Epistola Pascasini* is edited in Krusch, *Studien*, I, pp 245–50.

Being unable to do the essential calculations on his own, Dionysius went searching for texts that would provide him with a ready-made solution. He had learned from Paschasinus' letter that the Alexandrians divided the 19-year cycle into two periods, the first eight years (called *ogdoas*) and the final eleven years (called *hendecas*). As none of the texts available to him discussed the construction of the 19-year cycle, Dionysius must have felt some excitement and satisfaction when reading Quintus Julius Hilarianus' *Expositum de die paschae et mensis*, which explains the cyclic character of an 8-year lunar cycle.[25] Dionysius copied Hilarianus' passage as proof that the *ogdoas*,[26] the first eight years of the 19-year cycle, are cyclic, not realizing the difference between an 8-year lunar cycle and the division of the 19-year lunar cycle into two parts. By doing this, Dionysius ran into all sorts of problems. Having proven the cyclic character of the first eight years of the 19-year lunar cycle, he was left with the task to prove the same for the final eleven years, the *hendecas*. As 8 and 11 do not share a common divisor, the *hendecas* could be cyclic only if one year was cyclic, but one lunar year obviously did not have the same number of days as one solar year. Thus, Dionysius was left with an impossible task. He conceded his incompetence by simply stating that the *hendecas* was almost cyclic.[27]

Seventh-century Irish computists, therefore, not only had no source available to them to explain the construction of the 19-year cycle, which is *the* central feature of this reckoning, but the evidence they had was full of mistakes while, at the same time, considered to be authoritative. In a slow process, they managed to find an acceptable solution.[28] This was only one of the problems Irish computists faced when introducing the Dionysiac reckoning. Here is not the place to list all such features or even explain them. A few of these will suffice to get a rough idea.

As a lunar year was eleven days shorter than a solar year, extra lunar months of 30 days (embolisms) had to be intercalated at appropriate places in the 19-year cycle to make up for this difference. Dionysius had specified the years, but not the exact placement within the Julian calendar year.[29] Until Bede's became standard in the ninth century, various systems existed. The first, however, to have raised the question and to have proposed exact placements were seventh-century Irish computists.[30]

A feature more commonly known is the solar bissextile day. Today, an extra calendar day is added at the end of February. This, however, was not medieval practice. From the texts available to seventh-century Irish scholars, it was not quite clear where this extra calendar day had to be intercalated, or rather which

25 Quintus Julius Hilarianus' *Expositum de die paschae et mensis* is edited in *PL* 13, 1105–14.
26 Cf. fig. 3.1. 27 Dionysius, *Epistola ad Bonifatium et Bonum* (Krusch, *Studien*, II, p. 83).
28 See Warntjes, *Munich Computus*, pp cxxxiv–cxxxvi, cxliv–cxlv, cxcvi–cxcvii, 256–67.
29 Dionysius, *Epistola ad Bonifatium et Bonum* (Krusch, *Studien*, II, pp 82–6). 30 See Warntjes, *Munich Computus*, pp xlviii, n. 111, 268–9, with further literature.

Julian calendar date had to be counted twice (*bis*). The name itself, *bis sextus*, implies that a sixth day had to be reckoned twice. Macrobius suggested the sixth calends of March, while Isidore preferred the sixth nones of March. Irish computists were not convinced by either option. If the world was created on 21 March, as it was commonly believed, then time-reckoning began on that day. If time-reckoning began on that day, then the 4-year bissextile cycle also started on 21 March and ended immediately before it. As exactly four years are needed to accumulate a full bissextile day, it had therefore to be implemented on 21 March. All three options were seriously considered by Irish intellectuals of the time, with Macrobius' option of placing it on the sixth calends of March winning the day, not only in seventh-century Ireland, but subsequently in the entire Latin West throughout the Middle Ages.[31]

A final example is the 40-day period of the Lenten fast, called *quadragesima*. It was not quite clear how those forty days were to be reckoned. Did the fast end at the last supper or the resurrection? Were Sundays to be included or excluded in the count?[32] By the seventh century, the custom had been established to begin the *quadragesima* forty-six days before Easter Sunday, on Ash Wednesday. Accordingly, the fast was broken on Holy Saturday – the Sundays in the Lenten period were not counted as days of fasting. In the liturgical calendar, the date of the beginning of the fast obviously was of prime importance. Dionysius and all other authors in the seventh century, however, remain silent about this issue. The mathematical relation between the beginning of Lent and Easter Sunday had to be fixed, calendrical algorithms for establishing the right date and lunar age had to be invented. This was first done by Irish scholars.[33]

What Irish scholars of the seventh century achieved, therefore, was a comprehensive understanding of the Easter reckoning, which was to become the unanimously accepted system for the calculation of Easter, from the ninth century onwards, for the rest of the Middle Ages and in the Orthodox Church to the present day. Their knowledge they put into writing, compiling, for the first time, comprehensive textbooks on the reckoning of time. The most famous computistical textbook and, undoubtedly, the most influential, was composed by the Anglo-Saxon scholar Bede in 725.[34] He has often been credited with being the first to do so, but the three decades before the publication of his work saw the production of at least three such texts by Irish scholars. The oldest of these has only very recently been discovered in the Benedictine monastery of

31 Warntjes, *Munich Computus*, pp cxliii–cxliv, cxcviii, 130–1. 32 For early medieval approaches to this question, see Winfried Böhne, 'Beginn und Dauer der römischen Fastenzeit im sechsten Jahrhundert', *Zeitschrift für Kirchengeschichte*, 77 (1966), 224–37; Immo Warntjes, 'The Munich Computus and the 84 (14)-year Easter reckoning', *PRIA*, 107C (2007), 31–85 at 51–4. 33 See Warntjes, *Munich Computus*, pp 230–41, 330–1. 34 Bede, *De temporum ratione* (Jones, *Bedae opera*, pp 140–291, 329–91; Jones' later edition in CCSL, 123, pp 241–544 is inferior to his first, not least because it lacks Jones' brilliant commentary; note, however, that the final chapters of this work, Bede's *Chronica maiora*, is only included in this later CCSL edition).

Einsiedeln in the Swiss Alps. It probably dates to around 700.[35] The second text has been known since its introduction into modern scholarship by Bruno Krusch in 1880. Fortunately, it contains an unambiguous dating clause corresponding to 719. Since it survives in only one manuscript (formerly in the Irish foundation of St Emmeram in Regensburg, but since the secularization of the early nineteenth century in the Staatsbibliothek in Munich), it has become known as the Munich Computus.[36] A third Irish textbook on time-reckoning, the most sophisticated of the three and in its clarity and precision rivalling Bede, has been discovered and published by Dáibhí Ó Cróinín in the 1980s, who termed it *De ratione conputandi*.[37]

One of the interesting features of the earliest two textbooks, the *Computus Einsidlensis* and the Munich Computus, is that they contain a considerable number of Old Irish terms and phrases embedded in the otherwise Latin text, which reveals that these early eighth-century Irish scholars were the first to put the respective concepts into writing. There were no Latin models that Irish computists of this time could have copied or followed to express the definitions, concepts and algorithms they deemed important. Accordingly, they had the tendency to switch to the mother tongue of author and audience for didactic purposes whenever an explanation required absolute precision, as they presumably would have done in the classroom. A good example of this can be found in the *Computus Einsidlensis* (pl. 2).[38] In a, for its time, rather complex calculation,

35 Einsiedeln, Stiftsbibliothek, 321 (647), pp 82–125, which is now avaibale online: www.e-codices.unifr.ch/de/list/one/sbe/0321 (10 June 2013). There is no edition available at present, but a substantial part of this text is transcribed in the *apparatus fontium* of Warntjes, *Munich Computus*, pp 1–317. The discovery of this text is announced in Immo Warntjes, 'A newly discovered Irish computus: *Computus Einsidlensis*', *Peritia*, 19 (2005), 61–4. It has been studied since in Immo Warntjes and Jacopo Bisagni, 'The early Old Irish material in the newly discovered *Computus Einsidlensis* (ca. AD700)', *Ériu*, 58 (2008), 77–105; Warntjes, *Munich Computus*, pp cxxxiii–clii; L.A. Holford-Strevens, 'Paschal lunar calendars up to Bede', *Peritia*, 20 (2008), 165–208 at 201. These studies have demonstrated that the text, which lacks a dating clause, was composed in the period 689–719; the discovery of a prologue to the Victorian Easter table of 699 makes it plausible that the *Computus Einsidlensis* was not written before that year: see Immo Warntjes, 'A newly discovered prologue of AD699 to the Easter table of Victorius of Aquitaine', *Peritia*, 21 (2010), 255–84. 36 Munich, Bayerische Staatsbibliothek, Clm 14456, fos 8r–46r, which is now available online at: http://daten.digitale-sammlungen.de/~db/0004/bsb00046449/images/. The Munich Computus is edited by Warntjes, *Munich Computus*. The scholarship on this text is conveniently summarized in Warntjes, 'Irische Komputistik', 10–11. 37 Brussels, Bibliothèque Royale, 5413–22, fos 77v–107v; an abbreviated version of this text can be found in Vatican, Biblioteca Apostolica, Reg. lat. 1260, fos 87r–99v. From these two manuscripts, the text is edited by Dáibhí Ó Cróinín in Maura Walsh and Dáibhí Ó Cróinín (eds), *Cummian's letter 'De controversia paschali' and the 'De ratione conputandi'* (Toronto, 1988), pp 99–213. The first thirteen chapters are also transmitted in Paris, Bibliothèque Nationale, Nouvelle acquisition latin 456, fos 189v–190v (cf. Arno Borst, *Das Buch der Naturgeschichte: Plinius und seine Leser im Zeitalter des Pergaments* (Heidelberg, 1995), p. 97, n. 48), while others are copied separately elsewhere. For analyses of this text, see Dáibhí Ó Cróinín, 'A seventh-century Irish computus from the circle of Cummianus', *PRIA*, 82C (1982), 405–30, repr. in

the author introduces the fraction 5/12. Generally, the fractions commonly used were ½, ⅓ or ¼, and multiples thereof. More complicated fractions needed explanation. The calculation stretched over three pages, with the fraction 5/12 occurring half-way through. Thus, the author of this text had to make absolutely sure that this fraction was understood, otherwise his further argumentation could not have been followed. Therefore, he defines it precisely by the Irish phrase *máa triun laigu leut* ('more than a third, less than a half'), as 5/12 is the exact mean value of ⅓ and ½ (⅓ = 4/12 < 5/12 < 6/12 = ½) (pl. 2.1).[39]

Another example is the famous list of Old Irish numerals in the *Computus Einsidlensis* (pl. 3; 4, conjectural representation).[40] The purpose here for employing Old Irish is to clearly distinguish between the two parallel columns, the first representing a consecutive count of days, the other the Julian calendar dates of January. Accordingly, those two columns have to be read line by line as 'the first day (of January) is termed calends' (*oin Kl*), 'the second day (of January) is termed the fourth nones' (*ailiu IIII n*), etc. Had the first column been written in Latin also, a misreading could have occurred already in the first entry, as *primus Kl* could have been simply read as 'the first calends'.

By the time of composition of *De ratione conputandi*, however, the methods invented one or two decades earlier had become standard, and they had been refined and clarified, so that they could easily be expressed in Latin. The Irish element was gone; but so also were clumsy definitions, garbled concepts and diffuse algorithms. This is also reflected in the reception of these texts. Interest in the ideas expressed in the *Computus Einsidlensis* and the Munich Computus drew to an end in the first half of the ninth century, while some of the concepts of *De ratione conputandi* still enjoyed some popularity in the tenth.[41] The Irish computistical textbooks could not rival Bede's in popularity and influence. Still, it was seventh-century Irish computists rather than Bede who first outlined a

idem, *EIHC*, pp 99–130; Arno Borst, *Die karolingische Kalenderreform* (Hannover, 1998), pp 180–1; idem, *Schriften zur Komputistik im Frankenreich von 721 bis 818* (3 vols, Hannover, 2006), cross-references to *Scottus Rat.*, pp 1503–4; Holford-Strevens, 'Paschal lunar calendars', pp 198–201; Warntjes, *Munich Computus*, pp cxci–cci and cross-references to *De ratione conputandi* on pp 386–7; Mc Carthy, 'Study and use of numbers', pp 234–5. 38 For this passage, see Bisagni and Warntjes, 'Early Old Irish material', 102–4; Warntjes, *Munich Computus*, pp cxxxvi–cxxxviii. For the calculation: ibidem, pp 272–7 and cross-references there. 39 Note that the need to explain this fraction was felt by the author of this text because he and his audience were unfamiliar with the Roman fractions as outlined in the *Calculus* of Victorius of Aquitaine, where 5/12 is termed *quincunx* in the *praefatio*: see Alison M. Peden, *Abbo of Fleury and Ramsey: commentary on the Calculus of Victorius of Aquitaine* (Oxford, 2003), p. 3. 40 For this passage, see especially Bisagni and Warntjes, 'Early Old Irish material', 96–102. Before the recent uncovering of the Irish context of this text, Gallus Morel, 'Handschriften der Klosterbibliothek zu Einsiedeln', *Archiv der Gesellschaft für ältere deutsche Geschichtskunde*, 8 (1843), 736–49 at 743 believed these to be German numerals. 41 For the reception of the Munich Computus, see Warntjes, *Munich Computus*, pp clix–cci. The reception of the *Computus Einsidlensis* still needs to be studied. That *De ratione conputandi* enjoyed a considerably longer popularity is illustrated by the fact that some of its chapters were still copied in a codex from thorough understanding for time-reckoning and with this shaped, if not founded, the science that was to dominate the early Middle Ages before the introduction of Arabic texts.

Irish as well as later computists of the early Middle Ages did not invent new reckonings or reform old ones, as the Gregorian reform of 1582 would do. Their field of activity was confined by the prevailing system. Within these confines, however, they constantly revised old concepts, invented new algorithms, constructed ground-breaking tools, and acquired unparalleled skills. They were creative and innovative, not an ignorant bunch stagnant in ideas and only concerned with transmitting concepts of old without critical analysis, as renaissance scholars had portrayed them. It was this flexibility of mind and interest in scientific innovation that made the reception of Arabic science possible, as will be seen further below.

SCIENTIFIC RATIONALITY

As the early Middle Ages are generally perceived as stagnant, modern scholars have not considered the analysis of the scientific mentality of early medieval intellectuals a fruitful field of research. The general view in this question has, as so often, been shaped and clouded by a focus on Bede's scientific works. Faith Wallis, in her excellent translation of Bede's *De temporum ratione*, has drawn particular attention to the fact that for Bede science was a *doctrina christiana*.[42] Two aspects central to this *doctrina christiana* have, however, to be clearly distinguished, namely approach and argumentation. Surely, the approach to science was decidedly Christian throughout the early Middle Ages, whether by Bede or any other scholar of the time. This only changed with the introduction of Arabic science and, more forcefully, with the shift of learning from monastic schools to secular universities. This, however, does not mean that the argumentation was generally Christian, or rather theological-dogmatic. Bede's method of argumentation is that he silences any potential debate or criticism either by a Biblical quote or by references to Christian tradition or ecclesiastical authority. But Bede was only one of many scholars. Computistical literature of the period 500–1000 is abundant, collected in several thousands of manuscripts, and to reduce the arguments about early medieval scientific mentality to a single or a handful of authors does not do the period any justice.[43]

Auch of *c.*1000 (the Paris manuscript mentioned in n. 37); cf. also the ninth-century interest in *De ratione conputandi*'s discussion of the *lunae abortivae*, as outlined in Warntjes, *Munich Computus*, pp lxx–lxxi, clxv–clxvi, 248–9, 333–6; Warntjes, 'Irische Komputistik', 16–17. 42 Faith Wallis, *Bede: the reckoning of time, translated, with introduction, notes and commentary* (Liverpool, 1999), pp xviii–xxxiv. 43 This is the approach of Nadja Germann, *De temporum ratione: Quadrivium und Gotteserkenntnis am Beispiel Abbos von Fleury und Hermanns von Reichenau* (Leiden, 2006), esp. pp 29–176.

The Irish textbooks introduced in the previous section are a case in point. With the broad reception of Bede's writings in the Carolingian Empire from the early ninth century onwards, discussion about the correct method for the calculation of Easter ceased. From the sixth to the eighth century, however, three reckonings competed for acceptance. The first was invented by Sulpicius Severus *c.*410, called *latercus* by its author as well as most recent scholarship.[44] It was based on an 84-year luni-solar cycle and promoted Julian calendar limits for Easter Sunday of 26 March to 23 April and, more importantly, lunar limits of *luna* 14 to 20 on that date. Some fifty years later, in 457, the papal curia, at this time following yet another system, the *Supputatio Romana*, commissioned the Aquitanian scientist Victorius to establish a reckoning that would sort out the difference in observance between the Roman and the scientifically superior Alexandrian church.[45] Victorius created a 532-year Easter cycle with Julian calendar limits for Easter Sunday of 22 March to 24 April and lunar limits of *luna* 16 to 22. This, however, did not solve the problems Rome had with Alexandria. Victorius, by making some rather unconventional choices, even added insult to injury. A final attempt by the papal curia to bring their own Easter dates into line with the Alexandrians was initiated in 525 by commissioning Dionysius Exiguus to simply translate the Alexandrian system into Latin, as we have seen above.[46] The Dionysiac reckoning, as it became known, introduced Julian calendar limits for Easter Sunday of 22 March to 25 April and *luna* 15 to 21 on that date. Even though a 532-year Easter cycle also underlay this method, Dionysius only established a table of 95 years, which had to be extended in the third decade of the seventh century.

This is the situation Irish scientists faced in the seventh century, with three competing systems available. In the first half of the fifth century, the *latercus* was introduced into Ireland and remained the undisputed reckoning until the early seventh century.[47] With Columbanus' missionary activity on the Continent and

44 The *latercus* could only be positively reconstructed with Dáibhí Ó Cróinín's discovery of an Easter table of this reckoning (Padua, Biblioteca Antoniana, I 27, fos 76r–77v); for the discovery, see Daniel Mc Carthy and Dáibhí Ó Cróinín, 'The "lost" Irish 84-year Easter table rediscovered', *Peritia*, 6–7 (1987–8), 227–42, reprinted in Ó Cróinín, *EIHC*, pp 58–75; for the reconstruction Daniel Mc Carthy, 'Easter principles and a fifth-century lunar cycle used in the British Isles', *Journal for the History of Astronomy*, 24 (1993), 204–24. This reckoning can conveniently be accessed through Bonnie Blackburn and Leofranc Holford-Strevens, *The Oxford companion to the year* (Oxford, 1999), pp 870–5. Sulpicius' authorship is proven by Daniel Mc Carthy, 'The origin of the *latercus* paschal cycle of the Insular Celtic churches', *CMCS*, 28 (1994), 25–49; further arguments are provided by Warntjes, '84 (14)-year Easter reckoning', 34–7. For the technicalities of this reckoning see, most recently, Mc Carthy, 'The study and use of numbers', pp 228–33; Holford-Strevens, 'Paschal lunar calendars', 178–87. 45 Victorius' *computistica* are edited in Krusch, *Studien*, II, pp 4–52. Modern scholarship on this reckoning is conveniently listed in Warntjes, *Munich Computus*, p. xxxviii, n. 82; to these studies add Holford-Strevens, 'Paschal lunar calendars', 192–6. 46 See nn 9, 10. 47 For the introduction of the *latercus* into Ireland, see Daniel Mc Carthy, 'On the arrival of the *latercus* in Ireland' in Warntjes

the Roman mission to Anglo-Saxon England, the Irish felt some pressure to deal with the Victorian reckoning, which closed in both from the south and from the east. The southern Irish submitted to the pressure, the northern Irish clergy stuck to what they considered as the right method of old, the *latercus*. This led to a controversy between the two parts of Ireland for the rest of the century. In the end, the schism was solved when both the southern and the northern Irish churches accepted the third reckoning, the Dionysiac, in a slow process between *c.*680 and 720.[48]

The way in which these reckonings were handled and how their introductions came about reveals a great deal about the scientific mentality of the intellectuals involved in the process. Columbanus, in his letter to Pope Gregory written *c.*600, states that the *antiqui philosophi* and *sapientissimi computarii* of his homeland ridiculed rather than embraced the Victorian system.[49] This dismissal of Victorius' system was done despite it being unanimously accepted by the neighbouring Frankish clergy and the papal curia, which had commissioned it in the first place. Blind adoption of a system promoted by other authorities was not an option. The system had to convince by its details, not by the authority attached to it. This the Victorian table did not do, as Columbanus outlined in some detail: lunar ages 21 and 22 were unacceptable, as the moon on these dates was more covered than illuminated, darkness had won over light, which was hardly appropriate for celebration of the Lord's resurrection; the observance of the Easter full moon before the spring equinox was equally unacceptable.[50]

Even more revealing is one of the central documents of seventh-century Ireland – the Paschal letter of the southern Irish cleric Cummian to the northern Irish abbot of Iona, Ségéne, and Béccán the hermit, in 632. Cummian relates

and Ó Cróinín, *Easter controversy*, pp 48–75. For what follows in this paragraph, see Warntjes, *Munich Computus*, pp xxxvi–xli, with further cross-references. **48** The most recent treatments of the Insular paschal controversy are Thomas M. Charles-Edwards, *Early Christian Ireland* (Cambridge, 2000), pp 391–415; Caitlin Corning, *The Celtic and the Roman traditions: conflict and consensus in the early medieval church* (New York, 2006). Rather irritatingly, both studies do not consider the *computistica* of the period, which have been further increased by very recent discoveries; for these reasons, the period *c.*680–720, especially the last two decades of the seventh century, still awaits a thorough analysis taking the conflicts between adherents of Victorius and those of Dionysius seriously. Though outdated, the accounts by Bruno Krusch, 'Die Einführung des griechischen Paschalritus im Abendlande', *Neues Archiv der Gesellschaft für ältere deutsche Geschichtskunde*, 9 (1884), 99–169; Joseph Schmid, *Die Osterfestberechnung auf den britischen Inseln vom Anfang des vierten bis zum Ende des achten Jahrhunderts* (Regensburg, 1904) are still worth studying. **49** Columbanus, *Epistulae*, I, 4 (G.S.M. Walker (ed.), *Sancti Columbani opera* (Dublin, 1957; repr. 1997), p. 6): *Scias namque nostris magistris et Hibernicis antiquis philosophis et sapientissimis componendi calculi computariis Victorium non fuisse receptum, sed magis risu vel venia dignum quam auctoritate.* **50** Columbanus, *Epistulae*, I, 3 (Walker, *Sancti Columbani opera*, pp 2, 4); cf. also Columbanus, *Epistulae*, II, 5 (Walker, *Sancti Columbani opera*, p. 16). Columbanus presents one more argument against Victorius' reckoning, his listing of double dates: Columbanus, *Epistulae*, II, 7 (Walker, *Sancti Columbani opera*, p. 18): *Victorium nuper dubie scribentem et, ubi necesse erat, nihil definientem.*

that a delegation had been sent to Rome to inquire about the correct date for celebrating the most important Christian feast. In Rome they became aware that the whole world celebrated Easter on the same date, except for the Irish.[51] On the basis of this passage it may be argued that there was no scientific method involved in the introduction of the Victorian reckoning into Ireland. The southern Irish simply bowed to Roman authority. This, however, is only half the story. Cummian also mentions that he had studied all texts available to him dealing with the problem for a full year and the delegation to Rome had been sent, first and foremost, to gather more information.[52]

Certainly the three Irish textbooks are witnesses to the Irish scientific rationality in the seventh century. In not a single instance is a problem solved, or an argument silenced, by reference to a biblical, patristic or contemporary ecclesiastical authority. The argument is always based on all evidence available, and the solution greatly expands from this transmitted knowledge. This is especially true for the comparisons of the prevailing Easter reckonings, especially the Victorian and the Dionysiac, to which these textbooks and other treatises from the same intellectual milieu bear witness. Epacts, embolisms, *saltus*, Easter limits, initial years, Lenten and resurrection data, dating systems and other features are not only compared in more detail than ever before, in almost all instances such comparisons are actually made for the first time. Seventh-century Irish scholars tried to get a handle on the known reckonings by comparing the meagre information of one with the more detailed data of the other. Also, and more importantly, by comparing these systems, they checked the mathematical and biblical accuracy. Accordingly, the scientific method applied by these seventh-century Irish scholars – comparison of all available information – appears very modern and certainly owes its approach to scientific rationality.

One example may suffice to illustrate that the comparison of systems was not a trivial operation. All three reckonings employed different dating systems. In the *latercus*, a year was dated by the lunar age and weekday on 1 January, and it was this dating system that was first introduced into the Irish annals.[53] The same

51 Cummian, *De controversia paschali*, lines 259–88 (Walsh and Ó Cróinín, *Cummian's letter*, pp 90, 92, 94). For a detailed discussion of Cummian's letter, see Maura Walsh, 'Cummian's letter: science and heresy in seventh-century Ireland' elsewhere in this volume. 52 Cummian, *De controversia paschali*, lines 14–17 (Walsh and Ó Cróinín, *Cummian's letter*, p. 58): *Deinde apostolorum interrogans dicentem: Omnia probabte, quod bonum est tenete, antequam gustarem non fastidiui. Hinc per annum secretus sanctarium Dei ingressus, hoc est scripturam sanctam, ut u<a>lui inuolui, deinde historias, postremo cyclos quos inuenire potui.* 53 The weekday and lunar data of 1 January constitute the first two columns in the *latercus*: cf. the facsimile in Warntjes, '84 (14)-year Easter reckoning', 80–2 as well as Mc Carthy, 'Easter principles', 218–19. For this being the earliest dating system of the Irish annals, see especially Daniel Mc Carthy, 'The chronology of the Irish annals', *PRIA*, 98C (1998), 203–55, and now Daniel Mc Carthy, *The Irish annals: their genesis, evolution and history* (Dublin, 2008; repr. 2010), pp 8–10, 111–52 (with the more debatable argument that kalend + ferial data was first introduced into chronicling by Rufinus in a now lost work).

system was employed by the Victorian reckoning.[54] Still, the lunar data differed considerably, so that an alignment of the two tables was extremely problematic. The Dionysiac table also provided lunar and weekday data, though for different dates, the lunar age for 22 March, the weekday data for 24 March, and this was not even explicitly stated.[55] In order to compare the Dionysiac system with the earlier two, a seventh-century computist first had to translate the Dionysiac data to 1 January. And even when this was accomplished, again the lunar data differed from the other two.[56] The easiest way of aligning two tables was by comparing their linear dating systems and working out the relation between them. This, however, sounds simpler than it was. The *latercus* had no linear dating system. Victorius, for his part, had employed years from the passion (*anni passionis* = AP), Dionysius years from the incarnation of Christ (*anni domini* = AD).[57] How these two timelines were to be aligned was not clear. From the Gospels it was known that Christ was crucified either in the 30th or in the 33rd year of his life. Neither the 30th nor the 33rd year of the Dionysiac table could, however, in any way be reconciled with Victorius' first year. A solution could only be found by comparing the lunar and weekday data. When the two tables were finally aligned by this painstaking method, this revealed that the relation between AP and AD was, in fact, AP = AD-27.[58]

The comparisons especially of the Victorian and the Dionysiac reckonings continued right into the Carolingian era. In the tenth century, received wisdom was fundamentally challenged on chronological grounds, no doubt owing to the problems revealed by seventh-century Irish scholars and then popularized by Bede. Faced by the question of why Dionysius did not provide data for the crucifixion and resurrection of Christ in accordance with either the Synoptic Gospels or Johanine tradition, Bede simply answered with sarcasm and the authority of

54 See the fourth and fifth columns of the Victorian Easter table (Krusch, *Studien*, II, pp 27–52). 55 See the third and fourth columns of the Dionysiac Easter table (Krusch, *Studien*, II, pp 69–74). 56 The problems seventh- and eighth-century Irish computists had in comparing the lunar data of these three tables are well documented in the *Computus Einsidlensis* (Einsiedeln, Stiftsbibliothek, 321 (647), pp 109–11), the tract *De comparatione epactarum Dionysii et Victorii* (Warntjes, *Munich Computus*, pp 322–6), and esp. the Munich Computus, chs 49, 50, 52 (Warntjes, *Munich Computus*, pp 168–87, 194–203). Facsimiles of the three tables that were produced in this context are printed in Warntjes, *Munich Computus*, pp 319–21, 328; the two best of these, from the *Computus Einsidlensis* and *De comparatione epactarum Dionysii*, are reproduced here in figs 3.2, 3.3. 57 AP and AD constitute the first column in the Victorian and Dionysiac Easter table respectively (Krusch, *Studien*, II, pp 27–52, 69–74). 58 The earliest document to prove that an alignment of AP with AD had been achieved, the newly discovered prologue of 699 to the Victorian Easter table, is in all likelihood of Irish origin: cf. Warntjes, 'Prologue of AD699', 271–4 (ed. and trans. of the text, in which the alignment of AP, AD and AM – *anni mundi* – is omnipresent), 263–8 (Irish origin), 281–2 (context). It is an outstanding witness for the synchronization of chronological systems in the seventh century. That this was not a trivial operation is demonstrated by the only known manuscript that transmits the Victorian and Dionysiac Easter tables in parallel (Leiden, Universiteitsbibliotheek, MS SCA 28, fos 2r–11r), as the first years of the Dionysiac table are noted in the wrong place; cf. fig. 3.4.

tradition.[59] This, however, did not silence the critics. The debate continued, though often below the tangible surface, and reached its height when Abbo of Fleury toward the end of the tenth century revolutionized Dionysiac chronology.[60] Again, this was done by comparison of all the chronological data available, biblical and other, and very much resembles not only the reformation of the world era (*anni mundi*) that was popularized (but not invented) by Bede in the eighth century,[61] but also the approach of the great Joseph Scaliger, lauded as one of the greatest Renaissance scientists. We do not know enough about computistics in the intervening period, the ninth and tenth centuries. What is known, however, like the transformation of this science from a text-based to a highly visual one and especially the reception of the first glimpses of Eastern science in the West, suggests that scientific rationality was constantly applied throughout the early Middle Ages. There certainly have been numerous intellectuals like Bede preferring a theological-dogmatic approach. The scholars applying more scientifically verifiable methods (and by this scientific rationality should be defined), on the other hand, appear to have existed at any given time in the early Middle Ages. The difference between the early and the late Middle Ages lay not in scientific mentality, but simply in the texts available.

PRECONDITIONS FOR THE ADOPTION OF ARABIC SCIENCE

In the previous two sections, it has been argued that early medieval scientists kept an agile and innovative mind throughout this period and that their methods were not less scientific than those of later centuries. Still, it is obviously not easy to draw a direct connection between the themes discussed in the early and the later period, and to introduce seventh-century Irish science into the equation. The biggest difference between pre-Arabic science and that influenced by the Arabic world was not simply that better texts became available, leading to an increase in knowledge unimaginable only a few decades earlier. Rather the intro-

59 Bede, *De temporum ratione*, 47, 61 (Jones, *Bedae opera*, pp 265–8, 281–3). **60** Abbo set the tone for numerous attempts between the end of the tenth and the mid-twelfth century to re-calculate the base year of the incarnation era; for these, see now Peter Verbist, *Duelling with the past: medieval authors and the problem of the Christian era (c.990–1135)* (Turnhout, 2010); C. Philipp E. Nothaft, *Dating the passion: the life of Jesus and the emergence of scientific chronology (200–1600)* (Leiden, 2012), pp 103–12. **61** Cf. esp. Daniel Mc Carthy, 'Bede's primary source for the Vulgate chronology in his chronicles in *De temporibus* and *De temporum ratione*' in Warntjes and Ó Cróinín, *Computus and its cultural context*, pp 159–89; Mc Carthy, *Irish annals*, pp 118–52. For the reception of this recalculated AM era in the Carolingian period, see Richard Landes, 'Lest the millennium be fulfilled: apocalyptic expectations and the pattern of Western chronography, 100–800CE' in Werner Verbeke et al. (eds), *The use and abuse of eschatology in the Middle Ages* (Leuven, 1988), pp 137–211 at 174–203; James Palmer, 'Calculating time and the end of time in the Carolingian world, *c*.740–820', *English Historical Review*, 126 (2011), 1307–31.

duction of scientific instruments, especially the astrolabe, revolutionized the field. Science in the early Middle Ages was principally and almost exclusively based on mathematical models. After the introduction of Arabic science, astronomical observations (or at least astronomical tables) won the day. Thus, there was a major shift from non-observational to observational science, from mathematical structures and algorithms to empirical analysis.

This can well be illustrated by one specific problem, the length of a lunar month, or, in modern terminology based on astronomical observations, the moon phase. And in this instance, the story certainly begins in seventh-century Ireland. A synodic lunar month consists of 29.53059 days or 29 days, 12 hours, 44 minutes and 3 seconds. In order to press these into a simple mathematical model without fractions, lunar months were, since antiquity, reckoned as alternating between 30 and 29 days, called full and hollow. This created the perception among early medieval computists that an average lunar month consisted of 29½ days. Still, the mathematical model of the 19-year lunar cycle did not simply consist of lunar months alternating between full and hollow lunations of 30 and 29 days respectively, as we have already seen. First, so-called embolismic lunar months of 30 days length were introduced at appropriate places within the cycle, leading to two consecutive 30-day lunar months. Also, a lunar bissextile day was introduced in hollow February every four years, leading to three consecutive 30-day lunar months. That early medieval computists were well aware of these phenomena is demonstrated by an interesting text of 816, revealingly composed in the Irish foundation of St Emmeram in Regensburg.[62] In that year, as the astonished author relates, full January was followed by bissextile February, then full March, followed by an embolism, a total of four consecutive 30-day lunar months in a row. Such phenomena certainly created a general awareness that the average of 29½ days for a lunar month was simply an approximation. This awareness was further increased by an algorithm invented in seventh-century Ireland. The seven embolismic lunar months of the 19-year cycle all consisted of 30 days – that is they were 12 hours longer than the average. These in turn accumulated to an error of approximately one lunar day, which had to be subtracted at the end of the 19-year lunar cycle.[63] In medieval terminology, one lunar day was leaped over, so that the subtracting of this day was called *saltus lunae*, 'leap of the moon'. Irish computists were struck by this feature and tried to explain it in various ways. One way of doing this was to define it, quite appropriately, as an error accumulating linearly throughout the 235 lunar months of a 19-year cycle. In order to assign a numerical value to this monthly increment of the error, the 24 hours of the *saltus lunae* had to be divided by the 235 lunar months of a 19-year lunar cycle. The result was $^{24}/_{235}$ =

62 This text is edited under the abbreviated title *Arg. Aquens.* in Borst, *Schriften*, pp 1350–66.
63 Cf. p. 50, above.

0.102127659557447 hours, or four *momenta*, 1/12 of a *momentum*, and 1/47 of 1/12 of a *momentum* in the terminology of seventh-century Irish scientists, with 40 *momenta* equalling an hour.[64] The importance for the discussion here is that this calculation explicitly assigned an average excess to each lunar month – and it is the only early medieval calculation to do so – creating awareness that a lunar month could not simply consist of 29½ days, but that its length was more complex and had to be painstakingly computed. Bede, then, in a rather vague statement in his discussion of the *saltus lunae*, argues that the calculated lunar month often appears longer than the natural moon phase.[65] With this Bede also drew attention, at least indirectly, to the fact that a better mathematical approximation of the lunar month is needed.

Both Bede's *De temporum ratione*, as well as the Irish calculation of the monthly increase of the *saltus*, remained extremely popular throughout the early Middle Ages. The matter was eventually taken up in a tract entitled *De saltu lunae*.[66] In the manuscripts transmitting this text, it is ascribed to Columbanus, but the content renders his authorship impossible.[67] Unfortunately, the text itself does not provide a dating clause, so that only the transmission and reception of this work can shed some light on the time of its composition. The earliest codices containing *De saltu lunae* were composed towards the end of the ninth century,[68] but it is cited

64 For this algorithm and its transmission in seventh- and eighth-century Irish and later Continental texts, see Warntjes, *Munich Computus*, pp 272–7, with cross-references. 65 Bede, *De temporum ratione*, 42 (Jones, *Bedae opera*, p. 255). Likewise, in ch. 11 of *De temporum ratione*, Bede (Jones, *Bedae opera*, p. 204) argues that the length of a lunar month is 29½ days, save for the calculation of the *saltus lunae*, therewith implying that a lunation is, in fact, shorter rather than longer than this average value. 66 Pseudo-Columbanus' *De saltu lunae* was first edited by Gabriel Meyer, 'Die sieben freien Künste im Mittelalter, Teil 2', *Jahresbericht über die Lehr- und Erziehungsanstalt des Benediktiner-Stiftes Maria-Einsiedeln der Studienjahre 1886/87* (Einsiedeln, 1887), pp 3–36 at p. 30, from only one manuscript, St Gall, Stiftsbibliothek, 250, pp 112–14. Walker, *Sancti Columbani opera*, pp lxiii, 212–15 consulted only one further manuscript for his more recent edition. A critical edition based on all eight manuscripts and including a comprehensive source analysis is a pressing *desideratum*. 67 Columbanus had always been a staunch adherent of the *latercus*, which was based on an 84-year luni-solar cycle, while the author of *De saltu lunae* favours the 19-year lunar cycle as employed by Victorius and Dionysius. Cf. Bruno Krusch, *Ionae Vitae sanctorum Columbani, Vedastis, Iohannis* (Hannover, 1905), p. 32; Jones, *Bedae opera*, p. 376; and especially Dáibhí Ó Cróinín, 'The computistical works of Columbanus' in Michael Lapidge (ed.), *Columbanus: studies on the Latin writings* (Woodbridge, 1997), pp 264–70, repr. in Ó Cróinín, *EIHC*, pp 48–55 at pp 52–3. 68 Columbanus' *De saltu lunae* is transmitted in the following manuscripts: St Gall, Stiftsbibliothek, 250, pp 112–14 (St Gall, *c*.889); St Gall, Stiftsbibliothek, 459, pp 125–6 (St Gall, late ninth century); Zürich, Zentralbibliothek, C 62, fo. 217r–v (St Gall?, tenth century); Zürich, Zentralbibliothek, Car. C 176, fos 174v–175v (St Gall, early eleventh century); Munich, Bayerische Staatsbibliothek, Clm 14569, fos 26r–28r (probably St Gall, then St Emmeram in Regensburg, probably second third of eleventh century); Malibu, Getty Museum, Ludwig XII.5, fo. 108v (Reichenau?, then England, early twelfth century); Munich, Bayerische Staatsbibliothek, Clm 10270, fos 12v–13r (Upper Rhine, first quarter of twelfth century); Munich, Bayerische Staatsbibliothek, Clm 22307, fo. 191r–v (Tegernsee?, middle of twelfth century); Karlsruhe, Landesbibliothek, Ac 132 (destroyed in WWII).

even earlier, in a tract entitled *De lunae cursu* in the Bobbio Computus of 827, as Arno Borst has established.[69] Interestingly, none of the three major Carolingian computistical encyclopediae of 793, 809 and 818 cites either *De lunae cursu* or the original *De saltu lunae*. Still, the results of Pseudo-Columbanus' calculation found their way, both directly and indirectly through *De lunae cursu*, into the curious Cologne Computus of 805.[70] Accordingly, *De lunae cursu* and more importantly Pseudo-Columbanus' *De saltu lunae* must have been composed before this date. Because of the rather limited reception of the Pseudo-Columban treatise, it seems unlikely that *De saltu lunae* was compiled long before this time, so a date of composition of *c*.800 seems most plausible. Just who the author was, must remain speculative. The transmission and eleventh-century reception of this text shows a strong concentration in the Lake Constance region, especially around St Gall.[71] This, like its early reception in a Cologne text and a Bobbio codex, certainly reveals an Irish connection. The content itself is highly original, even provocative for the traditionalists. The first two decades of the ninth century saw a considerable number of original minds, the foremost among computists being the Irishman Dicuil.[72] Still, the authorship of *De saltu lunae* cannot easily be ascribed to Dicuil or his contemporary Dungal, but pinning the authorship of this treatise down to an Irishman working in the Carolingian Empire around 800 does not stretch the limits of hypothesis too far.

Pseudo-Columbanus presents the best mathematical approximation of the length of a lunar month to his day. He argues that the average lunar month is $29\frac{1}{2}$ days long. Within the 19-year lunar cycle, however, this average is frequently exceeded, namely by the seven embolisms, which are always 12 hours longer than the average, and by the extra $4\frac{3}{4}$ lunar bissextile days. These add up to 7×12 hours for the embolisms plus $4\frac{3}{4} \times 24$ hours for the lunar bissextile days, a total of 198 hours. Of these, the 24 hours of a *saltus lunae* have to be subtracted, which leaves 174 hours. The task, therefore, was to distribute these 174 hours equally among the 235 lunations of the 19-year lunar cycle. In truly early medieval mathematical fashion, Pseudo-Columbanus first converted 174 hours into 348 half hours, of which he assigns one to each of the 235 lunations with a

69 Bobbio Computus 47 (*PL* 129, col. 1298; the passage in question can also be found in Paris, Bibliothèque Nationale, Lat. 6400B, fo. 291v). This quote is first noted by Arno Borst in his still unpublished edition of Notker of St Gall's *De quatuor quaestionibus compoti* (cf. n. 73, below). 70 *Comp. Col.* III 7B, IIII 4 (Borst, *Schriften*, pp 906, 913–15). For the details, see figs 3.5, 3.6. 71 For the manuscript transmission of this tract, see n. 68, above; for the eleventh-century reception, see below. 72 Dicuil's *Liber de astronomia* is transmitted in only one manuscript, Valenciennes, Bibliothèque Municipale, 404 (386), fos 66r–118r, which is available online at http://gallica.bnf.fr/ark:/12148/btv1b8452582n/; the text is edited by Mario Esposito, 'An unpublished astronomical treatise by the Irish monk Dicuil', *PRIA*, 26C (1907), 378–446, repr. in idem, Michael Lapidge (ed.), *Irish books and learning in medieval Europe* (Aldershot, 1990), article VII. A new edition by Werner Bergmann is in preparation. For this text, see the often problematic analyses by Alfred Cordoliani, 'Le comput de Dicuil', *Cahiers de civilisation médiévale*, 4 (1960), 325–37; Werner Bergmann, 'Dicuils Osterfestalgorithmus im *Liber de astronomia*'

remainder of 113 half hours. If those are converted into quarter hours, the result is 226 quarter hours, which almost equals the number of lunar months in a 19-year lunar cycle (235). As a quarter-hour equals 10 *momenta* in early medieval terminology, Pseudo-Columbanus concludes that a lunar month consisted of 29½ days, half an hour, and almost 10 *momenta*.

As the manuscript context has already revealed, the reception of Pseudo-Columbanus' work did not considerably stretch beyond the Lake Constance, upper Rhine region, and this is fully confirmed by the few authors quoting *De saltu lunae*. First, Notker of St Gall refers to this text in his own treatise on the 19-year cycle, written early in the eleventh century.[73] Through Notker's writing, Hermann the Lame, around 1040, got acquainted with Pseudo-Columbanus' treatise.[74] He was not impressed, as Pseudo-Columbanus had refrained from finalizing the calculation by use of smaller fractions. Hermann continued where Pseudo-Columbanus had given up, calculating the lunar month as consisting of 29½ days, half an hour, 3 *puncta* and 33 *particulae*, with Hermann's smaller units of time differing considerably from those of Pseudo-Columbanus. But Hermann did not stop there. He used his result to establish a mathematical model to predict solar and lunar eclipses.[75] In this he failed. But by applying traditional *computistica*, by developing centuries-old ideas further using

in Warntjes and Ó Cróinín, *Easter controversy*, pp 242–87. 73 Notker's Computus was first edited by Meyer, 'Die sieben freien Künste', pp 31–4 (the relevant passage at 34, col. 1), and then by Petrus Tax in James A. King and Petrus W. Tax, *Die Werke Notkers des Deutschen*, 7 (Tübingen, 1996), pp cv–cxvi, 315–28 (the relevant passage at 327), 7A (Tübingen, 2003), pp 163–71. The discovery of further manuscript witnesses since then (see especially Norbert Kruse, 'Eine neue Schrift Notkers des Deutschen: Der althochdeutsche Computus', *Sprachwissenschaft*, 28 (2003), 123–55) makes a new edition mandatory, which is in preparation; it will be published in 2014 posthumously under Arno Borst's name in the *MGH*, *Quellen zur Geistesgeschichte des Mittelalters*. 74 Hermann's *Epistola ad Herrandum* survives in only one manuscript: Paris, Bibliothèque Nationale, Nouvelle acquisition latine 229, fos 17r–19r; it was first edited by Meyer, 'Die sieben freien Künste', pp 34–6 at p. 35, col. 1, and then by Arno Borst, 'Ein Forschungsbericht Hermanns des Lahmen', *Deutsches Archiv für Erforschung des Mittelalters*, 40 (1984), 379–477, text of letter at 474–7; reference to Pseudo-Columbanus at 475: *Refert tamen Columbanum dixisse saepe dictum mensem supra praedictam quantitatem semissem horae et pene X momenta recipere. Pene dixit ille et non plene. Cumque illud cur ambiguum poneret diligenter requisissem, deprehendi tales CCXXXV menses V̄IDCCCCXXXVIIII dies, horas XX ac X momenta ita recipere, sicque duabus horis et X momentis iustum terminum excedere*; a new edition with substantial source analysis and commentary is in preparation; it will be published in 2014 posthumously under Arno Borst's name in the *MGH*, *Quellen zur Geistesgeschichte des Mittelalters*. For analyses of this text, see Borst, 'Forschungsbericht', pp 395, 407–26; Alfred Cordoliani, 'Le computiste Hermann de Reichenau', *Miscellanea Storica Ligure*, 3 (1963), 165–90 at 187–8; Germann, *Quadrivium*, pp 185–99. 75 Hermann's *Prognostica de defectu solis et lunae* is edited by Germann, *Quadrivium*, pp 341–50 from three of the eight known codices; a critical edition based on all manuscript witnesses, including a thorough *apparatus fontium* and substantial commentary, is in preparation; it will be published in 2014 posthumously under Arno Borst's name in the *MGH*, *Quellen zur Geistesgeschichte des Mittelalters*. For analyses of this text, see Borst, 'Forschungsbericht', pp 436–41; Cordoliani, 'Hermann de Reichenau', pp 180–6; Germann, *Quadrivium*, pp 219–32.

centuries-old mathematical methods, he defined the areas in which the application of the new astronomical tables and instruments were considered most fruitful. He exemplifies how early medieval Western scientific thought laid the ground for the reception of Arabic science, how old science merged into new science. The problems had always been there and had been constantly discussed since the time of seventh-century Irish computists. But only now, in the eleventh century, did texts, tables and instruments become available to present new solutions, based on astronomical observation rather than mathematical approximation.

This example shows that the questions posed in the early Middle Ages were more satisfactorily answered by astronomical observations in the later period. Still, the interest and curiosity had always been there, and early medieval scientists proved to be extremely progressive within the limits imposed by the knowledge available. Their achievements laid the foundation for the reception of Arabic science. Seventh-century Ireland helped shape these minds.

Quintus Julius Hilarianus, *Expositum de die paschae et mensis*, 12–13 (*PL* 13, col. 1113):	Dionysius, *Epistola ad Bonifatium et Bonum* (Krusch, *Studien* II, p. 83):
aut enim undecim dies minus habet ratio lunae a ratione solis et facit annum communem, aut x et ix transcendit, et facit embolysmum. Inde per annos singulos hac diversitate dies annorum lunae cum diebus solis, qui in trecentis sexaginta quinque et quadran conficit annum, non sibi concordare videntur. Proinde etiam hoc ostendimus, ut appareat omnibus aequales eos invicem dies habere, et ab initio cursus eorum isto ordine cucurrisse, et damna communium annorum eadem luna in annis embolysmis compensasse. Octo annorum rationem, quae perfecta fore videtur, **in medium proferamus.**	Denique in annis communibus ad rationem solaris anni XI dies lunae deesse cernunt. In embolismis vero X et VIIII diebus eundem annum videtur solarem luna transcendere. Quapropter ogduadis et endicadis annos iuxta praefati circuli ordinem **in medium proferamus**, et liquido probavimus per octo annos et XI lunae cursum cum sole contendere: quando tot dies illa collegit quot ille cucurrit.
In octo scilicet **annos** luna **quinque annos** habet **communes** et **tres embolysmos. Ergo quinquies CCCLIV** faciunt **<M>DCCLXX, et ter CCCLXXXIV faciunt MCLII; fiunt simul dies duo millia DCCCCXXII**. Sic deinde et solis **octo** annorum summam in unum colligamus **octies CCCLXV quadran. duo millia DCCCCXXII.** Ita igitur congregavit solis ac lunae cursus cum diebus suprascriptis.	**In** ogduade diximus **V annos** esse **communes, tres embolismos. Quinquies ergo trecenteni quinquageni quaterni** fiunt **$\overline{I}$DCCLXX et ter trecenteni octuageni quaterni, $\overline{I}$CLII**, ac per hoc **simul fiunt $\overline{II}$DCCCCXXII**. Similiter **octo** anni solares, si in summam redigantur, id est, **octies trecenteni sexageni quini** et **quadrantes**, faciunt simul **$\overline{II}$DCCCCXXII**. Simil modo et endicadis annos, qui sunt communes VII et quattuor embolismi, si in summam ea, qua diximus, supputatione congesseris, tantundem paene repperies, quantum XI. solares anni conficiunt hoc est $\overline{IIII}$XIIII. Haec est igitur embolismorum, sicut praediximus, ratio, ut incrementis suis communium annorum detrimenta compensent.

3.1 Dionysius Exiguus' dependency on Quintus Julius Hilarianus' *Expositium de die paschae et mensis* (verbatim citations are highlighted in bold font).

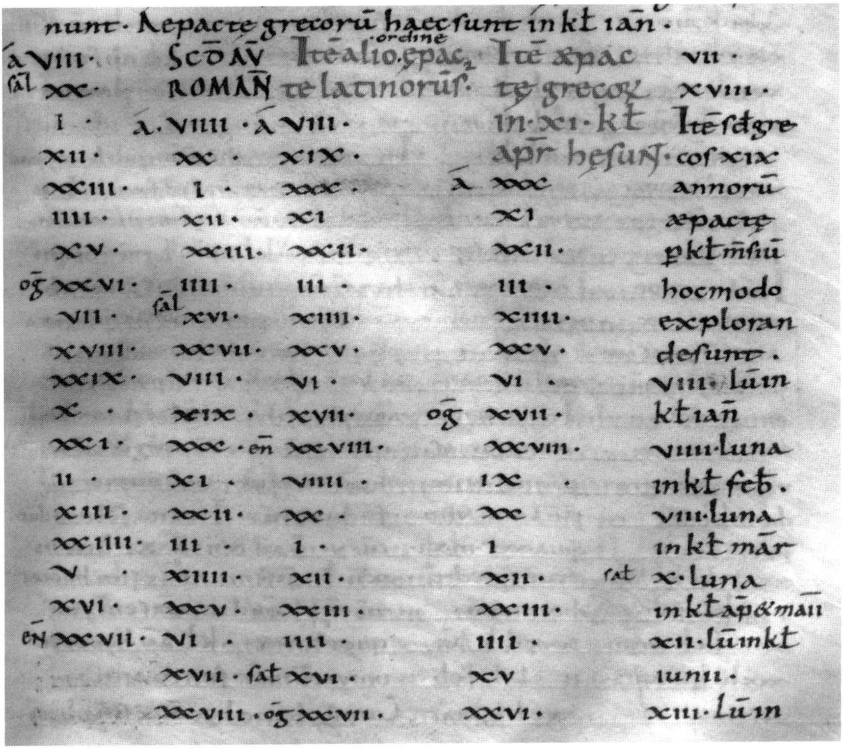

3.2 *Computus Einsidlensis*, Einsiedeln, Stiftsbibliothek, 321 (647), p. 109 (© Einsiedeln Stiftsbibliothek). The columns of this table contain the following data: 1. Dionysiac epacts (lunar age) of 1 January, starting with the *ogdoas* (first eight years of the *cyclus decemnovenalis*); 2. Victorian epacts of 1 January, starting with the *ogdoas*; 3. Victorian epacts of 1 January, starting with the *hendecas* (the final eleven years of the *cyclus decemnovenalis*); 4. Dionysiac epacts of 22 March, starting with the *ogdoas*.

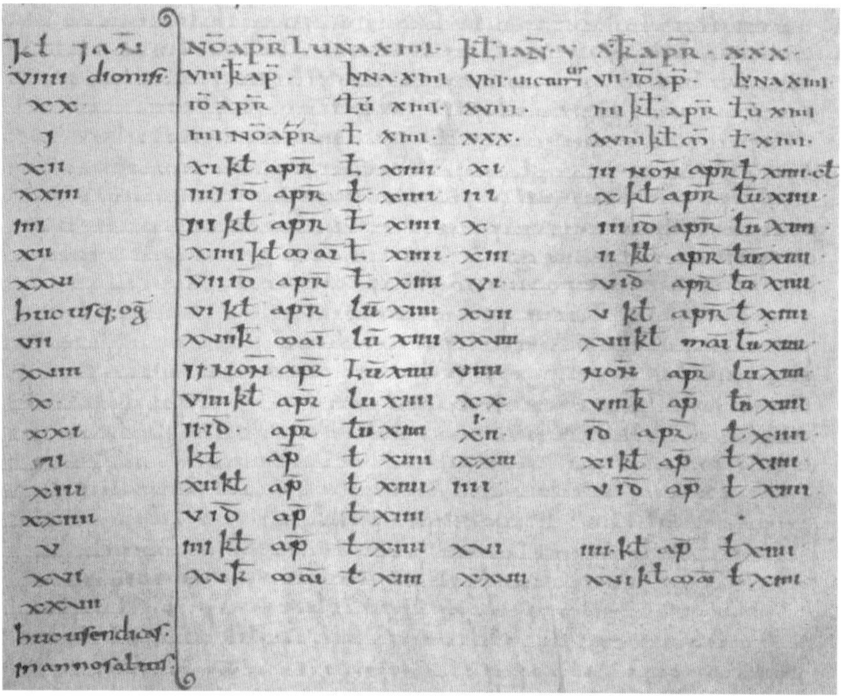

3.3 *De comparatione epactarum Dionysii et Victorii*, Cologne, Dombibliothek, 83–II, fo. 176v (image courtesy of Dombibliothek, Cologne). The columns of this table contain the following data: 1. Dionysiac epacts of 1 January, starting with the *ogdoas*; 2. Julian calendar dates of the Dionysiac Easter full moon, starting with the *ogdoas* (followed by a descriptive column, indicating that the dates refer to *luna* 14); 3. Victorian epacts of 1 January, starting with the *hendecas*; 4. Julian calendar dates of the Victorian Easter full moon, starting with the *hendecas*, followed by a descriptive column, indicating that the dates refer to *luna* 14. An edition of this table can be found in Warntjes, *Munich Computus*, p. 323.

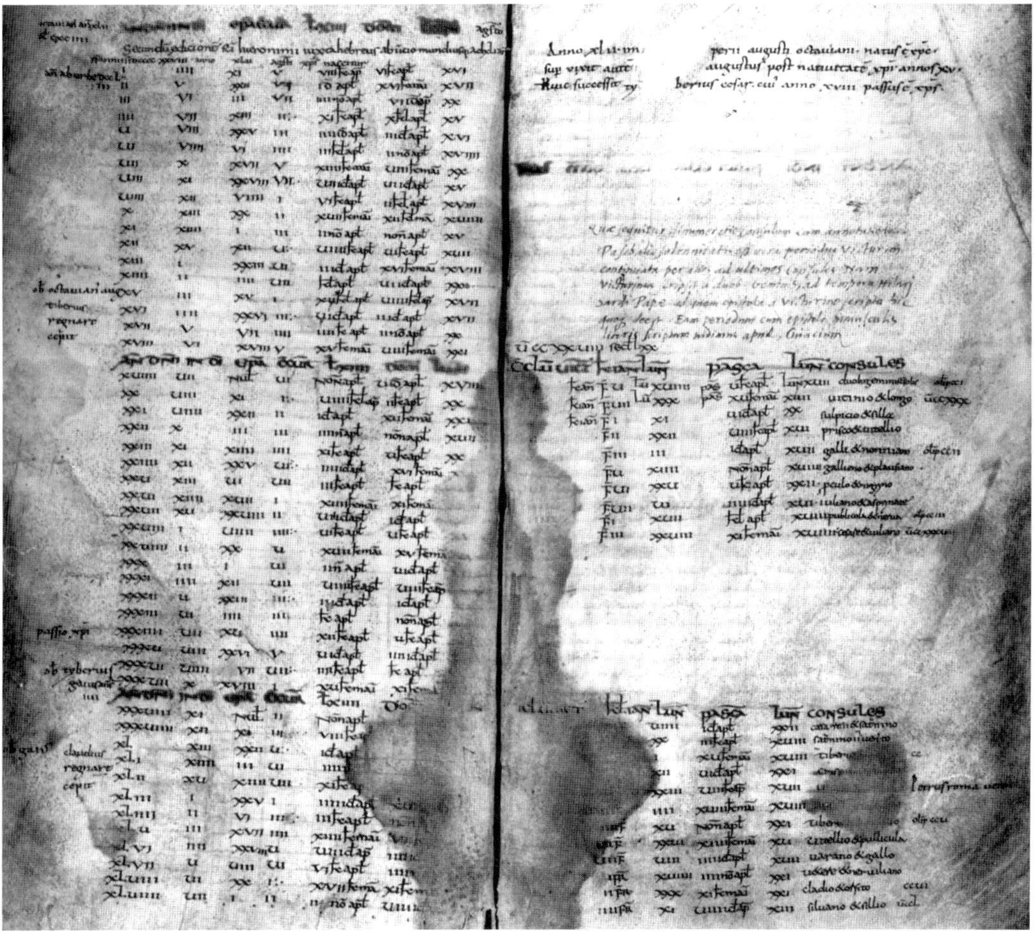

3.4 Leiden, Universiteitsbibliotheek, MS SCA 28, fos 2v–3r (reproduced with permission from the Universiteitsbibliotheek of Leiden), alignment of Victorius' (left) and Dionysius' (right) Easter tables.

Pseudo-Columbanus' *De saltu lunae* (Walker, *Sancti Columbani opera*, pp 212, 214):	*De lunae cursu* (Bobbio Computus 47; *PL* 129, col. 1298; Milan, Biblioteca Ambrosiana, H 150 inf., 19v):	*Comp. Col.* III 7B (Borst, *Schriften*, p. 906):
Omnis igitur lunaris cursus secundum hebraeorum & egyptiorum supputationem potest facere per singulos menses dies XXVIIII & semissem [...] Hoc ideo dicimus ut omnem lunarem cursum dies naturales habere comprobemus **XXVIIII et semissem et horae dimidium & X pene momenta**.	**Omnes igitur lunaris cursus secundum Hebreorum Aegyptiorum** quae [recte que] **subputationem potest facere** iuxta naturalem cursum **per singulos menses dies XXVIIII et semis et dimidium horae et decim poene momenta**. [...]	
	In quo **manifeste** dicere videtur **in XXVIIII diebus et semis**, id **est VI horis, et dimidi**um **horae**, id **est XX momenta, luna**rem **cursum finire, additis tamen**, ut supra dictum est, **X poenae momenta**, et tunc ad alteram lunam incipit prima.	Igitur **manifest**um est **in viginti novem diebus et sex horis**, hoc **est semisse, et horae dimidi**o, hoc **est viginti momentis, luna**m **cursum** suum **finire, additis tamen decem paene moment**is, ut alii adfirmant.

3.5 Dependency between Pseudo-Columbanus' *De saltu lunae*, *De lunae cursu* (Bobbio Computus 47) and the Cologne Computus of 805 (*Comp. Col.* III 7B) (verbatim citations are highlighted in bold font).

Pseudo-Columbanus' *De saltu lunae* (Walker, *Sancti Columbani opera*, pp 212, 214):	*Comp. Col.* IIII 4 (Borst, *Schriften*, pp 913–15):
utrum in XX & VIIII diebus ac semisse tantum luna semper suum cursum peragat, an tunc aliqua momenta retineat; quae aliquot lunas quadam supputandi difficultate neglecta, nunc tandem in unum congregata, quodam calculandi compendio reddita videantur. Omnis igitur lunaris cursus secundum hebraeorum & egyptiorum supputationem potest facere per singulos menses dies XXVIIII & semissem, quia nimis errant, qui lunam peragere cursum sui circuli XXX dierum spaciis aestimant; cum diligens inquisitio veritatis ostenderet in duobus lunae circulis non LX dies sed LVIIII debere computari. Ex quo accidit ut nunc XXX nunc XXVIIII computari videatur. Unde et in communibus annis VI lunae XXX & VI XXVIIII peractae videntur; quibus dies CCCLIIII efficiuntur. Si ergo XXVIIII & semisse luna naturali quodam cursu mensem peragere comprobatur, quare luna embolismi XXX computatur, cum nec ab altera luna semissem, hoc est XII horas habeat? Necnon et in februario luna, quae semper XXVIIII ex more computari solet, cum bissexti dies advenerit, XXX calculatur; ita ut in illa diem integrum, hoc est XXIIII horas, addi videamus. Per ogdoadem ergo et endecadem VII embolismi & V bissexti repperiunt expleti, excepto uno quadrante, hoc est VI horis, quae de uno bissexto restare videntur. In VII embolismis, VII semisses supputantur, hoc est tres dies & XII horae; ubi LXXX & IIII horas	utrum in viginti novem diebus ac semisse tantum luna semper suum cursum peragat, an tunc aliqua momenta retineat, quia per aliqua, sed lunam (MS) quadam supputandi difficaltate neglecta, nunc tandem in unum concreta, quodam calculandi compendio reddita videntur. Omnis igitur lunarum cursus secundum Ebraeorum et Egyptiorum supputationem potest facere per singulos menses dies viginti novem et semis. Quia nimis errant, qui lunam peragere cursum sui circuli triginta dierum spatiis estimant; cum diligens inquisitio veritatis ostenderit, in duobus lunae circulis non sexaginta dies, sed quinquaginta novem debere numerari. Ex quo accidit, ut non tricesimae, sed vicesimae nonae peractae videatur, in quibus dies trecenti quinquaginta quattuor efficiantur. Si viginti novem et semis naturali quodam cursu mensem semper peragere comprobatur, quare luna embolismi tricesima computatur, cum nec ab altera luna semissem, hoc est duodecim horas, habeat necnon et in Febroario luna, quae semper vicesima nona ex more computatur, cum byssextus dies advenerit, triginta dierum, ut alii adfirmant, ita ut in illa diem integrum, hoc est viginti quattuor, addi videamus? Per ogdoadum ergo et endegadum septem embolismi et quinque byssexti repperiuntur impleti, excepto uno quadrante, hoc est tribus horis, que de quinto byssexto restare videntur. In septem embolismis

computando repperies. In bissextis, **CXIIII horas** invenies. **Quibus omnibus in** unam summam **redactis, invenies horas CXCVIII.** [...]	**septem semisses** numerantur, **hoc est tres dies et duodecim horae. Ubi octoginta quattuor horas computando reperies? In byssext**o vero quinto dies quattuor et octodecim horae conplentur, in quibus **centum quattuordecim hor**is supputabis. **Quibus omnibus in una summa redactis, invenies horas centum nonaginta octo.**
Hoc ideo dicimus ut omnem lunarem cursum dies naturales habere comprobemus XXVIIII, et semissem, et horae dimidium, & X pene momenta; unde enim tantam horarum ac **momentorum multitudinem crevisse dicemus, si unaquaeque, ut supradictum est, luna XXVIIII dies tantum** & **semissem** possidet?	**Hec ideo dicimus, ut omnem lunarem cursum dies naturales habere conprobemus viginti novem et semissem et horae dimidium et decem paene momenta. Unde enim tantam horarum** et **momentorum multitudinem crevisse dicimus, si unaqueque, ut supra dictum est, luna viginti novem dies tantum** ac **semissem** habet?

3.6 The quotation in the Cologne Computus of 805 (*Comp. Col.* IIII 4) of Pseudo-Columbanus' *De saltu lunae* (verbatim citations are in bold).

From observation to scientific speculation in seventh-century Ireland

MARINA SMYTH

This essay will examine the working methods of some Irish monks in the middle of the seventh century, thereby giving a sample of the quality of intellectual activity in Ireland in the early Middle Ages, some two hundred years after Christianity reached the island. By then, a number of Irish scholars had acquired a very good command of Latin, and we are fortunate that some of the texts they wrote have survived. Thus, while we know little about the thoughts of the common man at that time, we can talk about the work of some Irish scholars raised in the Latin-language tradition of the church.

I will be paying particular attention to the work of two scholars each of whom left us a treatise informative of their views on the universe. One of them is now known as the Irish Augustine, for the simple reason that he was Irish and called himself *Augustinus*. His treatise *De mirabilibus sacrae scripturae* 'On the marvels of Holy Scripture'[1] [= IA] sets out to show that after God's initial great act of Creation, nothing new was ever created in the universe. He argues that even miracles – which would seem to be totally novel events – do not contravene the natures created by God. We would now talk of the immutable laws of nature underlying all scientific investigation. Demonstrating his thesis for all the marvels recounted in both the Old and the New Testament not only required that the Irish Augustine reveal his assumptions on the natural world, but also forced him to carefully investigate the surrounding physical world for further information.[2] The other work is an anonymous treatise known as *Liber de ordine*

1 An edition of both the long and the short versions of this text can be found in F.P. [Gerard] MacGinty, 'The treatise *De mirabilibus sacrae scripturae*' (PhD, UCD, 1971), but the most accessible edition remains that in J.P. Migne, *PL*, 35, cols 2149–200. Passages from this text pertaining to the present discussion have been translated by M. Smyth in *Understanding the universe in seventh-century Ireland* (Woodbridge, 1996). For the translation of some additional passages, see 'Selections from Augustinus Hibernicus, *On the miracles of Holy Scripture*' in J. Carey, *King of mysteries: early Irish religious writings* (Dublin, 2000), pp 51–74. 2 I was glad to see the Irish Augustine described recently as a true precursor of modern scientists due to 'his recording direct observations of nature and drawing conclusions from them': see C. Moriarty, 'The early Irish naturalists' in J.W. Foster and H.C.G. Chesney (eds), *Nature in Ireland: a scientific and cultural history* (Dublin, 1997), pp 71–90 at pp 71–3. See M. Smyth, 'Seventh-century Irish zoologists?' in A. Handy and B. Ó Conchubhair (eds), *The language of gender, power and agency in Celtic Studies* (Dublin, 2014), pp 59–74.

creaturarum 'On the ordering of creation'[3] [= DOC], a grand survey of the entire universe both spiritual and material, with time and the history of salvation thrown in for good measure. This treatise is clearly influenced by the previous work, but it breaks away from a strictly biblical context, so that statements first formulated by the Irish Augustine to explain a particular miracle have acquired independent standing as part of the description of reality. The rational, methodical approach of *De ordine creaturarum* is striking, with the physical universe deliberately described in descending order – from the upper reaches of the sky down to Hell below the earth – rather than in the order of the seven days of creation as one might have expected at that time, given the tradition of hexaemeral literature.[4] Several shorter works surviving from mid-seventh-century Ireland will be introduced when they throw additional light on the state of scientific knowledge at that time.

We may not accept or even recognize some of the underlying assumptions on which scientific thought rested in mid-seventh-century Ireland, but I should hasten to add that studying the assumptions of our ancestors has led me to wonder: what will scholars a few hundred years from now think of assumptions we take for granted? Will future developments make our present scientific endeavours look merely naive? It seems to me that what really matters in the scientific process is whether a scholar is curious and open-minded, willing to seek information and apply logical reasoning to the available information in order to formulate a rational conjecture. I propose to show that this scientific enterprise was alive and well in mid-seventh-century Ireland.

THE FOUR-ELEMENTS THEORY

The authors under consideration were heirs to the culture and learning developed on the continent of Europe in the waning days of the Roman Empire. We know that this learning was transmitted to them in a surprisingly wide range of texts composed by Christian scholars such as St Augustine, St Jerome and St Gregory. Those Fathers of the Church are rightly better known for their contri-

3 Edited with a Spanish translation by M. Díaz y Díaz, *Liber de ordine creaturarum: un anónimo irlandés del siglo VII* (Santiago de Compostela, 1972), but more readily available in the early edition in J.P. Migne, *PL*, 83, cols 913–54. Translated in M. Smyth, 'The seventh-century Hiberno-Latin treatise *Liber de ordine creaturarum*: a translation', *Journal of Medieval Latin*, 21 (2011), 137–222. 4 This is all the more remarkable given the author's apparent familiarity with St Augustine's magnificent treatise in the hexaemeral tradition, the *De Genesi ad litteram* 'On the literal meaning of Genesis' – or at least with selections from that treatise. Apart from the early edition in J.P. Migne, *PL*, 34, the only edition currently available for St Augustine's treatise is the 1894 edition by J. Zycha in 28:1 of the series *Corpus scriptorum ecclesiasticorum latinorum* (hereafter, CSEL). There is however a convenient annotated translation by J.H. Taylor SJ, *St Augustine: the literal meaning of Genesis* (2 vols, New York, 1982).

butions to Christian thought than for their understanding of science, but having been brought up in the Roman system of education, they were familiar with the basic tenets of ancient cosmology and scattered these concepts throughout their works.

A fundamental assumption in Late Antiquity, and indeed throughout the Middle Ages, was that all matter is composed of some combination of four elements – earth, water, air and fire. This knowledge was transmitted as a given in early Christian works, and it is not surprising that Irish Christian scholars also took it for granted.[5]

Thus, for example, it is the presence of *earth* in the air surrounding us which gives it the firmness necessary for bearing the weight of flying birds,[6] not to mention clouds full of water.[7] This earthy component also explains how the air above could produce manna, a substance solid enough to nourish the Israelites in the desert.[8] There is also *water* in the air around us, since minute particles of water are drawn up from the land and the sea by the clouds, in a process similar to that of sponges absorbing water.[9] And of course there is *fire* in the air, since fire by its very nature tends to rise to the upper regions of the sky, as was held to be obvious from watching a flame burning upwards.[10] This belief that all matter contains one or more of the four elements would have accustomed our scholars to the idea that any material object might contain components that are not at all obvious. One favourite approach to explaining a miracle was then to argue that if a component or a property can be shown to exist within some created thing, God is not interfering with the nature of that thing when he radically increases the volume of that component or intensifies that property. Think of Lot's wife, for instance, who was transformed into a pillar of salt. That miracle is explained as follows: the saltiness of sweat and tears shows that there is salt in the human body, and to perform that miracle, God simply radically increased the volume of that salt component.[11] Similarly, there must be earth in water since it has sufficient solidity to support the weight of a swimmer or of ships, and can even acquire the firmness of ice.[12] It was therefore nothing really new when God, who daily controls the flow of the tide, pushed back the water of the Red Sea into frozen walls between which the Israelites could safely cross – for such is the explanation given for the crossing of the Red Sea.[13]

5 See Smyth, *Understanding the universe*, pp 47–87. 6 IA 1.23; DOC 6.2, 7.4. 7 IA 2.1; DOC 6.2, 7.4. 8 IA 1.23. See Smyth, *Understanding the universe*, pp 190–2. 9 IA 1.7, 1.22, 2.18; DOC 7.5. 10 IA 2.31; DOC 4.6. 11 IA 1.11; Smyth, *Understanding the universe*, p. 302. J.F. Kenney cited this particular explanation as an example of the reasoning of the Irish Augustine: see *The sources for the early history of Ireland: ecclesiastical. An introduction and guide* (Dublin, 1997 [1929]), pp 275–7, §104. 12 IA 3.8. Nor should we be surprised that God could allow the water of the Jordan to become so dense as to push an axe that had fallen into it back up to the surface (4 Kings 6:5–6), not to mention Jesus and Peter walking on water (IA 2.25 and IA 3.8). 13 IA 1.20. Some early Latin versions of the Bible clearly suggested that the water of the Red Sea froze at that time; Ambrose used such a version, as is evident from *Hexaemeron* 2.11: *gelauerunt* [...]

THE FORMATION OF ROCKS

Another miracle associated with the Exodus from Egypt called for a sophisticated sustained argument to show that it was not contrary to the nature of things that clear drinking water sprang from a rock when Moses struck it with his staff (Exodus 17:6). How could water spring from a rock, unless water was somehow hidden within the rock? If it could be shown that there was indeed water within the rock, despite all appearances to the contrary, then the miracle could be explained since the power of God is such that he would make the presence of this water obvious for all to see, if needs be increasing its volume or perhaps rather reversing the process of formation of the stone. When seeking to explain this miracle, the Irish Augustine therefore searched about in his books and in his experience for examples demonstrating the presence of water in very hard, solid substances.[14]

In his readings he had come across the commonly held view that snow frozen over a very long period transforms into rock-crystal,[15] thereby showing that water can become very solid under certain circumstances. Indeed, to a Christian scholar such as the Irish Augustine, marine life provided further examples of water itself becoming very solid, hard matter. This is because the Book of Genesis asserts that birds and fishes and all other dwellers in the sea were produced from water at the time of Creation. It was therefore reasonable to assume that the hard shells of sea turtles originated in water. This must be what the Irish author had in mind when he said 'Several marine reptiles are covered all their life with shells which are clearly much harder than many stones'.[16] From examining the occasional sea turtle stranded dead or alive on Irish beaches (most likely after being carried from the Gulf of Mexico by the Gulf Stream), the Irish Augustine could have observed first hand that these shells can be very hard. I am not talking about the more common Leatherback Turtle (*Dermochelys coriacea*), but two types of smaller sea turtles with hard shells can still be found sometimes along the Irish coast – the Loggerhead (*Caretta caretta*) and the Kemp's Ridley (*Lepidochelys kempii*), now both endangered species.[17]

fluctus, 'the waters froze' (K. Schenkl (ed.), *Sancti Ambrosii opera* (3 vols, Vienna, 1897–1902), CSEL, 32:1, p. 49). The records for Exodus 14:2 and 15:8 in the *Vetus Latina* database published by Brepols provide additional evidence for this version. See Smyth, *Understanding the universe*, p. 212, n. 127. **14** He does not appeal to the idea that *all* material objects contain *all four* elements, so he apparently did not know that aspect of the classical four-element theory. He rather builds his argument based on the data available to him. See Smyth, *Understanding the universe*, pp 76, 81–7. **15** This was commonplace knowledge in Late Antiquity (Smyth, *Understanding the universe*, p. 74, n. 123.). See, for example, St Augustine's comment on Psalm 147:17 (E. Dekkers and J. Fraipont (eds), *Enarrationes in Psalmos CI–CL* (Turnhout, 1990 [1956]), CCSL, 40, pp 2160–1), which I translate as follows: 'What is crystal? It has become very hard, it has frozen a great deal; and it cannot melt easily like snow. Snow that has hardened over many years and centuries is called crystal.' **16** IA I.24. **17** R.S. Penrose and L.R. Gander, *British Isles & Republic of Ireland turtle strandings & sightings – Annual Report* 2011, Marine

Whales are now also endangered animals, more especially the Northern Right Whale (*Eubalaena glacialis*), also known as the Nordcaper, which was once plentiful in the North Atlantic but became a favourite target of whalers who referred to it as 'the right' whale to hunt for its oily blubber and baleens, especially since it did not sink when killed. These whales are characterized by several large protuberances on their skin, patches of very rough skin usually infested with sea lice. In particular, the callosity at the top of the head of a Northern Right Whale is referred to as the 'bonnet'. While these callosities attract barnacles, and thus become quite prominent, they are an intrinsic part of the body of the whale and are present at birth.[18] The Irish Augustine must have seen for himself – or had someone trustworthy report to him – these callosities on the body of a stranded whale. This is what he says, 'Upon those whales which the sea casts ashore from time to time, we can see rough stony lumps (*roccae nagarba*), which we can show upon close inspection not to have been raised from the bottom of the sea, but rather to have been solidified thus from the very quality of sea water and from the extremely hardened skin of their bodies'.[19] Apart from providing the first written evidence for the stranding of marine animals on Irish beaches,[20] the Irish Augustine makes it clear that he understood these rough growths on the body of whales to be another example of the transformation of water into something very hard.[21]

The author of the treatise on miracles also reflects on the presence of pebbles in unexpected places – within the soil accumulating at the top of a rocky crag for instance or even in the fork of the branches of a large tree. He argues that the soil in such elevated places must originate from the decomposition of earlier vegetation, mosses or leaves for instance – a reasonable surmise, especially if he

Environmental Monitoring, available at www.strandings.com/Graphics%20active/2011%20 Turtle%20Strandings%20Report.pdf or http://ebookbrowse.com/2011-turtle-strandings-report-pdf-d334984059 (Sept. 2012). These rare turtle strandings are often reported in the press – for example, G. Burrows, 'Turtle found on beach in Co. Dublin', *Irish Times* (2 Jan. 1969); C. Moriarty, 'July and August peak months for turtles', *Irish Times* (26 July 1983); M. Viney, 'The perilous tale of a travelling turtle', *Irish Times* (25 Aug. 1990) – mostly about the Leatherback, but the author says he has found specimens of the two other types on the Mayo coast; G. Deegan, 'Turtle leaves west Clare for wilds', *Irish Times* (7 July 2001); A. Lucey, 'Loggerhead turtle found on Kerry beach nursed back to health', *Irish Times* (29 Apr. 2002). **18** J. Fairley, *Irish whales and whaling* (Belfast, 1981), pp 35–6, and various websites sponsored by conservation groups, such as the *Canisius Ambassadors for Conservation* at www.conservenature.org/learn_about_wildlife/marine_mammals/northern_southern_right_whale.htm (Sept. 2012). **19** IA 1.24. See Smyth, *Understanding the universe*, pp 84–5. **20** The next earliest record occurs in the annotations by H^2 in the Annals of Ulster for the year 752, referring to the earlier, rather mythical, stranding of a whale with three gold teeth 'at the time of Fiachna son of Áed Rón, king of Ulaid' (S. Mac Airt and G. Mac Niocaill (eds), *The Annals of Ulster (to AD1131)*, pt I: text and translation (Dublin, 1983), pp 208–9). **21** A stranded Humpback Whale (*Megaptera novaeangliae*) might also have warranted this observation, though the tubercles – that is, the protruding knobs along the edges of the flippers and on the head and lower jaw of this species of whale – are somewhat softer. Numerous photographs of the popular Humpback Whale are available on the internet.

had occasion to examine a sod of turf. Indeed, vegetation can often be seen on the erratic boulders – remnants of the Ice Age – scattered in parts of Ireland, in the Burren for example. But the Irish Augustine wonders further how it can be that one sometimes finds pebbles in this soil which is so far above the ground? I find myself thinking of children or tourists throwing pebbles onto the top of dolmens, but that is apparently not what the Irish scholar had in mind. He argues quite reasonably that rainwater falling from above is the only other possible component in these pebbles and he therefore conjectures that the combination of rain with the decomposed vegetation must have gradually generated the stones.

The Irish Augustine then points out that constantly flowing water will harden soils rich in clay, but the most astounding argument is drawn from his observation that one often sees smaller stones imbedded in rock. Conglomerate rocks are abundant in Ireland, so that this observation could have been made in many locations. To give but a few examples from various parts of the island: there are sandstone conglomerate cliffs at Dunmore East, Co. Waterford, alternating sandstone and conglomerate cliffs at Harrylock Bay, Co. Wexford, quartzite pebbles in conglomerate at Cushendun, Co. Antrim, not to mention the conglomerate boulders on Renvyle Beach, Co. Galway.[22]

The author quite correctly deduced that this rock must at one time have been soft – that is, it clearly was moist and thus contained water. I am not aware of anyone saying that before him, nor indeed for quite some time after, although I was fascinated to read recently that in the eighteenth century there was a dispute among scholars as to the origin of the basalt pillars of the Giant's Causeway in Co. Antrim – were the pillars secreted from the sea, that is, 'precipitated from water' or were they rather formed by volcanic activity? That was the dispute between the 'Neptunists' or 'watermen' and the 'Vulcanists' or 'firemen'.[23]

But to get back to the Irish Augustine, he points out that the hollow sea shells (*concavos ostreos*), presumably fossils, imbedded in rock tell the same story, that is – the rock was originally soft. While he might have borrowed his awareness of such fossils from Christian texts available to him, it seems more likely that he observed them himself since rocks bearing remnants of their earlier marine surroundings are not uncommon in Ireland, where Brachiopod fossils are found in carboniferous sandstone (as in Co. Galway) or carboniferous limestone (as in Co. Fermanagh).[24] I should mention here that earlier Christian authors had seen the presence of remnants of sea life at high altitudes as proof of the historicity of the Great Flood.[25] Our Irish scholar, however, was clearly a cut above enter-

22 Images of these rock formations are available on Google. 23 P.N. Wyse Jackson, 'Fluctuations in fortune: three hundred years of Irish geology' in Foster and Chesney (eds), *Nature in Ireland*, pp 91–114 at pp 93–4. 24 Images are available on Google. 25 For example, Tertullian, *De pallio*, §2 (*PL*, 2, cols 1032–4). Indeed, the Irish Augustine appears to have been familiar with the *Historiae aduersum paganos* of Orosius, in which that author also remarks on this

taining such considerations. Indeed, after assembling all this evidence, he paused to summarize what he had established so far – that the hardness of stone can originate from water alone or from the combination of water and earth.

The Irish Augustine then approached his problem from a different angle, seeking examples of the opposite process – very hard substances becoming soft. Metals are an obvious example, as is the 'stone' – *lapis* mentioned in the Book of Job (Job 28:2), which melts and turns into brass when exposed to heat. Indeed, he notes that metals are examples of substances capable of switching back and forth between hardness and liquidity. Salt is another such substance – it crystallizes in salt pans or as rock salt, but dissolves in liquids. And so observation reinforces his hunch that even very hard substances can contain water, and that human or natural agency can sometimes reverse the process, transforming very hard substances into liquid.

His next step is straightforward – God was not in any way interfering with the natural order when he reversed the process of formation of the rock at Horeb and had water gush out from it to quench the thirst of the Israelites. What is truly remarkable is how the mental gymnastics necessary to explain away miracles based on the immutable laws of nature, that is, the laws God created once and for all 'in the beginning', forced the Irish Augustine back to observation of the natural world around him. This, in turn, led him to explaining natural phenomena, in this case the formation of rocks. It goes without saying that the Irish Augustine was a true precursor of the eighteenth- and nineteenth-century geologists James Hutton and Charles Lyell in showing the slow progress of rock formation from sediment to hard rock. Moreover, this passage is a wonderful example of using the scientific process to solve a problem – find examples of similar phenomena and see if you can identify the underlying principle.

THE NATURE OF WATER

Our mid-seventh-century Irish scholars have a very interesting theory about water, namely that saltiness and sweetness are both intrinsic properties of water, since salt water and sweet water can readily be transformed one into the other. This theory first arises in the context of the discussion of the miracle of the brackish water turned into drinking water at the touch of Moses' staff (Exodus 15:23–5).[26] Following his practice of dismissing outlandish explanations, the Irish Augustine first eliminates the idea that some property of the staff itself somehow effected the change. He argues that even if the staff were somehow

'proof' of the historicity of the Flood (C. Zangemeister (ed.), *Pauli Orosii Historiarum adversum Paganos Libri VII* (Vienna, 1882), 1.3, CSEL, 5, p. 41). **26** IA 1.22; Smyth, *Understanding the universe*, pp 229–36.

'sweet' it could not transform the water into drinking water, just as adding a honey-comb will not make a bitter liquid sweet. I rather think he had tried this himself:

> Could it be that the nature of this wood was of such great sweetness that no sooner was it dipped into the most bitter water of the river than this became sweet and savoury? And yet, if someone were to throw a honey-comb into bitter water, would not the honey-comb become bitter rather than the water sweet?[27]

He adduces instead examples to show that the transformation of salt water into sweet water is not at all unusual; in fact, he affirms that it happens all the time since the clouds transform the seawater they have sucked up into themselves into soft rain to benefit the crops. A few years later, the anonymous author of the treatise on the universe would modify this statement slightly, affirming that the transformation occurs as the salt water falls from the clouds through the air[28] – further evidence (if such were needed) that these scholars were independent thinkers and did not merely copy from their sources. In any case, either the clouds or the air are the agents of a fundamental transformation. It is not just a matter of filtering out the salt – as some claimed in Antiquity. Both Irish authors point out that humans can also bring about this transformation, since sailors strain sea water through earth, indeed 'through the best earth' (*per humum optimae terrae*, according to the Irish Augustine), to make it drinkable by transforming its bitterness into sweetness – a totally plausible observation, as I am told by people who understand much better than I do the process of deion-izing salt molecules.[29]

While the argument so far is sufficient to explain the miracle recounted in Exodus, the Irish Augustine does not stop there. Now that he has started thinking about sweet and salt water, it occurs to him that daily life offers examples of sweet water transformed into salt water through the addition of the ashes of plants. This would have occurred routinely in Ireland with the practice of preserving food in brine prepared in this way, or in the creation of the local organic shampoo, which was well known to be salty – as was to be expected due to the ashes providing the alkali necessary to make a soapy liquid.[30]

And thus the Irish Augustine proved to his satisfaction the total mutability of salt and sweet water, concluding that salt and sweet water are essentially the same. In his view, the external agents triggered a change *within the water itself* – a conception that explains the complete solubility of salt in water (until saturation is reached of course), and, in the opposite direction, the deionization of the

27 IA 1.22. 28 DOC 7.8, 9.2. 29 Smyth, *Understanding the universe*, pp 227–36 and 'The seventh-century Hiberno-Latin treatise', 187, n. 106. 30 Smyth, *Understanding the universe*, pp 263–5.

salt molecules within the water. These early Irish scholars did not understand why, but they realized that this was not a simple matter of removing the salt component from salt water.

COSMOLOGY IN MID-SEVENTH-CENTURY IRELAND

Some scholars in Antiquity had formed the very clear picture of a spherical Earth at the centre of a spherical universe, with the sun, the moon and the other five visible planets all circling the Earth, but this was by no means common currency, and I do not believe this scheme was accepted in Ireland until possibly the end of the seventh century, that is, after the texts I am considering were written. In fact, there was considerable confusion in these texts on the meaning of all words associated with roundness.[31] Scholars in mid-seventh-century Ireland visualized the Earth as a flat stretch of land surrounded by the Ocean, with the vault of the sky above and a watery abyss below. This is nothing very surprising really; why should they think otherwise? This was the cosmology presented by Scripture. The author of the *Liber de ordine creaturarum* does allow for the possibility that the firmament is surrounding the Earth 'as the shell of an egg', but he borrowed this concept from the grammatical tradition of learning,[32] and he does not in any way embrace it. In fact, he considers it equally plausible that the sky is a vault or a flat disc above the Earth – the same hesitation he found in a treatise by St Augustine.[33] I have found no trace in Ireland at this early time of any awareness of the notion of the seven visible planets circling the Earth, as had been identified in Antiquity. For these early Irish authors, the sun, the moon and the stars are the only inhabitants of the upper regions of the sky, where part of their course can be observed. Some people in Ireland might even have assumed the heavenly bodies were animate, as a number of earlier Christian writers had asserted.[34] Under those circumstances, regular phenomena such as the changing phases of the moon remained mysterious, as clearly stated by the Irish Augustine – 'Who will ever be able to understand the causes of the full alternations of the moon, waxing for fifteen days and waning for the very same length of time?'[35]

Having inherited no theoretical basis for a possible explanation of the moon's behaviour, observation was the only way forward – and this is a good thing from

31 For example, the word *hemisphaeriae* does not mean 'hemispheres', but seems to refer rather to meteorological phenomena such as rain, snow, thunder etc. (Smyth, *Understanding the universe*, pp 188–9). Again, in *De ordine creaturarum*, 5.9, the moon is referred to as possibly a *rota*, 'a wheel', suggesting a flat circular object. 32 Smyth, *Understanding the universe*, p. 109. 33 Zycha, *Augustine, De Genesi ad litteram*, bk 2, 9.20–2, pp 45–7; trans. Taylor, *St Augustine: the literal meaning of Genesis*, i, pp 58–60. 34 DOC 5.3, 5.11; Smyth, *Understanding the universe*, pp 173–5. 35 IA 1.7.

a scientific point of view. So what did they observe? First of all they observed something we tend to forget nowadays, that the moon is often visible during the day, especially in the morning and evening light. From watching over time the relative location of the sun and the moon in the sky, these early Irish scholars realized that there was a pattern underlying the changes in the appearance of the moon. We understand today that on account of the daily eastward spin of the Earth, both the sun and the moon appear to the observer on Earth to be travelling daily from east to west, and as the moon becomes more full it becomes visible in the daytime sky, where it appears to be trailing the sun more and more every day. Conversely, as the moon is waning, it rises in the east later and later during the night; that is, closer and closer to sunrise, and one might well say that it then precedes the sun. The author of *De ordine creaturarum* describes this:

> Through the daily movements of its ministration [to the needs of man], in as much as the moon follows the sun, so it increases its light, and in as much as it precedes the sun, so it decreases its brightness.[36] [...] Or is it a disc illuminated by the rays of the sun, which, whenever it comes close to the sun, either in front or behind it, is illuminated as though at the rim by a beam of light? When however, it is seen to recede further and further, it gradually increases its light more and more from the splendour of the sun, until it stands face to face with the sun with fully equal orb, and can hold the full image of the sun within itself.[37]

He observes but he cannot explain. This author had read at least part of St Augustine's treatise on the *Literal meaning of Genesis*, and it is clear from the second passage quoted above that he was familiar with the African bishop's speculations on the nature of the moon – whether it was a half-bright and half-dark sphere slowly spinning with respect to the Earth, or was it illuminated by the rays of the sun so that whenever it comes close to the sun 'it can appear only in the shape of a horn'?[38] The Irish author's concise description of the moon preceding or following the sun suggests that he had gone to the trouble of figuring out for himself through observation what St Augustine was talking about.

The Irish Augustine had earlier considered the possibility that even though both the sun and the moon halted in the sky at the time of Joshua's battle (Jos 10:13),[39] the moon might have continued changing phase. He dismissed the idea,

36 DOC 5.8; Smyth, 'The seventh-century Hiberno-Latin treatise', 176; eadem, *Understanding the universe*, pp 157–9. 37 DOC 5.9. 38 Zycha, *Augustine, De Genesi ad litteram*, bk 2, 15.31; trans. Taylor, *St Augustine: the literal meaning of Genesis*, i, pp 68–9. 39 For one consequence perceived in the early Middle Ages for the pause of the sun during that battle, see Smyth, 'Once in four: the leap year in early medieval thought', *Proceedings of the Third International Conference on the Science of Computus in Ireland and Europe, Galway, 16–18 July 2010* (Turnhout), forthcoming.

but not because he believed the phases of the moon are dependent on the moon's position in the sky relative to the sun. His approach is radically different – using a *reductio ad absurdum* type of argument, he argues that the phases of the moon cannot have changed during that miracle since there was no disruption at that time in the course of the great 532-year cycles he perceived in the pattern of periodicity of the relative courses of the sun and the moon.

One might wonder where the Irish Augustine got the notion of such a lunar-solar cycle, which was obviously not something he could observe in the sky. Recall that the church was wrestling at that time with the vexed question of determining the correct date for Easter. Everybody agreed by then that Easter must be celebrated on the first Sunday after the first full moon after the spring equinox, but how to work this out in practice ahead of time, so that one would know, for instance, when to start the forty days of the Lenten fast? Several methods had been implemented over the centuries and discarded, but the latest solution known in Ireland seemed to have settled the problem. The Irish Augustine was acquainted with the 532-year table which Victorius of Aquitaine had created in the mid-fifth century for determining the date of Easter.[40] He knew that the data repeated after 532 years and believed the table was flawless. He overlooked, however, the rather artificial construction of Easter tables – which are only loosely tied to astronomical reality, given the need to accommodate the convention of a seven-day week, not to mention additional technical adjustments. The Irish Augustine therefore *incorrectly* inferred from his *correct observation* that the Victorius Easter table was cyclical, that after 532 years the sun and the moon are back in the very same configuration in the sky as they were 532 years previously. He then posited that this 532-year cycle in the relative behaviour of the sun and the moon is part of the created order of the universe, valid ever since creation. If the moon had continued to change phases during Joshua's miracle, while remaining in the same location in the sky, this lunar increment beyond the course of the sun would have disrupted the pattern of the 532-year cycle.

The conviction of the permanence since Creation in the duration of this great 532-year unit of time, combined no doubt with technical concerns associated with the dating of Easter, must be at the root of the matter-of-fact way in which these texts assume that the equinoctial hour is the norm.[41] The units of time are absolutes for these Irish scholars, independent of the seasons, and surviving computistical works originating just a little later in seventh-century

40 The standard edition is in B. Krusch, *Studien* II, pp 17–52; for additional bibliography see I. Warntjes, *The Munich Computus: text and translation. Irish computistics between Isidore of Seville and the Venerable Bede and its reception in Carolingian times* (Stuttgart, 2010), p. xxxviii, n. 82. See elsewhere in this volume, Warntjes, 'Seventh-century Ireland: the cradle of medieval science?'
41 As opposed to the changing daytime and night-time hours normally used in Late Antiquity.

Ireland will normally begin with extensive discussion of the units for the division of time.[42]

THE TIDES

Discussion of the origin of the waters of the great Flood – the superabundance of which certainly qualifies as miraculous – reminds us of the overall view of the material universe which scholars in seventh-century Ireland inherited from their Christian sources. It is a given that there are waters above the firmament, as affirmed in the Genesis account of Creation (Gen 1:7; see also Ps 148:4), and the Irish Augustine cannot decide whether the water pouring down for forty days (Gen 7:11–12) was dropped through celestial trapdoors from that vast reservoir above the firmament, or whether it was simply a matter of more rain than usual.[43] As for the waters rising from the abyss according to the Genesis account (Gen 7:11), that is exactly where he thinks these waters most likely originated, that is, from the great abyss below the Earth – to which they withdrew as the land became dry again. This is completely consistent with Genesis: 'All the springs of the great abyss were burst open'.[44] Indeed, for the Irish Augustine this biblical text provides a possible account of the mechanism of tides.

Tides were a striking natural occurrence for areas bordering on the Atlantic, but little had been written about them by ancient scholars since tides are practically non-existent in the Mediterranean Sea, so that Irish scholars had to be creative on this topic. The Irish Augustine speculated that the water of the ebbing tide rises from the abyss and returns there as the tide recedes. This makes perfect sense – he pointed out that the water of the rising tide must come from somewhere, and it must return to that same place as the tide recedes, just as happened at the time of the great Flood. But do note that this theory assumes that the level of the sea changes uniformly, so that the sea would be seen to rise simultaneously and then subside simultaneously on all coastal areas. The same result could have been achieved if God simply increased the volume of water in the sea. While the Irish Augustine himself could not decide between these two options for the origin of the rising waters of the tide,[45] his successor, the anonymous author of *Liber de ordine creaturarum*, clearly thought that the volume of water remains constant in the sea, and his account of tides implies that this water is simply moving back and forth across the surface of the Earth:

> And because it is impossible that it should not be leaving some place – though it be a place we do not know – when it is flooding the land in our

42 Warntjes, *The Munich Computus*, pp cxli–cxliii. 43 IA 1.6; Smyth, *Understanding the universe*, pp 98–101. 44 Genesis 7:11 and also Job 38:8: 'Who shut up the sea with doors, when it broke forth as issuing out of the womb?' 45 Smyth, *Understanding the universe*, pp 244–9.

regions, and similarly filling that place when it retreats from us – so that what is flow here is ebb there, and what is ebb here is flow there – we are forced to acknowledge that we know nothing of half the ministry of the sea.[46]

The phrase 'what is flow here is ebb there, and what is ebb here is flow there' strongly suggests that the anonymous author is not thinking of water rising from and receding back into the abyss, but rather of water swishing back and forth over the surface of the Earth.

The early eighth-century Northumbrian scholar Bede has been credited with the major scientific achievement of discovering the true explanation of tides – namely, that the mass of water on the surface of the Earth is swinging to and fro, dragged by the moon as it circles the Earth.[47] We would talk about the gravitational pull of the moon. Would Bede have come to that conclusion had he not read the anonymous treatise written in Ireland more than twenty-five years earlier?

And yet, while the author of *De ordine creaturarum* gives a reasonable account of the rising and ebbing of the tide, he does not explain *why* the tide ebbs and flows twice in twenty-four hours, each day with high tide some fifty minutes later than the previous day. Nonetheless, the observation of a *monthly pattern* in the intensity of tides was pointing our Irish authors in the right direction, as they noted that the intensity of tides varied with the phases of the moon. More precisely, they observed that tides tend to be stronger in the week surrounding the new moon and the full moon than in the intervening weeks surrounding the half-moon phase.[48] They called these the *malinae* tides and the *ledones* tides, respectively.[49]

This appears to have been a common distinction in Ireland, since these terms occur not only in both texts under consideration, but also in the *Hisperica Famina*, those extraordinarily complex short essays composed in Ireland at that time. The section devoted to the sea assumes that the *malina* tides are especially strong and stormy, while the intervening *ledones* are more subdued.[50] I am inclined to believe that this common awareness of periods of higher and lower high tides may be related to coastal salt production in Ireland, since salt pans

46 DOC 9.7; Smyth, 'The seventh-century Hiberno-Latin treatise', 189. 47 P. Duhem, *Le système du monde: histoire des doctrines cosmologiques, de Platon à Copernic* (10 vols, Paris, 1913–59), iii, p. 20; W.M. Stevens, *Bede's scientific achievement* (Jarrow, 1986); F. Wallis, *Bede: the reckoning of time, translated, with introduction, notes and commentary* (Liverpool, 1999), pp 82–5 for the translation of the text of Bede, *The reckoning of time*, §29, and pp 307–12 for commentary. 48 Smyth, *Understanding the universe*, pp 251–62. 49 M. Smyth, 'The date and origin of *Liber de ordine creaturarum*', *Peritia*, 17–18 (2003–4), 1–39 at 8, n. 43; eadem, 'The seventh-century Hiberno-Latin treatise', 188–9. 50 *Hisperica Famina* A399–411, ed. and trans. M.W. Herren, *The Hisperica Famina: I. The A-Text* (Toronto, 1974), pp 94–7; Smyth, *Understanding the universe*, p. 255.

need to remain moist and different procedures will have to be followed to ensure this, depending on the intensity of tides. Watching the phases of the moon would have been a practical way of determining when to initiate the appropriate adjustments. It should be noted, however, that there is a lag between the phases of the moon and their effect on the magnitude of tides, so that the tidy symmetrical pattern described by the Irish Augustine cannot be the result of personal observation, though it could be a deliberate simplification for didactic purposes.

While Irish scholars were unable to establish the causal relationship between the course of the moon and the tides, the Irish Augustine saw the observed association between the intensity of tides and the phases of the moon as part of the order of the created universe. He therefore had no qualms in adding details to scriptural episodes involving unusual lunar behaviour – since the moon stood still at the time of Joshua's battle, the pattern in the intensity of the tide must have stopped also.[51] Again, since the 532-year cycles were not disrupted when the sun went backward at the time of the prophet Ezechiel, it must be that the moon went backward also, so that the pattern in the intensity of tides must also have reversed temporarily during that miracle.[52] Observation brought these Irish authors so very close to figuring out the explanation of tides! But they lacked the astronomical concept of the moon circling around a spherical Earth, so that it was impossible for them to resolve the question.

THE FORMATION OF ISLANDS

But when they limited their attention to the surface of the Earth, observation served them rather better, witness the confident assertion by the Irish Augustine on the origin of islands. Far from asserting that all islands were created in the original act of Creation, he adopts a sophisticated geological approach. The question arises in the course of the discussion of a conundrum that had preoccupied earlier Christian authors – since all land animals must be descended from those Noah brought into his ark, how come there are wild animals on islands which are clearly so far removed from the mainland that animals could not possibly swim to them? Were these wild animals transported to these islands in ships?[53] The Irish Augustine dismisses that notion as ridiculous and offers a far more rational explanation – islands were once attached to the mainland but got separated over time. Indeed, he affirms that Ireland itself was severed from the mainland.[54] Reflecting on the origin of the numerous sea stacks dotting the coast

51 IA 2.4. 52 IA 2.28. 53 See, for example, St Augustine, *The City of God*, 16.7; Smyth, 'Zoologists in seventh-century Ireland?', pp 62–5. 54 IA 1.7. To be sure, the explanation offered by the Irish Augustine is not in full agreement with that offered by present-day geologists. The rising levels of the sea as the Ice Age was ending account for the separation from the

of Ireland, he must have conjectured correctly that they had become separated from the mainland. The sharply severed tip of Loop Head, Co. Clare, would be a striking example, or the sea stack in Dingle Bay, Co. Kerry, as well as the sea arches such as can be seen near Rossbeg or at Horn Head (both in Co. Donegal), which one can imagine to erode and eventually collapse, giving rise to new sea stacks. The Irish Augustine even claims that older people in his generation had seen such islands break away, which is quite possible if we think of the Bridges of Ross, Co. Clare, where such an arch is supposed to have crumbled some 150 years ago.[55] And then there is Dún Briste in Co. Mayo, which appears to have broken off in the fourteenth century, and still holds the remnants of medieval buildings.[56]

NUMBERS AND ARITHMETIC

Let us now turn to numbers, another essential component of the order of the universe, not least because they allow the measurement of time. In the mid-seventh century, two important chronological tables were available to Irish scholars, at least in the southern part of Ireland. I already mentioned the 532-year Easter table of Victorius of Aquitaine, which begins with the year of the Passion. I will now briefly describe the *World chronicle of Eusebius/Jerome*, so-named because St Jerome translated into Latin the tables originally designed in Greek by Eusebius of Caesarea in the fourth century. The fifth-century manuscript from Italy, Oxford, Bodleian Library, MS Auct. T. 2. 26, is an early copy of the *Chronicle of Eusebius/Jerome* and gives a good idea of the manuscripts of these synchronizing tables that would have been available in Ireland. I will explain the innovative layout in such manuscripts with reference to that early Bodleian manuscript – images of which are available on the internet.[57] The goal was to synchronize the history of all known peoples, and the result is page upon page of columns of roman numerals, each column recording for each one of the various civilizations known in Antiquity the years in the life

continent of Europe first of the combined land-mass of Britain and Ireland, and later for the separation of Ireland from Britain. See, for example, J. Feehan, 'The heritage of the rocks' in Foster and Chesney (eds), *Nature in Ireland*, pp 3–22 at pp 17–18. **55** Once more, Google Images is very helpful. **56** An RTÉ radio documentary about a visit by various scholars to this spectacular severed promontory was first broadcast on 21 Mar. 1981. It is available at www.rte.ie/radio1/doconone/radio-documentary-saint-patrick-was-a-mayoman.html. **57** Digital images of significant pages and openings from this manuscript are available on the web at the LUNA site of the Bodleian Library, http://bodley30.bodley.ox.ac.uk:8180/luna/servlet, where a simple search for the word *Eusebius* will then bring up these images. There is an early monochrome facsimile of this manuscript: *The Bodleian manuscript of Jerome's version of the Chronicles of Eusebius*. Reproduced in collotype, with an introduction by J.K. Fotheringham (Oxford, 1905).

or the rule of some eminent leader or for major historical occurrences. Eusebius was of the opinion that detailed historical information was not available before the birth of Abraham, and he therefore chose to start his chronicle at that point in time, numbering the years in his concordance consecutively from the birth of Abraham. To facilitate counting, he also decided to clearly indicate every decade, duly recording in the left-hand margin of each table the appropriate number for the year corresponding to each decade since the birth of Abraham. Thus, it can easily be seen that the Fall of Troy occurred 835 years after the birth of Abraham (figs 4.1 and 4.2, fos 44v–45r), and that Solomon started to build the Temple in Jerusalem 984 years after the birth of Abraham, that is, four years after the marginal notation for the year 980 (figs 4.3 and 4.4, fos 51v–52r). For this later period, the sequential numbers in the columns correspond to the years of the reign of the named rulers of the Assyrians, the Hebrews, the Athenians, the Latins, the Lacedemonians, the Corynthians and the Egyptians. This means that the reader is faced with seven vertical columns of roman numerals, with only occasional commentary mentioning major events. Such an emphasis on numbers must have made a big impression on the viewer.

Despite his uncertainties on the dates of events before the Flood, Eusebius gave his estimate of 2,242 years from Adam to the Flood and 942 years from the Flood to the birth of Abraham. In the introductions to their tables, both Jerome and Victorius relayed this information provided by Eusebius. A good manuscript of the Eusebius/Jerome tables provides a great deal of numerical information clearly setting important events within the overall history of the world. Thus, the page recording the birth of Jesus in Bethlehem (fo. 109v in the fifth-century Bodley manuscript) states explicitly that 'From Abraham until the birth of Christ, there are 2,015 years' (fig. 4.5). This can easily be verified by noting the 2020 mark in the left-hand margin and working backwards through the individual years. The Eusebius/Jerome table will also tell you (fo. 111v) that when Jesus began his public life in the fifteenth year of the Emperor Tiberius (Lk 3:1), this was 2,044 years after the birth of Abraham, which occurred 942 years after the Flood, which occurred itself 2,242 years after Adam. The table also spells out that 548 years had elapsed since the rebuilding of the Temple under Darius, 1,060 years since the building of the Temple by Solomon, and 1,539 years since the departure of the Israelites from Egypt, so that the event is clearly located within the history of the world (fig. 4.6).[58]

For the period following the Passion, if one was so inclined, one could get some additional historical information by consulting the Easter tables of Victorius, which provided the names of the Roman consuls in those years for

58 See fo. 111v in the fifth-century Bodley manuscript, which can be viewed at the LUNA site of the Bodleian Library. All this data can be seen more clearly on fo. 127r of the ninth-century manuscript Oxford, Merton College, MS Coxe 315 (probably from Reichenau), digitized at http://image.ox.ac.uk/show?collection=merton&manuscript=ms315.

which they were available. Irish scholars in the mid-seventh century apparently did exactly that, continuing the *Eusebius/Jerome chronicle* with the Easter tables of Victorius. Evidence for this can been seen in fos 112v–113r[59] of the manuscript Oxford, Bodleian Library, MS Bodley 309 (a later copy of a collection of computistical material which would have been available to Bede from Irish sources)[60] where a much-simplified Eusebius/Jerome table is suddenly followed by consular data drawn from the tables of Victorius.[61]

The Irish Augustine apparently did something similar when he set out to insert sequences of 532-year cycles into the chronology of the world since Creation. After some initial arithmetic, adding up several 532-year cycles until the sixth cycle ended eight years after the birth of Abraham, all he needed to do was carefully map the 532-year cycles against the sequence of years in a manuscript of the *Eusebius/Jerome chronicle*, availing of the numbers in the margin marking the decades. He did not even need to perform multiplications. When he reached the cycle including the year of the Passion, that is the *tenth* 532-year cycle since Creation, he started referring to the consuls listed in the table of Victorius for the year of the Passion and thereafter.[62] I have often thought that after this methodical but tedious process, the Irish Augustine must have been very disappointed that his careful reckoning failed to reveal that events in the life of Christ corresponded to some momentous, or at least significant, numerical milestone in the unfolding of time.

This elaborate comparison of the alleged recurrence of the 532-year lunar-solar cycles with the information contained in the *Eusebius/Jerome chronicle* is not the only evidence in mid-seventh-century Ireland for a fondness, perhaps even an obsession, for locating current events in the history of the world. The evidence suggests that the influx of facts and figures associated with the tables of Eusebius/Jerome and of Victorius went to the head of Irish scholars, who then went to town performing calculations of their own based on the information provided in these two chronological tables.

A first example is the Latin poem *Versus de annis a principio* 'On the years since the beginning', with incipit *Deus a quo facta fuit* and dated internally to AD642 or thereabouts.[63] The eighteen strophes of this poem give much relative dating information for major events listed in the *Chronicle of Eusebius/Jerome*,

59 This opening is unfortunately not available on the LUNA site of the Bodleian Library. 60 As shown by C.W. Jones, 'The "lost" Sirmond manuscript of Bede's computus', *English Historical Review*, 52 (1937), 204–19. See the recent bibliography in n. 36 on p. xxi of Warntjes, *The Munich Computus*. 61 This was also noted by C.P.E. Nothaft who gave an extensive analysis of this transition at the *Fourth International Conference on the Science of Computus, Galway, 13–15 July 2012*: 'The date of the Passion in early medieval computistics: the strange case of the Victorian-Dionysiac "hybrid" table in the Sirmond manuscript (Oxford, Bodl. Libr., Bodley 309)', to be published in the Proceedings of that conference. 62 See D. Ó Cróinín, 'Early Irish annals from Easter tables: a case restated', *Peritia*, 2 (1983), 74–86 at 81–2. 63 K. Strecker (ed.), *Poetae latini aevi Carolini*, 4, fasc. 2, *MGH*, Poetae (Berlin, 1896), pp 695–7.

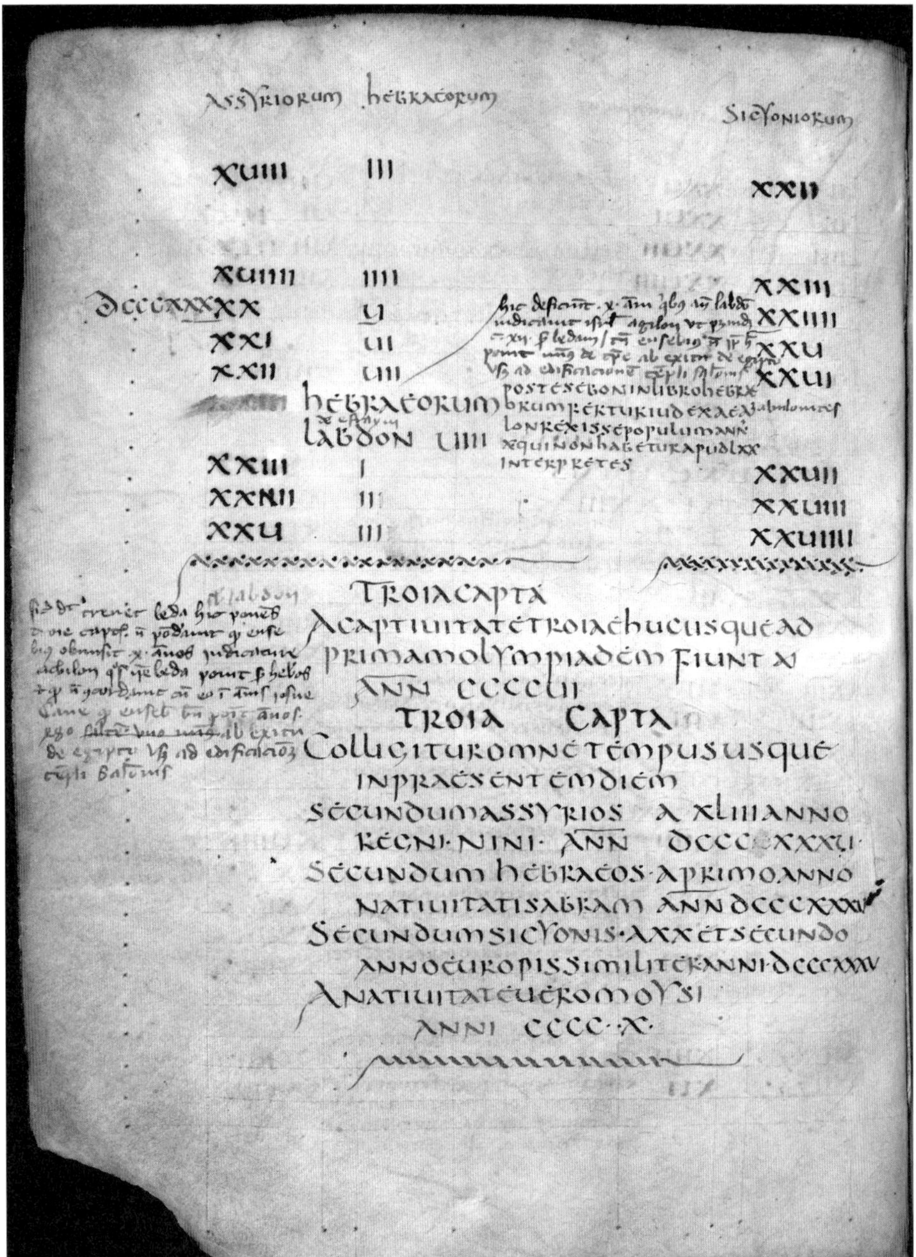

4.1

4.1 and 4.2 Oxford, Bodleian Library, MS Auct. T.2.26, fos 44v–45r (reproduced with permission from the Bodleian Libraries, University of Oxford). This mid-fifth-century manuscript of *The Chronicle of Eusebius/Jerome* shows the layout and impressive array of numbers characteristic of this synchronizing chronicle. At least one such manuscript must have been available in Ireland in the seventh century. Note the number *DCCCXXX* (= 830) in the left margin of this opening, marking the horizontal row of events believed to have occurred 830 years after the birth of Abraham in year 1. The Fall of Troy, identified by the

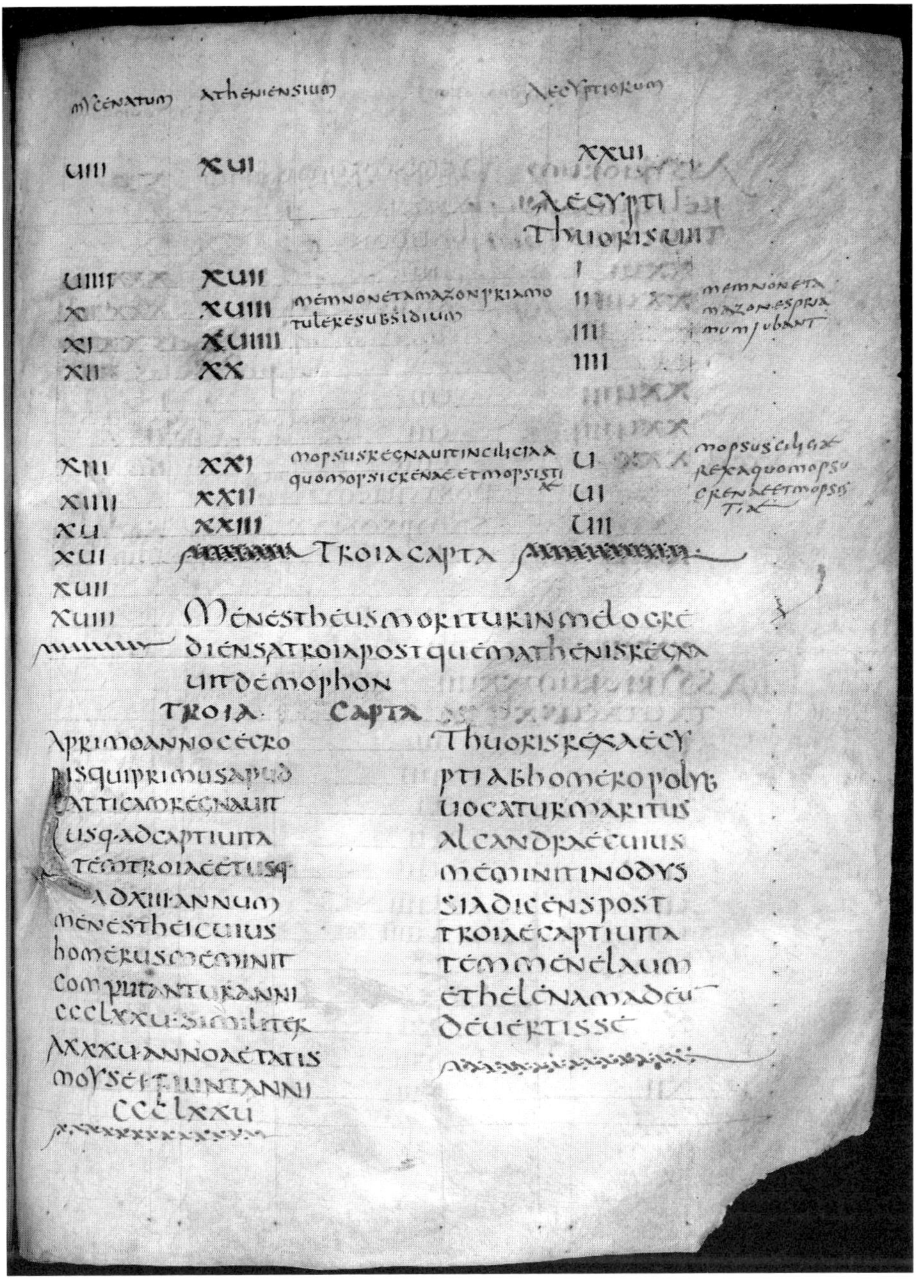

4.2

statement *TROIA CAPTA* (in red ink), can thus easily be seen to have occurred 835 years after the birth of Abraham, as is explicitly reported in one of the statements on fo. 44v: *SECUNDUM HEBRAEOS A PRIMO ANNO NATIUITATIS ABRAM ANNO DCCCXXXU* (there is a small hole in the manuscript next to the *U*): 'according to the Hebrews, 835 years from the first year of the birth of Abraham'. Numerous later comments were added throughout this manuscript.

including the beginning of Jesus' public life. The poem says that the death of an Irish king 'Domnal' occurred 618 years after the baptism of Jesus, information that must have been drawn from Easter tables/annals based on the calculations of Victorius of Aquitaine. The *Annals of Ulster* record the death of Domnall mac Áedo, said to be *rex Hibernie*, for *anno domini* 641 (= AD642), and this is likely the 'Domnal' of the poem.[64] Like Prosper of Aquitaine (*c*.390–*c*.463), for whom simple addition of the intervals listed in the Eusebius/Jerome tables demonstrated that the public life of Jesus started 5,228 years after Creation,[65] the author of the poem apparently assumed the baptism marking the beginning of Jesus' public life took place 5,228 years after Creation (2,044 + 942 + 2,242 = 5,228). This would explain why he sets the death of king 'Domnal' 5,845 years after Creation (5,228 + 617 = 5,845, avoiding double-counting of the year in which Jesus was baptized). It was then commonly believed that the world we know would last no longer than six thousand years after Creation.[66] The subtraction is actually made, yielding the additional information that this king died 155 years before the end of the world. The very existence of such a poem based on simple addition and subtraction suggests that it was considered a great accomplishment to be able to perform such operations.

The second example, the 'computistical bit' recorded on fos. 95v–95bis r in the manuscript Oxford, Bodleian Library, MS Bodley 309, and in other related manuscripts,[67] is more closely tied to the numerical data drawn from the Prologue to the Easter table of Victorius of Aquitaine, where we find that the first Passover took place a full 3,689 years after Creation, that is, in *anno mundi* 3690, and that Christ was crucified after 5,228 years had been completed since Creation, that is, in *anno mundi* 5229.[68] Subtraction then shows that there were 1,539 years between the first and second Paschal events. Here also the death of an Irish ruler, Suibne mac Commáin – the ruler of the Déssi who died in AD658 according to the *Annals of Inisfallen* – is located within the chronology of the

64 See Ó Cróinín, 'Early Irish annals', pp 79–81. 65 Prosper of Aquitaine, *Epitoma Chronicon* in Th. Mommsen (ed.), *Chronica minora saec. IV, V, VI, VII*, *MGH*, *Auct. Antiq.*, 9 (Berlin, 1892), pp 343–499, at 409: *Ab Adam itaque usque in quintum decimum annum Tiberii sunt anni V̄CCXXVIII.* See C.P.E. Nothaft, *Dating the Passion: the life of Jesus and the emergence of scientific chronology (200–1600)* (Leiden, 2012), pp 73–4. 66 R. Landes, 'Lest the millennium be fulfilled: apocalyptic expectations and the pattern of western chronography, 100–800 CE' in W. Verbeke et al. (eds), *The use and abuse of eschatology in the Middle Ages* (Leuven, 1988), pp 137–211, esp. pp 137–96, and most recently 'Computistical and chronological activity at the approach of the millennial years 6000/800 and 1000', presented at the *Fourth International Conference on the Science of Computus, Galway, 13–15 July, 2012*, forthcoming. 67 Edited by Ó Cróinín in 'Early Irish annals', pp 82–3; trans. in M. Smyth, 'Isidore of Seville and early Irish cosmography', *CMCS*, 14 (1987), 69–102 at 95–8. 68 Prologue §9: Krusch, *Studien* II, p. 24. Note the inconsistency with the data in the *Chronicle of Eusebius/Jerome*, which implied that the public life of Jesus started 5,228 years after Creation and that the Crucifixion occurred three years later; see Nothaft, *Dating the Passion*, p. 74.

history of the world, and this ruler is said to have died 141 years before the expected end of the world.

Another short piece, known by modern scholars of the *computus* as the pseudo-Dionysian *Argumentum* 16,[69] gives us a good idea of how early Irish scholars actually carried out their calculations using roman numerals. Immo Warntjes has recently shown that this piece was very likely composed in Ireland some time before 675.[70] Curiously, *Argumentum* 16 sets out to demonstrate that the leap year day results from an annual increment of 3 hours in the difference between the calendar year of 365 days and the actual solar year. This annual increment of 3 hours is then said to add up to a leap year day of 12 hours every four years.[71] The argument explaining this is based on the significance of the number 7 in the Christian tradition – and thus presumably in the order of the universe, as suggested by the explicit reference to God's rest on the seventh day in some versions of the text. We are first told that if you divide by 7 the hours in a day, the remainder is 3, and this causes the leap year day. Note that the author is assuming a 24-hour day.

The second corroborating argument is even more surprising, though it does indicate awareness of the importance of the 365-day calendar year in the matter of the leap year. The idea here is to find the number of hours in a year and then divide that by 7. This division is also shown to have remainder 3, and these three hours are said to add up every four years to a leap year day of 12 daytime hours. Transferring the roman numerals used in the early Middle Ages into arabic numerals, this is how it was done:

> To find the number of hours in a year:
> 12 hours × 365 = (10 + 2) × (300 + 60 + 5) hours
> (10 × 300) + (2 × 300) = 3,600 hours,
> (10 × 60) + (2 × 60) = 720 hours,
> (10 × 5) + (2 × 5) = 60 hours.

Adding up gives 4,380 daytime hours in a year. Adding an equal amount of night-time hours gives a total of 8,760 hours in a year.

The author now divides 8,760 by 7. Division is done by approximation through multiplication – the very same process as in our algorithms, but it is trickier to follow the reasoning because the steps are described in prose:

69 The Latin text of all genuine and pseudo-Dionysian *Argumenta* as edited by Krusch, *Studien* II, pp 75–81 can be found with an English translation at the site of M. Deckers on the Dionysian 19-year cycle: http://hbar.phys.msu.su/gorm/chrono/paschata.htm. 70 I. Warntjes, 'The *argumenta* of Dionysius Exiguus and their early recensions' in I. Warntjes and D. Ó Cróinín (eds), *Computus and its cultural context in the Latin West, AD300–1200, Proceedings of the First International Conference on the Science of Computus in Ireland and Europe, Galway, 14–16 July, 2006* (Turnhout, 2010), pp 40–111 at pp 92–5. 71 Smyth, 'Once in four', where it is shown that it was not unusual in the early Middle Ages to think of the leap-year day as having only twelve hours, probably because of a false etymology of the word *bissextus*.

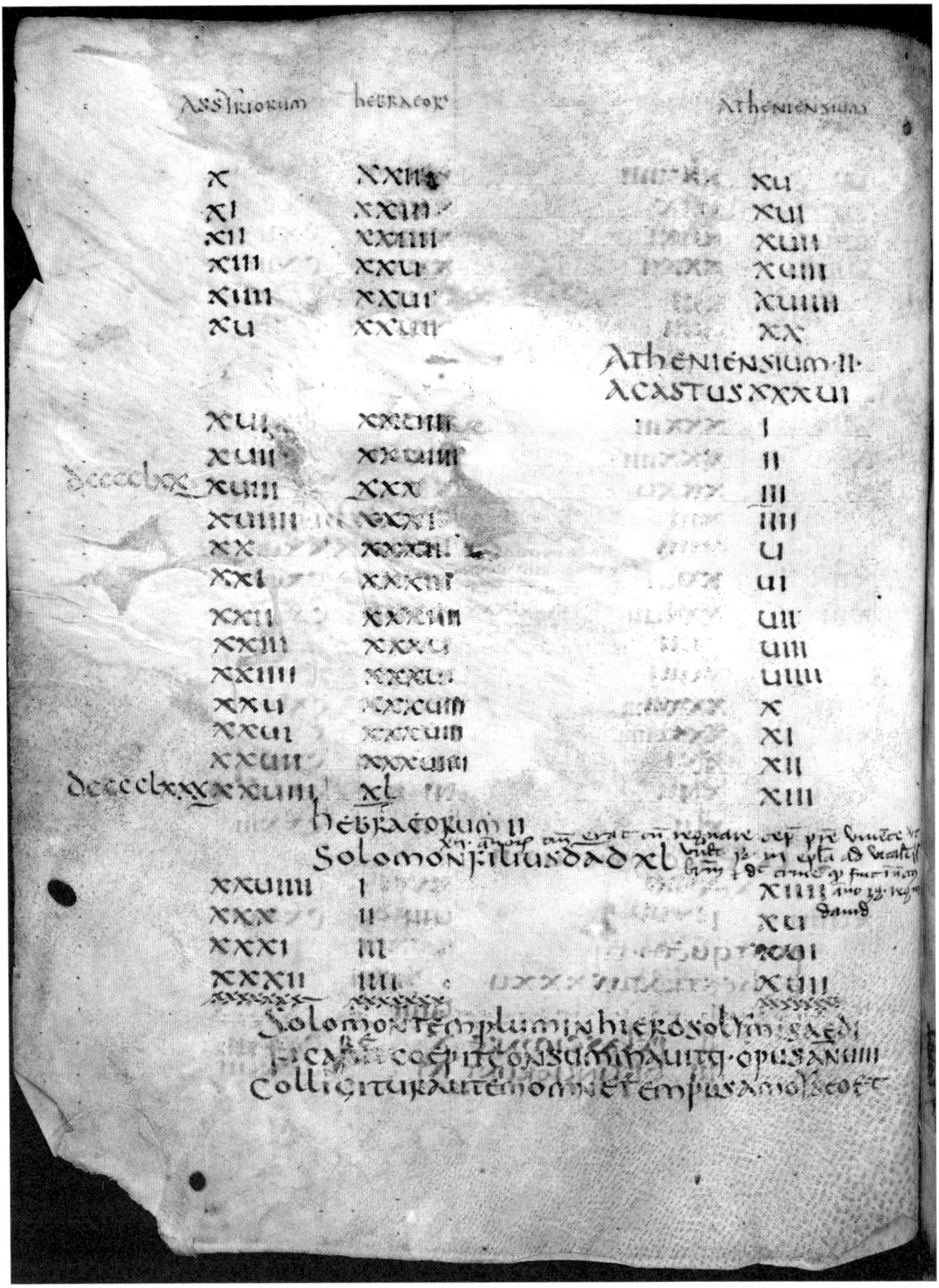

4.3

4.3 and 4.4 Oxford, Bodleian Library, MS Auct. T.2.26, fos 51v–52r (reproduced with permission from the Bodleian Libraries, University of Oxford). Again, note the *DCCCCLXXX* (= 980) in the left margin, marking the row of events associated with the 980th year since the birth of Abraham, as recorded in the seven columns corresponding to the rulers of the Assyrians, the Hebrews, the Athenians, the Latins, the Lacedemonians, the

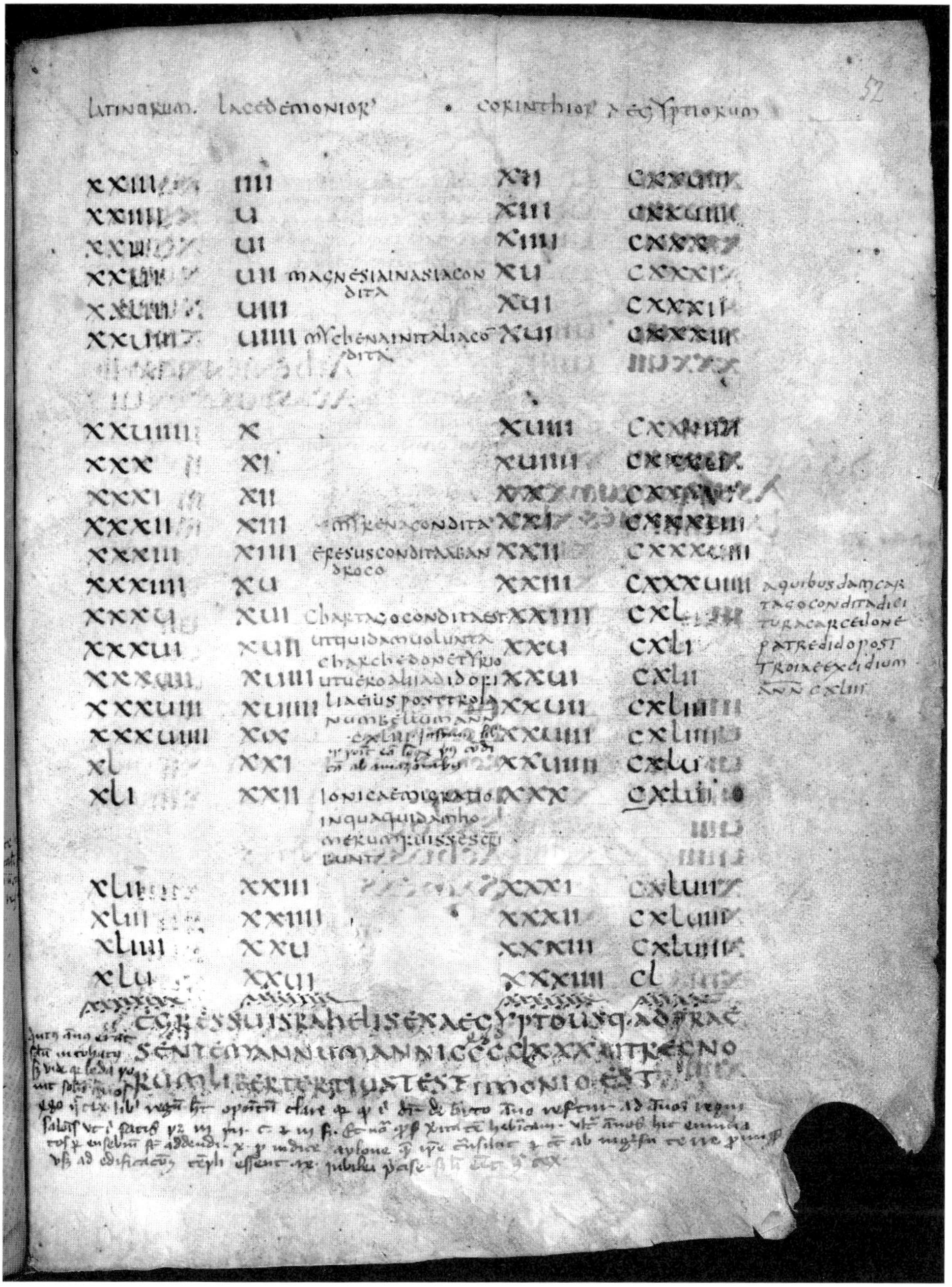

4·4

Corynthians and the Egyptians. The statement (in red ink) at the bottom of fo. 51v makes it clear that four years later, Solomon started building the temple in Jerusalem and that this construction lasted *UIII* (= 8) years. The text then goes on to state (still in red ink) that this occurred *CCCCLXXX* (= 480) years after the Israelites left Egypt (bottom of fo. 52r).

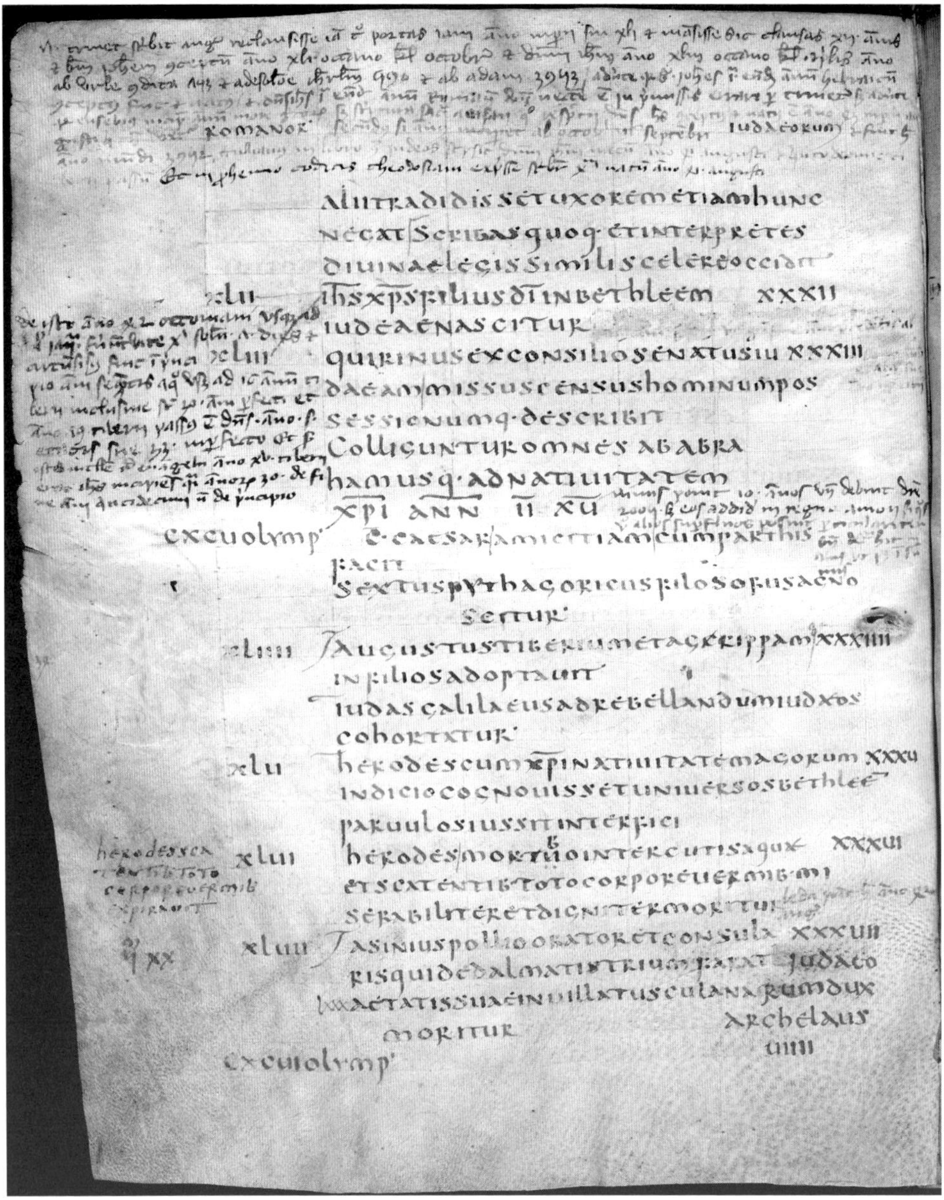

4.5 Oxford, Bodleian Library, MS Auct. T.2.26, fo. 109v (reproduced with permission from the Bodleian Libraries, University of Oxford). The birth of Jesus is recorded in red ink on lines 4 and 5: *IESUS CHRISTUS FILIUS DEI IN BETHLEEM IUDEAE NASCITUR*, followed a little later (again in red ink) with the statement that 2,015 years went by from Abraham to the birth of Christ: *COLLIGUNTUR OMNES AB ABRAHAM USQUE AD NATIUITATEM CHRISTI ANNI ĪĪXU*. Note the symbol for *ĪĪXX* (= 2020) in the left margin near the bottom of the text. Working backwards through the years associated with the Roman Empire makes it clear that Jesus was indeed born 2,015 years after the birth of Abraham.

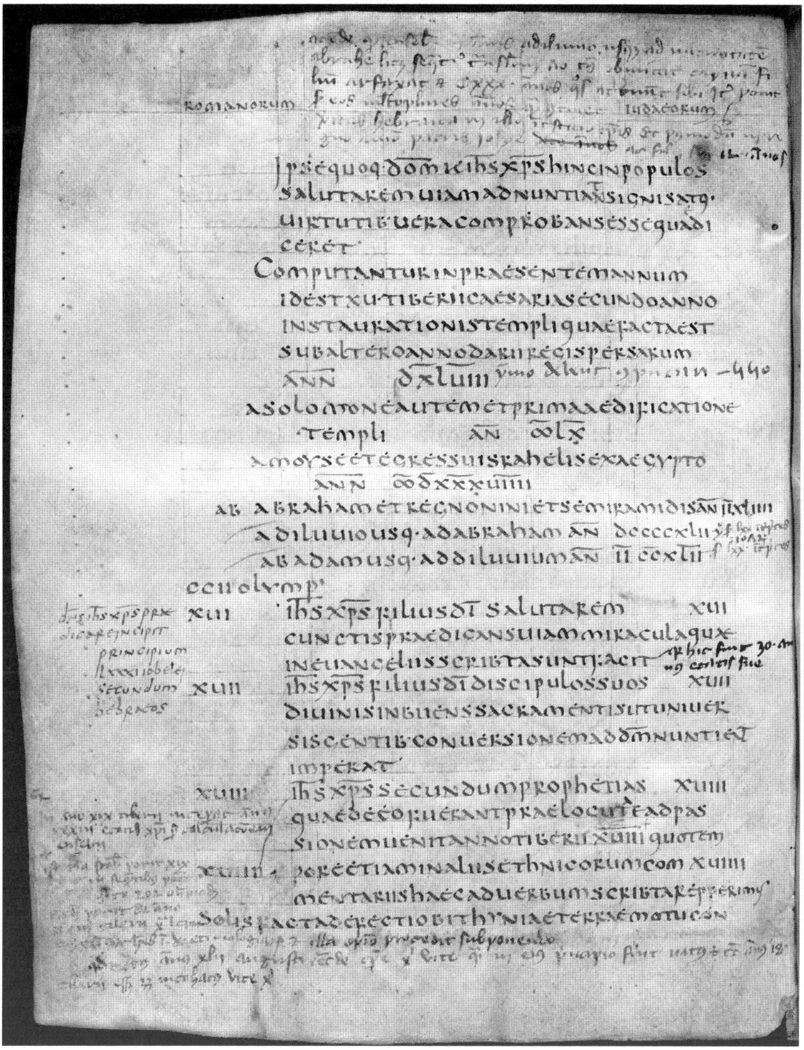

4.6 Oxford, Bodleian Library, MS Auct. T.2.26, fo. 111v (reproduced with permission from the Bodleian Libraries, University of Oxford). The text at the beginning of this page is the end of the entry corresponding to the fifteenth year of Emperor Tiberius, when Luke 3:1 states that Jesus was baptized and commenced his public ministry. This significant fifteenth year of Tiberius is then carefully situated within the history of the world: 548 years after the rebuilding of the Temple under Darius, 1,060 years since the building of the Temple, 1,539 years since the departure from Egypt and 2,044 years after the birth of Abraham (which was itself 942 years after the Flood, which was in turn 2,242 years after Adam). These numbers are all in red ink, and tend to be further emphasized by their location in the middle of a line. *The Eusebius/Jerome Chronicle* is explicit about the duration of Jesus' public life, spelling out activities by Jesus for each one of the years corresponding to the sixteenth and seventeenth year of Tiberius, and setting the Passion in the eighteenth year of Tiberius: *IESUS CHRISTUS SECUNDUM PROPHETIAS QUAE DE EO FUERANT PRAELOCUTAE AD PASSIONEM UENIT ANNO TIBERII XUIII.*

7 × 1,000 = 7,000, which leaves us with 1,760 hours,
7 × 200 = 1,400, which leaves us with 360 hours,
7 × 50 = 350, which leaves us with 10 hours,
7 × 1 = 7, which leaves us with 3 hours.

These 3 hours are said to build up each year toward the *bissextus*, the leap year day! Note how the distributive method is used for multiplication. Note also that it was considered much simpler to multiply by 12 and then double the result, than to multiply directly by 24 – a wise decision when using roman numerals. Division is done through approximation by multiplication – the very same thing we do with our algorithm, except that our technique lays it all out much more easily and we do not need to think about what we are actually doing – 7 times one thousand is 7,000, subtracting this from 8,760 leaves 1,760 hours, and so forth. These steps were not laid out neatly in the manuscripts, and the combination of prose, abbreviations and roman numerals made it look altogether more confusing.[72] Nonetheless, we thus have a good idea of how *relatively simple* arithmetic was carried out in Ireland in the second half of the seventh century. It is a bit clumsy, but these scholars understood what they were doing, which is more than most of us can claim nowadays!

72 As can be seen in the digital image of the section *De bissecto* in the ninth-century copy of this second paragraph of *Argumentum* 16 in Einsiedeln, Stiftsbibliothek, Codex 321 (647), p. 108. The image of this page is easily found on the internet by entering the elements of the shelf mark as keywords, then scrolling to p. 108; or try directly *http://*www.e-codices.unifr.ch/de/sbe/0321/108/small. Note that only the second paragraph of *Argumentum* 16 occurs in the *Computus Einsidlensis*, which Immo Warntjes ascribes to southern Ireland in the period between AD689 and AD719 (see *The Munich Computus*, pp cl–clii, where the relevant section of *Argumentum* 16 is edited from both the Einsiedeln manuscript and the later Munich Computus).

Cummian's letter: science and heresy in seventh-century Ireland

MAURA WALSH

In AD632 a cleric named Cummian wrote a closely reasoned and sometimes angry and emotional letter about the computation of Easter in response to a letter from two clerics named Ségéne and Béccán, who called Cummian a heretic. Why did Ségéne and Béccán call Cummian a heretic, and why is Cummian's response so impassioned?

Cummian's letter survives in one manuscript, which is now in London, British Library, Cotton Vitellius A.xii, fos 79r–80r.[1] It is part of a codex that gathered together a number of independent manuscripts written by late eleventh-century and early twelfth-century hands. The few folios in the codex that contain Cummian's letter also have another letter written in the same hand. Unfortunately, it is not Ségéne's letter to Cummian, which would have been rather wonderful. Alas, no copy of Ségéne and Béccán's letter to Cummian has yet come to light. Perhaps someday more folios from this manuscript with the missing letter will be discovered.

Although we can locate this compilation of independent manuscripts in southern England with some probability, we do not know where the folios containing Cummian's letter came from, or why they were copied by a twelfth-century English scribe. The world is full of mysteries, but for this one we are grateful!

The salutation of the letter identifies the writer as *Cummianus, supplex peccator* ('Cummian, a suppliant sinner'). Cummian addresses his letter to two people: Segienus and Beccanus.[2] Ségéne is the only one of these three that can be definitely identified. He was abbot of Iona, Columba's main foundation, from 623 to 652. Although Iona is a small island off the coast of Scotland, as the successor of Columba, Ségéne was a very powerful cleric on the island of Ireland, where there were a number of monastic communities that were part of the Columban paruchia. The monasteries in the Columban group formed a separate religious and political power-base in seventh-century Ireland.[3]

Cummian was a cleric and a scholar in a monastic community, most probably

1 For a complete description of the manuscript, see Maura Walsh and Dáibhí Ó Cróinín (eds), *Cummian's letter 'De controversia paschali' and the 'De ratione conputandi'* (Toronto, 1988), pp 51–3. 2 Ibid., p. 57; Latin text, p. 56, lines 1–5. 3 For a discussion on the structure of the early Irish church, see Dáibhí Ó Cróinín, *Early medieval Ireland* (London, 1995), pp 147–68.

in the south-central region of Ireland. It has been suggested that he may have been the bishop of Clonfert, Co. Galway, or of Durrow, Co. Offaly, but neither of these can be confirmed. There have been many articles that offer arguments for identifying Cummian with a number of Cummians who appear in other sources, such as the annals, poetry and saints' lives. However fascinating this question of identity is, it does not have a direct bearing on the theme of this essay, and as it is a complicated and disputed question, it will not be addressed further here.

Béccán cannot be positively identified either. Cummian refers to him as *caro carne et spiritu fratri* ('beloved in body and a brother in spirit'). He may be a brother or relative of Cummian, but it is possible that this is to be taken only in a religious sense. Again, there are a number of Béccáns in the contemporary sources and a number of scholarly conjectures about Béccán's identity, but again this will not be addressed here. Suffice to say that he joined Ségéne in disagreeing with Cummian.[4]

Although the letter was sent to Ségéne on Iona, the area where this particular dispute on the computation of Easter took place was in south-central Ireland. This identification of the location is based on the monastic representatives that Cummian mentions when he is describing the synod that is called to discuss the question. These are Ailbe of Emly, Co. Tipperary; Ciaran of Clonmacnoise, Co. Offaly; Brendan either of Clonfert, Co. Galway, or of Birr, Co. Offaly; Nessan of Mungret, Co. Limerick; and Lugaid of Cluain Ferta Mo-Lua, Co. Laois. The synod itself was held in Mag Léne, which may be identified with Moyleen, Co. Offaly.

The letter is dated by a combination of internal and external evidence. Firstly, we have dates for Ségéne's abbacy of Iona: 623–52.[5] Secondly, Cummian says that in the year in which the Irish delegation, which he sent out as we will see, was at Rome, the Irish and Roman Easter dates differed by a whole month. A comparison between the 19-year Easter table used in Rome and the 84-year cycle used in Ireland shows this occurring in 631. Thus, it is argued that Cummian wrote the letter in 632/3, as he says that the delegation that he sent out to Rome returned to Ireland in the third year after they left.

In his letter, Cummian is responding to Ségéne and Béccán and their party, who are very unhappy that Cummian and other clerics have decided to officially use an Easter table recently received from Rome. This Roman table was significantly different from the one that had been in use in the Irish churches from time immemorial and was one that had been rejected by Irish computists much earlier

4 See Walsh and Ó Cróinín, *Cummian's letter*, pp 7–15 for a discussion of the possible identities of Cummian and Béccán. The most recent suggestion for the identity of Cummian is Máirín Ní Dhonnchadha, 'The beginnings of Irish vernacular literary tradition', *L'Irlanda e Gli Irlandesi Nell'alto Medioevo, Spoleto, 16–21 aprile 2009* (Spoleto, 2010), pp 532–92. 5 Adomnán of Iona, *Life of Saint Columba*, trans. Richard Sharpe (London, 1995), pp 35–40.

on mathematical grounds. Ségéne and Béccán's letter (which we do not have) had apparently attacked this decision quite strongly, even going so far as to accuse Cummian and his party of heresy.

The 'Easter controversy' has at its root the complications of astronomy, mathematics, world chronology and theology that surround the calculation of the date of Easter for each year. The original Easter happened at the time of the Jewish Passover celebration, which begins with the sacrifice of the Paschal lamb on the fourteenth moon of the Jewish lunar month, Nisan, which starts on the first full moon of spring. However, when Christianity became the official religion of Rome, which used a solar calendar, the Christian church complicated the calculation by determining that Easter must also fall after the spring equinox and on a Sunday. This meant that the calculations for determining the proper date for Easter now had to reconcile the lunar year and the solar year, and these years and their months are not the same length. This introduced complications, which were the basis for disagreement not only in the British Isles, but also between the Eastern and Western churches. The goal of the church was that the whole of Christianity should be united in the date of its most important feast-day. Initially, the date of Easter in each year was determined in Rome and promulgated to the bishops throughout the Christian world. It quickly became apparent that sending out a letter from Rome with the proper date of Easter to the far corners of the world every year was not going to work. The letters often arrived too late or not at all. But what about a table of Easters for a number of years? This was a good idea, but nothing is easy. Tables were compiled giving the dates of Easter for a number of years. Unfortunately, the tables do not always agree on their dates because the mathematics that reconcile the lunar and solar year are used differently in each table. In addition, some tables used 25 March as the date of the spring equinox (the old Roman equinox), and some used 21 March (the Alexandrian equinox).

In the greater part of Ireland in the seventh century, the table that was used was an 84-year Easter table known as the latercus. Although modern scholars knew about the Irish form of the 84-year table, and although there existed one corrupt description that generations of scholars tried to interpret and reconstruct, it was not until 1985 that Dáibhí Ó Cróinín discovered a complete manuscript of this 84-year cycle that was accurate enough to reconstruct (the only known manuscript copy of the table).[6] This was a great step forward in the modern study of the science of medieval computus and our understanding of

6 This job was excellently done by mathematician and historian of mathematics, Daniel Mc Carthy: see Dáibhí Ó Cróinín and Daniel Mc Carthy, 'The "lost" Irish 84-year Easter table rediscovered', *Peritia*, 6–7 (1988), 227–42, repr. in Dáibhí Ó Cróinín, *EIHC* (Dublin, 2003), pp 58–75. Mc Carthy later revised the reconstruction on further information in 'Easter principles and a fifth-century lunar cycle used in the British Isles', *Journal for the History of Astronomy*, 24 (1993), 204–24.

the basis for the disagreements that are played out in the letter we are looking at. So, an 84-year cycle is pretty much the standard in Ireland and among the British (not the Anglo-Saxons in Britain) at the time of the letter. As a simplification for our purposes, this table observed lunar limits of xiv–xx and put the spring equinox on 25 March. This is only the tip of the iceberg, but these variations underpin the main theological, rather than scientific, ideas that Cummian is wrestling with.

In Gaul and among the Anglo-Saxons in Britain, however, another table was the standard. In Rome in 455 as a result of recurrent disagreements over the date of Easter, which had plagued the churches of East and West, Pope Leo I (440–61) asked Archdeacon Hilarus, soon to be Pope Hilarus (461–8), to do a thorough investigation of the Paschal problem and its resolution, a task that Hilarus passed on to Victorius. Victorius produced a 19-year cycle, which was expanded to twenty-eight 19-year cycles, at which point the data repeated – thus a perpetual table of 532 years. Seemingly, this was the answer to a prayer! But it was not without major problems. Victorius based his table on the Alexandrian 19-year cycle with the spring equinox on 21 March, but kept the Roman lunar limits of xvi–xxii. Because he knew that lunar limits xvi–xxii were not Alexandrian, he made a parallel set of calculations using the Alexandrian limits, xv–xxi, and these he placed side-by-side with the Roman dates in his table. In fact, to further confuse things, he did the Alexandrian calculation wrong. His view was that the pope should choose which of these double-dates should be used, but as the table spread with the double-dates included, there was much confusion about which date to use in years where there was a double-date. Victorius' double-dates were only the first of many problems with his table. However, in spite of this, Victorius' table was declared the official table for Gaul in 541 and was also used by the Anglo-Saxons in Britain. Again, very simply, this table had lunar limits of xvi–xxii for the Roman Easter dates and, where there were double-dates, lunar limits of xv–xxi for the Alexandrian dates. The spring equinox was set at 21 March to bring it into line with the Alexandrian table.

When the Victorian table first came to Ireland in the sixth century, the Irish were good enough computists to see the problems and inconsistencies in it and chose to continue using the 84-year latercus. As Columbanus said in his letter to Pope Gregory in about 600,

> You should know that Victorius has not been accepted by our teachers and by the old Irish experts and by the mathematicians most skilled in the calculation of the computus, but was considered more worthy of ridicule and pity than of authority.[7]

7 G.S.M. Walker (ed. and trans.), *Sancti Columbani opera* (Dublin, 1957), p. 7.

Things rested so for some time: the Irish continued to follow the 84-year latercus, which often gave dates for Easter that differed from the dates for Easter in the cycle that Rome, Gaul and Britain used. Trouble was brewing. In 628/9, the Irish churches received a letter from Pope Honorius, who warned them (as reported by Bede)

> not to imagine that their little community, isolated at the uttermost ends of the earth, had a monopoly of wisdom over all the ancient and new churches throughout the world, and he asked them not to keep a different Easter, contrary to the paschal calculations and synodical decrees of all the bishops of the world.[8]

Although Cummian does not specifically mention the letter of Pope Honorius (625–38), he does refer to the Irish as 'an insignificant group of Britons and Irish who are almost at the end of the earth, and, if I may say so, but pimples on the face of the earth'.[9] This may be an echo of Honorius' letter and this letter may have been the spur to the events that result in Cummian's letter. For Cummian notes that approximately four years before he wrote his letter, the cycle of 532 years began to be celebrated *a nostris* ('our party', 'our community', 'some of the Irish'?). This puts the beginning of the dispute at 628, which is the date that Bede has for the Honorius letter.

The debate on whether to adopt the Victorian 19-year table was hot and heavy in Ireland. Those who felt that the Irish should continue with the 84-year cycle did so on two main grounds. One was that they did not want to abandon the Easter table that was handed down to them by St Columba. The other was that their men of science and mathematics had shown convincingly that the version of the 19-year cycle that they first received, which was Victorius' table, was full of mathematical and astronomical error and was ambiguous because of the years in which there were double-dates. They really could not believe that Rome would approve such an error-ridden Easter table. On the other side, those who wished to adopt the Victorian table pointed out that the Victorian table was approved of by Pope Hilarus and, as far as they knew, was used throughout the universal church. They could also accuse those who used the 84-year cycle of falling into theological error, and even heresy. Cummian's letter did not convince Ségéne, Béccán and their party to change their Easter table. The Irish church continued to be split between the 'Roman' party and the 'latercus' party. Cummian describes the major disagreement in 630 that we are looking at. Bede describes another major ballyhoo at the Synod of Whitby in 664 when most of the Columban communities changed to a 19-year table, and the end of the split

8 Charles Plummer, *Baedae opera historica* (1896; repr. 1946); *Historia ecclesiastica gentis Anglorum*, Lib. II cap. xix, p. 122. 9 Walsh and Ó Cróinín, *Cummian's letter*, pp 73, 75; Latin text, p. 72, lines 109–10.

does not happen until 715, when Iona makes the decision to adopt the 'Roman' Easter.

Cummian tells us that when the 19-year table under discussion arrived (possibly with Honorius' letter), although some began to use it right away, he and others did not accept it immediately. He writes that he decided to do some research. He cloistered himself away for a full year and studied the Bible, the church Fathers, the canons of the church and any Easter cycles he could find. On the basis of his conclusions, he called a synod of bishops and abbots (or their representatives) to present the results of his study and to decide what route to take. This synod, after much discussion, decided to start using the 19-year table at the next Easter, but just as they were winding up with a prayer, a certain *paries dealbatus* ('whited sepulchre') got up and once again spread dissension. When Cummian describes this, his frustration and anger are palpable. The synod broke up in confusion. At this point, a delegation was sent to Rome *iuxta decretum synodicum* ('according to synodical decree') to see what Easter was celebrated there. Cummian reports on their journey:

> And they saw all things just as they had heard about them, but they found them more certain inasmuch as they were seen rather than heard. And they were in one lodging in the church of St Peter with a Greek, a Hebrew, a Scythian and an Egyptian at the same time at Easter, in which we differed by a whole month. And so they testified to us before the holy relics, saying: As far as we know, this Easter is celebrated throughout the whole world.[10]

As a result of this testimony, Cummian and those who agreed with him made their decision to use the 19-year table from then on. Representatives of the 'latercus, or 84-year' party at the synod apparently reported to Ségéne and Béccán on Cummian's presentation and decision subsequent to the return of the pilgrims, because they then sent to Cummian what, on the basis of Cummian's response, was a stinker of a letter, pointing out his lack of loyalty to Columba's Easter cycle, the computistical errors of the table he was supporting, and the theological heresy that he had fallen into. Although we do not have Ségéne and Béccán's letter to Cummian, all this is quite clear from Cummian's response. Cummian is outraged at their accusations, and in his letter, both defends his methodology for making his decision and tries to convince Ségéne and Béccán of the danger of the error they have fallen into.

Cummian reminds them that, first of all, he did not just jump on the bandwagon – he held back to study the situation. Cummian's initial reluctance to accept the Victorian table may reflect his judgment of the flawed mathematics

10 Ibid., pp 93, 95; Latin text, pp 92, 94, lines 280–5.

of that table, but he had what he felt were even more serious points to consider: whether the 84-year table was theologically sound and whether celebrating Easter on a date different to that of the rest of the church was a matter for excommunication or an action that would leave them open to an accusation of heresy.

And now we see the scholar at work: 'Hence, having cloistered myself for a year and having entered the sanctuary of God (that is sacred Scripture) I studied as much as I was able, then I examined the histories' (here he is referring to the writings of the Fathers of the church), 'and finally the cycles which I could find'.[11] This methodology is more than just good scholarship. The hierarchy of authorities that Cummian uses is one that is found in a canon ascribed to Pope Innocent I in the *Collectio Canonum Hibernensis*, which advises:

> first look in the Old Testament, then the four gospels and all the writings of the Apostles. Then in the lives of the saints. If the answer is not here, look in the histories of the church and the writings of the catholic doctors. If not here, inspect the canons of the apostolic seat.[12]

I think there is no doubt that Cummian is putting together not only an academic argument but also a legal argument, based on this canon.[13]

Cummian's biblical studies are concerned with ascertaining the correct lunar limits for the Pasch and Resurrection, based on the evidence of the description of Passover in the Old Testament and the Passion in the New. With the help of Jerome's commentaries on the Bible, he soon comes to believe that the Irish are in danger of being outside the church and heretical.

His first biblical texts are from Exodus, 12:2, 12:6, 13:3, 13:6, where the Passover and the days of unleavened bread are described. He finds that the sacrifice of the lamb happens on the fourteenth moon of the first month (Nisan). He follows this with a quote from 1 Cor 5:7: 'And I found that the Apostle spoke thus concerning the sacrifice of Christ, not concerning the Resurrection, saying: For Christ, our Paschal lamb, has been sacrificed'.[14] He then quotes Leviticus, Num 9:13: 'If anyone is clean and is not on a journey yet refrains from keeping the Passover, that person shall be cut off from his people, because he did not offer the Lord's offering at its appointed time'. And Cummian says *Et hoc timui* ('and I feared this').[15]

This gives us a good example of his methodology and a clue to one of the main points of his argument. If the Irish insist on continuing to follow their Easter cycle, they will be 'cut off from God's people because they do not offer

11 Ibid., p. 59; Latin text, p. 58, lines 15–17. 12 Hermann Wasserschleben (ed.), *Die irische Kanonensammlung* (Giessen, 1874/Leipzig 1885; repr. Aalen, 1966), p. 59. 13 Walsh and Ó Cróinín, *Cummian's letter*, pp 15–18 for a detailed look at the structure of the letter. 14 Ibid., p. 59; Latin text, p. 58, lines 23–5. 15 Ibid., p. 59; Latin text, p. 58, lines 27–9.

the Lord's offering at its appointed time'. Why? Because the Old Testament celebration of the sacrifice of the lamb on luna xiv parallels the New Testament's passion of Christ, not his resurrection, so that it is impossible that the Resurrection be on luna xiv. But the 84-year table has Easter limits of xiv–xx. *Et hoc timui!* ('And this I feared!').

Again: Jerome explaining the lunar limits of Easter says 'What more is necessary as testimony? The Law speaks, the Apostle proves, this remains: that the contradictor be cast out as a liar [...]' and Cummian says *Hoc timui et me perculit* ('I feared this and it disturbed me').[16] His conclusion of the biblical section of the letter makes the point once again: 'For, if the fourteenth moon were assigned to the Resurrection, as you do, then the thirteenth would fall on the Sepulture and the twelfth on the Passion, in an absurd order'.[17]

Having established to his satisfaction (but also to his horror!) that the 84-year cycle has uncanonical lunar limits, he turns to the writings of the Fathers of the church. His focus here is mainly the importance of the unity of the church as expressed by celebrating Easter on the same day throughout the Christian world. He finds much that disturbs him:

> In the canons of the synods [...] I found it written that 'those are to be excommunicated and driven from the church' and anathematized who go against the canonical statutes of the fourfold Apostolic See, namely that of Rome, Jerusalem, Antioch and Alexandria, which agree on the unity of Easter.[18]

He quotes the synod of Nicaea and the Synod of Arles and then quotes Jerome:

> Whence care should be taken [...] lest we eat the figural 'lamb' outside of the one house against the precept of the Law, that is outside of the universal church. From which it is manifest that Jews and heretics and all conventicles of perverse doctrines who do not eat the Lamb in the one church do not eat the flesh of the Lamb, but that of the dragon that was given, according to the Psalter, as food to the Ethiopian people.[19]

And Cummian says:

> *Et hoc uereor!* And this I am in fear of! But you ought to consider which are the conventicles of which I spoke, whether they are the Hebrews,

16 Ibid., p. 61; Latin text, p. 60, lines 47–51. **17** Ibid., p. 69; Latin text, p. 68, lines 84–5. **18** Ibid., pp 69, 71; Latin text, pp 68, 70, lines 86–93. The phrase *excommunicandos et de ecclesia pellendos* seems to come from the Prologue to the Dionysiac table, where it is quoted as a canon of the Council of Antioch. For a discussion of the source, see Walsh and Ó Cróinín, *Cummian's letter*, pp 70–1, n. 90. **19** Ibid., p. 73; Latin text, p. 72, lines 101–7. Here, Cummian is summarizing a passage from Jerome's *In Vigilia Pascha*, *CCSL*, 78:536.15–17.

> Greeks, Latins and Egyptians, who are united in their observance of the principal solemnities, or an insignificant group of Britons and Irish, who are almost at the end of the earth, and, if I may say so, but pimples on the face of the earth. Judge this for me. For with the judgment you pronounce, you will be judged, and this is the danger hanging over you. For you are the heads and the eyes of the people, and if they are led into error because of your obstinacy, you shall answer, according to Ezechiel, for the blood of each soul to the strict Judge.[20]

The Columban party accused Cummian of being a traitor to Columba by abandoning his Easter cycle and adopting instead the 19-year cycle in order to join the universal church. He defends himself and absolves the elders of the Irish church of any wrongdoing, but emphasizes that Ségéne and Béccán can no longer plead ignorance:

> Our elders, however, whom you hold as a cloak for your rejection, kept simply and faithfully without blame of any contradiction or animosity, that which they knew to be best in their day, and prescribed for their posterity thus, according to the Apostle: Test everything, hold fast what is good, abstain from every form of evil. *Sed vereor.* But I fear lest it happen that you should judge the unanimous rule of the Catholic church to be a form of evil, which heaven forbid! Need I say more?[21]

After which he goes on for several more pages.

In his research, Cummian has discovered quotes from practically every one of the Fathers of the church that paint a black picture for anyone who puts himself outside of the unity of the church. Note his reaction to what he found: *Angustiae michi undique!* 'Troubles on all sides of me! If I shout this with Jerome [...] I am opposed by you. If I do not cry out, I am excommunicated by the universal Catholic church'.[22]

Having quoted Cyprian on the unity of the church, he says *His perterritus*: 'Terrified by these things I turned to Augustine'. Here he found Augustine on church unity saying 'Whence it is written in the law: He who curses his father or mother shall be put to death'. And Cummian says 'What, then, more evil can be thought about Mother Church than if we say Rome errs, Jerusalem crrs, Alexandria errs, Antioch errs, the whole world errs; the Irish and British alone know what is right?'[23]

20 Walsh and Ó Cróinín, *Cummian's letter*, pp 73, 75; Latin text, pp 72, 74, lines 107–14. Here, we find what may be an echo of the letter that Honorius wrote to the Irish, reported by Bede in his *Historia ecclesiastica gentis Anglorum* in Plummer, *Baedae opera historica*, p. 122. 21 Walsh and Ó Cróinín, *Cummian's letter*, p. 75; Latin text, p. 74, lines 114–21. 22 Ibid., p. 79; Latin text, p. 78, lines 146–9. 23 Ibid., p. 81; Latin text, p. 80, lines 173–80.

He quotes Augustine, who says 'because whoever is outside of the unity, although he comes he cannot be healed', and Cummian says *Et hoc michi horret*, 'And this horrifies me, unless the trustworthy reasoning of your counsel should lead me to a more valid and more certain opinion'.[24]

Quoting Gregory on the church as protector, Cummian says *Et hoc timeo*, 'And I fear this. You consider carefully, brothers, whether it is with good reason or through cowardice'.

There is nothing dry and dusty here – no academic matter, this is a serious and quite emotional matter for Cummian. His fear of being outside of the unity of the church, excommunicated, anathematized or considered a heretic is palpable and he wants to convince Ségéne, Béccán and their party that they are in danger.

Now Cummian turns to the evidence of the computistical texts he has found. In his letter, Cummian identifies ten different Easter cycles that had been used by the church at some time. For some he has details, for others not. In fact, not all his Easter cycles are actually Easter tables. Some are writings about Easter that were never attached to tables. Each of these 'cycles' has a name attached to it, giving it some authority, and he notes that the Irish 84-year cycle is one 'whose author, place and time we are uncertain of'. The first Easter cycle Cummian notes is 'that which holy Patrick, Papa Noster ('our bishop'), brought and followed'. This is the earliest datable reference to St Patrick in Irish literature.[25] Cummian goes on to mention cycles by Anatolius, Theophilus, Dionysius, Cyril, Morinus, Augustine, Victorius, Pachomius (founder of the monasteries of Egypt), and finally 'the 19-year cycle of the 318 bishops [...] by which the feast of Easter can with certainty be found. Indeed, I found this cycle contrary in the calends, the bissextile, the epact, the fourteenth moon, the first month and the equinox' to the Irish 84-year cycle.[26]

So who are the heretics? Cummian's studies convince him that the 84-year cycle, with its lunar limits of xiv–xx, put the Irish in danger of being accused of the heresy of quartodecimanism, in which the Resurrection was celebrated on the fourteenth moon with the Jews. True quartodecimanism celebrates Easter on luna xiv every year. The Irish 84-year table allowed Easter to be celebrated on luna xiv when it fell after the spring equinox and on a Sunday, which would only happen sporadically. But this was enough for Cummian. Cummian also concludes that putting themselves outside the unity of the church, as they are doing, also puts them in danger of accusations of heresy and excommunication.

But, according to Ségéne and Béccán, Cummian and his party are the

24 Ibid., pp 81, 83; Latin text, pp 80, 82, lines 187–9. 25 Dáibhí Ó Cróinín has suggested to me that simply by mentioning a cycle brought by St Patrick to Ireland (even if it is no longer used by the Irish), he is reminding the Columban party that their championing of their cycle because it was Columba's cycle is not a valid reason to reject anything new (pers. comm.). 26 Ibid., pp 83, 85, 87; Latin text, pp 82, 84, 86, lines 204–22.

heretics. Quoting Paul and Augustine on the confusion that can arise between the Pasch on the fourteenth moon and the Resurrection, as we have seen, Cummian says:

> And this I ask: Consider diligently how you might excuse me or direct me to some better understanding by your words or with more valid writings and more certain proofs – if you have any – and I will gratefully accept it, as I have accepted this. If, however, you do not have any, be silent and do not call us heretics.[27]

On what basis were Ségéne and Béccán calling Cummian and his party heretics? The Victorian table used lunar limits of xvi–xxii for possible dates for Easter and a spring equinox of 21 March. These were found by Irish computists to be in grievous error. We can go to Columbanus to understand what Ségéne and Béccán most likely based their accusation on. Firstly, on the Victorian lunar limits, Columbanus says 'What then do you say about Easter on the twenty-first or twenty-second moon, which already [...] is proved by many computists to be no Easter, considering its darkness'. And he explains that these moons rose after midnight and hence darkness ruled over light, which could not be, since the feast of Easter was a celebration of light's victory over darkness.[28] The second problem with Victorius is even more grievous from the Irish computists' point of view. Again, Columbanus:

> We also read in the book of sacred dogma [Gennadius, *De dogm. eccl.* 87]: 'Easter, that is the festival of the Lord's resurrection, cannot be celebrated before the passing of the spring equinox, the beginning of the fourteenth moon', namely to avoid its preceding the equinox. Victorius has certainly broken this rule in his cycle and thus has long since introduced error into Gaul.[29]

Why did Columbanus and the Irish computists claim that Easter dates in Victorius preceded the spring equinox? Victorius used 21 March as the spring equinox, while the Irish used 25 March. It did happen, then, that in Victorius' table, Easter could occur before 25 March, which would, in their lights, indeed be heretical. Columbanus noted that the 21 March equinox had been roundly condemned by Anatolius, who had in turn been praised by Eusebius and Jerome. Thus, Ségéne and his party rejected 21 March and accused Cummian of heresy on two counts: using a 'dark' moon and celebrating before the equinox.

In conclusion, the letter of Cummian gives us an unusually personal glimpse

27 Ibid., p. 75; Latin text, p. 74, lines 129–32. 28 Walker, *Sancti Columbani opera*, p. 2.3. 29 Ibid., p. 4.3.

into the rifts between communities and people caused by the argument over the correct way to compute the date of Easter. It also shows us that, for Cummian, Ségéne and Béccán there is no dividing line between biblical exegesis and mathematics, each dependent on the other to argue their approach. This look at the structure and sources of Cummian's letter shows a seventh-century scholar in action: reading, researching and finally writing a beautifully structured argument, which throughout shows the emotional strength of his disagreement with the opposing party. It reminds us that the main purpose behind the mathematics of computus was to produce a table of Easters that supported and embodied the theological truths of the church. While computists continued to work on the nearly intractable mathematical problem of marrying the lunar and solar years, in fact, one of the most mathematically problematic Easter cycles became, for a while, the official table of the Western church. The choice for Cummian and the Irish was between unity of the church and scientific rigour. The Irish remained divided on this point for eighty-four more years after Cummian wrote his letter.

Music and the stars in early Irish compositions

DAVID HOWLETT

Among ancient and medieval men the *artes liberales*, the 'liberal arts', were studies appropriate to *homines liberales*, 'free men', and they were not discrete. There were seven of them, and they belonged together. Cicero in *De oratore* III 127 about 55BC, Vitruvius in *De architectura* between about 50 and 26BC, and the younger Seneca, 4BC–AD65, in *Epistula* LXXXVIII all wrote about the liberal arts, and in AD428 Augustine in the *Retractationes* referred, apparently for the first time, to the number seven. But for the Middle Ages the writings of four men of the sixth and seventh centuries established the canon of seven liberal arts divided into the *triuium* and the *quadruuium*. Martianus Capella wrote a prosimetron, *De nuptiis Philologiae et Mercurii*, to commend the union of the love of language, *philologia*, with intellect, represented by the god Mercurius, book I about the betrothal of Philology and Mercury, book II about their wedding, and the last seven about her dowry, the liberal arts. The senator and consul Anicius Manlius Torquatus Severinus Boethius, AD480–524, wrote explicitly in *De institutione arithmetica* I 1 and in *De institutione musica* II 3 about the coherence of the four mathematical disciplines of the *quadruuium*. The senator Cassiodorus, AD480–575, referring often specifically to Boethius, devoted the entire second book of his *Institutiones* to the seven liberal arts. Finally, Isidore bishop of Seville, AD570–636, made at the beginning of his encyclopaedic *Etymologiarum siue originum liber*, published in 640 by his successor Braulio, a definitive statement of the seven liberal arts.

All these authors and texts were known to early Insular writers.[1] Works of Cicero, Martianus Capella, Boethius and Isidore were known early among the Irish.[2] The oldest extant copy of Cicero *De oratore* survives written by the hand of Lupus of Ferrières, copied from a manuscript from an Insular centre lent between 829 and 835 by Einhard, biographer of the Emperor Charlemagne.[3] Alcuin, the architect of Charlemagne's educational reforms, had read Vitruvius.[4] He had also read the earliest extant Anglo-Latin author Aldhelm (d. 709), who

1 For accounts of individual writers, see L.D. Reynolds (ed.), *Texts and transmissions: a survey of the Latin classics* (Oxford, 1983; corr. repr. 1986). 2 D. Ó Cróinín, 'The Irish as mediators of antique culture on the Continent' in P.L. Butzer and D. Lohrmann (eds), *Science in Western and Eastern civilization in Carolingian times* (Basel, 1993), pp 41–52. 3 C.H. Beeson, *Lupus of Ferrières as scribe and text critic: a study of his autograph copy of Cicero's 'De oratore' with a facsimile of the manuscript* (Cambridge, MA, 1930). 4 E. Duemmler (ed.), *Epistolae Karolini aevi II* (Berlin, 1895), no. 308, pp 471–3 at p. 472.

cites material from Seneca's *Agamemnon* in *De pedum regulis* chapter 140,[5] and Bede (d. 735), who cites material from Seneca's *Epistula* XCIV in his Commentary on Genesis.[6] There are Hiberno-Latin commentaries on *De nuptiis*, notably by Iohannes Scottus Eriugena.[7] The first scholars to cite Martianus, Boethius and Isidore outside their original milieux are Irish.[8] Manuscripts of all the other texts survive from Anglo-Saxon England.[9]

The canon of the liberal arts as defined by Martianus Capella, Boethius, Cassiodorus and Isidore was divided into the *triuium* or 'three ways' and the *quadruuium* or 'four ways'. The former included the three human linguistic arts of grammar, rhetoric and logic, and the latter the four divine sciences of arithmetic, music, geometry and astronomy; arithmetic as static number, music as moving number, geometry as measurement of the static earth, and astronomy as measurement of the moving heavens, all considered as reflections of the Mind of God expressed in number. These seven liberal arts were early and often identified with the seven pillars that Wisdom hewed for her own house in Proverbs 9:1. All seven arts are present, three implicitly and four explicitly, in the cosmogony of Job 38:4–7.[10]

There is a recurrent insistence, not only that the precision of the divine mathematical sciences of the *quadruuium* be brought to bear on the human linguistic arts of the *triuium*, but that within each discipline practice be closely allied with theory. This is implicit in Cicero's affirmation that quadruvial music and trivial poetry were originally practised by the same artists. It is implicit also in Vitruvius' distinction between *fabrica* and *ratiocinatio*, and in Boethius *De consolatione philosophiae*, in which Philosophia appears in book I prose 1 in a garment of her own weaving, bearing both the letters Π for πρακτικη 'practice' and Θ for θεωρητικη 'theory', rent by her followers, but conceived and executed as one whole.

In each of what the early Irish considered *tres linguae sacrae* 'three sacred languages', Hebrew, Greek and Latin,[11] a single notational system served to record letters of the alphabet and numbers and musical notes. In the twenty-two-letter Hebrew alphabet א = 1, ב = 2, ג = 3; in the twenty-four-letter Greek

5 R. Ehwald, (ed.), *Aldhelmi opera* (Berlin, 1919; repr. 1961), p. 194. 6 C.W. Jones (ed.), *Libri quatuor in principium Genesis usque ad natiuitatem Isaac et eiectionem Ismahelis adnotationum*, *Bedae Venerabilis opera*, *Pars I: Opera exegetica* (Turnhout, 1967), pp 5–6. 7 M. Lapidge and R. Sharpe, *A bibliography of Celtic-Latin literature, 400–1200* (Dublin, 1985), nos 704–6, pp 188–90. See also no. 721, p. 194, and no. 1182, p. 316. 8 J.N. Hillgarth, 'Ireland and Spain in the seventh century', *Peritia*, 3 (1984), 1–16. D. Howlett, 'Seven studies in seventh-century texts', *Peritia*, 10 (1996), 1–70 at 1–6; idem, 'Hibero-Latin, Hiberno-Latin and the Irish foundation legend', *Peritia*, 19 (2005), 44–60 at 47–8. 9 H. Gneuss, 'A preliminary list of manuscripts written or owned in England up to 1100', *Anglo-Saxon England*, 9 (1981), 1–60; H. Gneuss, *Handlist of Anglo-Saxon manuscripts: a list of manuscripts and manuscript fragments written or owned in England up to 1100* (Tempe AZ, 2001). 10 D. Howlett, *The Celtic Latin tradition of biblical style* (Dublin, 1995), pp 6–8; idem, *British books in biblical style* (Dublin, 1997), pp 7–9. 11 D. Howlett, '*Tres linguae sacrae* and threefold play in Insular Latin', *Peritia*, 16 (2002), 94–115.

alphabet Α = 1, Β = 2, Γ = 3; and in the twenty-three-letter Latin alphabet A = 1, B = 2, C = 3. To this day we designate musical notes as ABCDEFG. In such a system every written word has a semantic meaning and a numerical value and a musical value.

In the Hebrew Old Testament a conspicuous example of gematria occurs in the account of the first of the Israelite Judges in Judges 3:7–11, in which the fifty-second word is כָּלֵב, of which the alphanumeric value is 52. In the Greek New Testament, an even more famous example occurs in Apocalypse 13:18, in which the number of the Beast is εξακοσιοι εξηκοντα εξ, 600, 60, 6. One expresses the written words of the number 666 from this Greek text as DCLXVI,[12] exhibiting the descending value of the Roman numerals for 500, 100, 50, 10, 5, 1. The name implied is the Latin NERO CAESAR, but spelled in Hebrew letters נרון קסר to produce 50+200+6+50+100+60+200 or 666. Within this passage one sees a multilingual exercise that includes all *tres linguae sacrae*, a Greek text, Roman numerals and a Latin name spelled in Hebrew letters. The most famous example of the technique in Latin literature is *De nuptiis*, in which at the beginning of book II Philologia calculates multilingual gematria on her name and that of Mercurius to determine their compatibility.[13]

MO-SINU MACCU MIN

The first text in the *Bibliography of Celtic-Latin literature*, Part B. Ireland, is a note, copied during the eighth century, preserved in Würzburg, Universitätsbibliothek, MS M.p.th.f.61, fo. 29.[14] In the text that follows, boldface represents *litterae notabiliores* and punctuation marks of the manuscript; italics mark rhyme; acute and grave accents mark rhythms of the cursus.

12 According to a treatise about Antichrist by Henry Harclay, a fourteenth-century chancellor of the University of Oxford, the name of Antichrist will be DICLVX, he who 'says' he is the 'light'. 13 For a tabulation of the Hebrew system, see *Gesenius' Hebrew grammar*, ed. E. Kautzsch [2nd Eng. ed. A.E. Cowley] (Oxford, 1910; repr. 1985), §5.2, p. 26. For evidence of knowledge of the shapes and names of letters of the Hebrew alphabet, see the seventh-century Old-Irish text *Auraicept na n-éces, the scholars' primer*, ed. & trans. G. Calder (Edinburgh, 1917; repr. Dublin, 1995), pp 86–7, 229–30. For early Insular explications of the Greek system, see Bede, *De temporum ratione*, ed. C.W. Jones, *Bedae Venerabilis opera Pars VI Opera Didascalica, II* (Turnhout, 1977), pp 272–3, and the seventh-century *Auraicept na n-éces*, pp 230–1. For actual calculation of Greek values, see the eighth-century Hiberno-Latin manuscript Milan, Biblioteca Ambrosiana F.60 sup., fo. 61rb. Howlett, *Celtic Latin tradition*, pp 29–32, *British books in biblical style*, pp 34–7. For a fourteenth-century Anglo-Norman poem that illustrates the phenomenon while discussing it, see D. Howlett, 'Gematria, number and name in Anglo-Norman', *French Studies Bulletin*, 27 (101) (2006), 90–2. 14 D. Ó Cróinín, 'Mo-sinu maccu Min and the computus at Bangor', *Peritia*, 1 (1982), 281–95. D. Howlett, 'Five experiments in textual reconstruction and analysis', *Peritia*, 9 (1995), 1–50 at 1–3; Howlett, 'Insular Latin writers' rhythms', *Peritia*, 11 (1997), 53–116 at 59–60.

Mo Sinu maccu *M*in scríba et *àb*bas B*é*nn*cuir* [space]	8	13	35
primus Hibernensium computum a Graeco quodam sapiente memor*áliter* d*í*d*icit* ..	9	27	65
Deinde Mo Cuoroc máccu Net Sémon .	6	11	27
quem Romani . doctorem totius múndi nominábant [space]	6	16	39
alumnúsque praefàti scríbae—	3	9	25
in insula quae dicitur Crannach Duin Lethglaisse hanc scienti*a*m *líter*is f*íxit*	11	24	67
ne *m*emória l*àb*er*étur* ;	3	9	18
	46	109	276

MS 2 hebernensium. compotem. dedicit. 3 maccumin. 6 insola.

'Mo-Sinu maccu Min, scribe and abbot of Bangor,
first of Irishmen learned by memory the computus from a certain Greek wise man.
Then Mo-Cuoróc maccu Net Sémon,
whom the Romans named teacher of the whole world,
and a pupil of the foresaid scribe,
in the island which is called Crannach of Downpatrick fixed this learning in letters,
lest the memory lapse.'

As Bangor was founded in the 550s and the death of Mo-Sinu is recorded in the Annals of Ulster *s.a.* 610, the oral learning of computus from a Greek occurred during the second half of the sixth century, about 120 years after the introduction of Latin learning to Ireland by the mission of Palladius from AD431 and the mission of St Patrick, according to tradition from 432. We know that in two traditions that survived for thousands of years, the Sanskrit in India and the Hebrew in Israel and the diaspora, the most intensely literate transmission coincided with intensely oral transmission. This note states explicitly that the beginning of the Irish tradition was both oral and literate.

In addition to making immediately comprehensible sense, the limpid prose of this text conforms with the rules of artistic composition by ending every line with a good Ciceronian clausula. The first line, a double spondee, is balanced by the last, a double trochee. The second line, a cretic-resolved trochee, is balanced by the penultimate, a cretic-trochee. The third, a cretic-trochee, is balanced by the antepenultimate, a molossus-trochee, around the central line, a double spondee and a molossus-double trochee. This coincides with the rhyme of the first line with the last and the rhyme of the second with the penultimate. The prose conforms also with the rules of the cursus, the first, fifth, and last lines ending in *velox*, the second in *tardus*, the third and sixth in *planus*, the fourth in *trispondiacus*.

In visual terms, the text exhibits spaces between words, a convention developed among the Irish perhaps during the sixth century, certainly by the seventh. It begins with a *littera notabilior* and the ends of lines are marked, the first and fourth by unusually large spaces, the second, third and seventh by punctuation points, and the fifth by a dash.

This text about historical events incorporates calendrical phenomena. There are seven lines, one for each day of the week. There are six times as many letters (276) as words (46), six representing perhaps the six days of the Hexaemeron, forty-six the number of words in the account of the perfection of Sabbath rest in Genesis 2.1–4,[15] 276 the number of days from the Annunciation, celebrated on the supposed anniversary of Creation, 25 March, to the Nativity, celebrated on 25 December.[16]

We know Mo-Sinu as Sinilis, tutor of Columbán, from Jonas' *Vita Columbani*,[17] and as Sinlanus, from the poem *Memoria abbatum nostrorum, Sancta sanctorum opera*, composed perhaps in AD683, extant in the Antiphonary of Bangor:[18]

Elegit sanctum Sinlanum	'He chose holy Sinlanus,
Famosum mundi magistrum.	famous master of the world'.

MO-CUORÓC MACCU NET SÉMON

Dáibhí Ó Cróinín has identified Mo-Sinu's pupil Mo-Cuoróc maccu Net Sémon with a man mentioned in Irish genealogical collections:[19]

Mo-Chuaróc ind ecnai	'Mo-Chuoróc of the wisdom
noéb ná dámair dígnae	a saint that endured not reproach'.

According to the Annals of Ulster, Domnall mac Ardhgair mic Lochlainn died on the feast of *Mochuaroc ind ecnai*, that is Wednesday, 9 February 1121. 'Little Cuor', patron saint of Cell Chuaráin, Kilcoran near Youghal, Co. Cork, is mentioned also in one manuscript of the commentary of the *Félire* as *Mochuaroc na nona* 'Mo-Cuoróc of the None'.

The poem that follows was the most widely disseminated computistical verse of the early Middle Ages, known by heart by every schoolboy, because, laden as it is with numbers and mnemonic devices, it:

15 Howlett, *Celtic-Latin tradition*, pp 33–45; idem, *British books in biblical style*, pp 37–50. 16 For other examples of this in Insular literature, see Howlett, *British books in biblical style*, p. 294, n. 59, D. Howlett, *Caledonian craftsmanship: the Scottish Latin tradition* (Dublin, 2000), pp 84, 89, 95, 159–60. 17 B. Krusch (ed.), *Ionae Vitae Sanctorum Columbani, Vedastis, Iohannis* (Hannover and Leipzig, 1905). 18 Howlett, *Celtic Latin tradition*, pp 187–9. 19 Ó Cróinín, 'Mo-sinu maccu Min', 286–7.

provided a formula for discovering the date of Easter in a given year without resort to complex calculations. Each line is formed from two parts, presenting first the Julian date for the full moon in the month in which Easter should fall within each year of the nineteen-year lunar cycle, and after this the ferial regular for each year, itself dependent on the solar cycle (indicating the difference between the weekday on 24 March and the weekday on which the paschal full moon appears).[20]

In this text, boldface marks alliteration at the beginning of each hemistich.[21] Italics mark rhymes at the end of each hemistich, the basis of rhyme understood here as a single letter, though complete syllables and entire words are italicized to illustrate the extent of rhyme in a composition that has been repeatedly affirmed to lack it. Underlining marks words for numbers that illustrate their value by their position. To the right of the text columns mark the regulars, numbers of lines, the rhyme scheme, and numbers of words, syllables and letters.

<table>
<tr><td>NONAE APRILIS</td><td></td><td></td><td></td><td>2</td><td>5</td><td>12</td></tr>
<tr><td>**No**nae *Aprilis* **no**runt *quinos* |</td><td>V</td><td></td><td>aa</td><td>4</td><td>9</td><td>24</td></tr>
<tr><td>**O**ctonae *Kalendae* **a**ssi*m* deprom*unt*</td><td>I</td><td></td><td>bc</td><td>4</td><td>11</td><td>30</td></tr>
<tr><td>**Id**us *Aprilis* **et**iam sex*is*</td><td>VI</td><td></td><td>aa</td><td>4</td><td>9</td><td>21</td></tr>
<tr><td>**N**onae | qua*ternae* **n**amque dipondio</td><td>II</td><td></td><td>bd</td><td>4</td><td>11</td><td>28</td></tr>
<tr><td>**It**em **un***denae* **am**biunt | *quinos*</td><td>V</td><td>5</td><td>ba</td><td>4</td><td>10</td><td>24</td></tr>
<tr><td>**Qu**atuor | *Idus* | **ca**piunt *ternos*</td><td>III</td><td></td><td>aa</td><td>4</td><td>10</td><td>24</td></tr>
<tr><td>**T**ernae *Kalendae* **t**itulant *senos*</td><td>VI</td><td></td><td>ba</td><td>4</td><td>10</td><td>27</td></tr>
<tr><td>**Qu**atuor*denae* **cu**bant in | quadr*is*</td><td>IIII</td><td></td><td>ba</td><td>4</td><td>10</td><td>27</td></tr>
<tr><td>**Sept**enas | *Idus* **sept**e*m* | *eligunt*</td><td>VII</td><td></td><td>ac</td><td>4</td><td>10</td><td>25</td></tr>
<tr><td>**S**enae *Kalendae* **s**ortiunt *ternos* |</td><td>III</td><td>10</td><td>ba</td><td>4</td><td>10</td><td>27</td></tr>
<tr><td>| **Den**ae | sep*tenae* **don**ant assi*m*</td><td>I</td><td></td><td>be</td><td>4</td><td>9</td><td>24</td></tr>
<tr><td>**Pr**idie Non*as* **p**o**r**ro *quaternis*</td><td>IIII</td><td></td><td>aa</td><td>4</td><td>10</td><td>25</td></tr>
<tr><td>**No**nae *Kalendae* **no**tantur **sept***enis* |</td><td>VII</td><td></td><td>ba</td><td>4</td><td>11</td><td>29</td></tr>
<tr><td>**P**ridie *Idus* **p**anditur quin*is* |</td><td>V</td><td></td><td>aa</td><td>4</td><td>10</td><td>24</td></tr>
<tr><td>**A**prilis *Kalendas* **e**xprimit unu*s*</td><td>I</td><td>15</td><td>aa</td><td>4</td><td>11</td><td>27</td></tr>
<tr><td>**D**uodenae namqu*e* **d**octe | *quaternis*</td><td>IIII</td><td></td><td>ba</td><td>4</td><td>11</td><td>28</td></tr>
<tr><td>**Spe**ciem quinta*m* **spe**ramus duo*bus* |</td><td>II</td><td></td><td>ea</td><td>4</td><td>11</td><td>28</td></tr>
<tr><td>**Qu**artanae *Kalendae* | **qui**nque conici*unt*</td><td>V</td><td></td><td>bc</td><td>4</td><td>12</td><td>32</td></tr>
<tr><td>**Quin**denae | tri*bus* **con**stant adept*is*</td><td>III</td><td>19</td><td>aa</td><td>4</td><td>10</td><td>30</td></tr>
<tr><td></td><td>42</td><td></td><td></td><td>76</td><td>195</td><td>504</td></tr>
<tr><td></td><td></td><td></td><td></td><td>97</td><td>236</td><td>640</td></tr>
</table>

20 S. Rankin, '*Si [...] memoraliter decantare consuescas*: melodies for computistical verse', *Anglo-Saxon*, 2 (forthcoming). 21 For earlier editions, see K. Strecker (ed.), *Poetae Latini aevi Carolini* (Berlin, 1923; repr. 1964), pp 670–1, Ó Cróinín, 'Mo-sinu maccu Min', 287, and Rankin, '*Si [...] memoraliter decantare consuescas*'.

'The Nones of April know five [regulars].
The eighth Kalends bring out one.
The Ides of April also with six.
The fourth Nones certainly with a double.
Likewise the eleventh [Kalends] embrace five.
The fourth [lit. 'four'] Ides take three.
The third Kalends give a title to six.
The fourteenth [Kalends] lie down in four.
The seventh Ides choose seven.
The sixth Kalends allot three.
The seventeenth [Kalends] grant one.
The day before Nones straightway in four.
The ninth Kalends are marked by seven.
The day before the Ides is opened by five.
The Kalends of April one expresses.
The twelfth [Kalends] certainly in four.
The appearance of the fifth Ides we hope for in two.
The fourth Kalends throw together five.
The fifteenth [Kalends] with three added.'

Let us note the patterns of numbers and division by ratios.

The seventy-six words divide by 1:1, 38:38, in the thirty-eight words from | *assim* 2 to *assim* | 11. Practising diminution, the thirty-four words from | *assim* 11 to the end of the poem divide again by the same ratio 1:1 at | *unus* | 15.

The seventy-six words divide by 2:1, 51:25, in the fifty-one words between *dipondio* | 4 and | *duobus* 17.

The seventy-six words divide into thirds at *ternae* | 7. There are thirty-three words between *ternos* | 10 and | *tribus* 19. The seventy-six words divide by 3:2 at 46 and 30. There are forty-six words from | *capiunt ternos* 6 to *duobus* | 17. Practising diminution the forty-six words divide again by the same ratio at 28 and 18. There are twenty-eight words from *ternos* | 10 to *duobus* | 17.

The seventy-six words divide into 1/4 and 3/4 at 19 and 57. There are nineteen words from | *quaternae* 4 to *quadris* | 8 and fifty-seven words from | *quaternae* 4 to *quartanae kalende* | 18. The seventy-six words divide by 4:3 at 43 and 33. There are forty-three words from *quatuor* | 6 to *quaternis* | 16. There are thirty-three words from | *quadris* 8 to *quaternis* | 16.

The poem contains five words for 5. The seventy-six words divide by 5 at 15. Between *quinos* | 1 and | *quinos* 5 there are fifteen words. The fifteenth word after *quinis* | 14 is *quinque* 18. The fifth word after *quintam* | 17 is *quinque* 18, after which there are five words to the end of the poem.

The seventy-six words divide by 7 at 11. From | *septenae* 11 to *septenis* | 13 there are eleven words. The seventy-six words divide by 7:6 at 41 and 35. Before | *septenae* 11 there are forty-one words. After *septem* | 9 there are forty-one words. The forty-one words divide in turn by 7:6 at 22 and 19. After *septenas* | 9 the nineteenth word is *septenis* 13. The nineteen words divide in turn by 7:6 at 10 and 9. After *septenas* | 9 the ninth word is *septene* 11.

The seventy-six words divide by the 17-part ratio 9:8 at 40 and 36. After *nonae* | 4 the thirty-sixth word is *nonae* | 13. From the beginning to | *Denae septenae* 11 there are forty words. The forty words divide by 1/8 and 7/8 at 5 and 35, at *octonae* | 2.

Let us note next the pattern of dates, three Nones, five Ides, eleven Kalends. There are neumes in this text, of which Susan Rankin has observed:

> Once the fact of repeating patterns in this melody is established, it becomes evident that this repetition was entirely regular: thus the melody for the whole is structured in short phrases matched to individual hemistichs, and then, in terms of nineteen full lines, as ABCBCDBCDBCDBCDBCD,

a pattern that coincides with the pattern of dates, thus:

A	B	C
Nones	Kalends	Ides
D	B	C
Nones	Kalends	Ides
D	B	C
Kalends	Kalends	Ides
D	B	C
Kalends	Kalends	Nones
D	B	C
Kalends	Ides	Kalends
D	B	C
Kalends	Ides	Kalends
D		
Kalends		

According to the Annals of Ulster, Mo-Sinu abbot of Bangor died in AD610, and his pupil Columbán of Bangor and Bobbio, the most forceful proponent of the Irish 84-year cycle, died in AD615. As the Würzburg note praises Mo-Cuoróc maccu Net Sémon *quem Romani doctorem totius mundi nominabant*, one infers that the *Romani* may be southern Irish churchmen who had accepted the Roman practice of calculating Easter before Cummian defended their position in his famous *Epistola de controuersia Paschali* to Ségéne abbot of Iona and Béccán the hermit in AD633.[22] If Mo-Cuoróc was a *Romanus*, his association with a church

22 Walsh and Ó Cróinín, *Cummian's letter 'De controversia paschali' and the 'De ratione conputandi'*. D. Howlett, 'The earliest Irish writers at home and abroad', *Peritia*, 8 (1994), 1–17

in Munster and his presence in the north is as easily accountable as the presence of that other southerner, the Leinsterman Columbán, in Bangor. But the note tells us not that Mo-Cuoróc was in Bangor, merely that he was a pupil of Mo-Sinu abbot of Bangor, and that he was on the island of Crannach of Downpatrick when he wrote. Two words of the title *Nonae Aprilis* and seventy-six words of the poem and nineteen regular numbers total ninety-seven, coincident with the alphanumeric value of MO-CUORÓC, 12+14+3+20 +14+17+14+3 or 97. Five syllables of the title and 195 syllables of the poem and thirty-six syllables of the regular numbers total 236, one more than the alphanumeric value of MO-CUORÓC MACCU NET SÉMON, 12+14+3+20+14+ 17+14+3+12+1+3+3+20+13+5+19+18+5+12+14+13 or 235. From the first words of the poem *Nonae Aprilis* to the last numeral of the regulars inclusive there are exactly 640 letters and spaces between words, which may provide a clue to the year in which Mo-Cuoróc maccu Net Sémon composed at Crannach of Downpatrick the poem that earned him the nickname *na Nóna* 'of the None' and the honorific title, in Latin *Sapiens* 'the Wise', and in Irish *ind ecnai* 'of the wisdom'. The poem with its calendrical learning and its neumes illustrates clearly the connection between music and the stars.

BISHOP ULTÁN MOCCU CONCHOBUIR OF ARD BRECCÁIN

Ultán bishop of Ard Breccáin near Navan in County Meath, tutor of Tírechán of Tirawley, is credited with authorship of the *Vita primitiua Sanctae Brigitae*, a work from which elements survive only in later Lives of Brigit that borrowed from it. To this Life he may have appended three abecedarian stanzas of his Brigitine hymn *Xpistus in nostra insula*, extant in three manuscripts.[23]

I				
Xpistus in nostra *i*nsul*a* que uocatur Hiber*nia*	a	7	16	39
ostensus est hom*i*n*ibus* maximis mirab*i*l*ibus*	b	5	16	38
que perfecit per fel*icem* celestis uite *uirginem*	c	7	16	41
precellentem pro mer*ito* magno in mundi circ*ulo*.	d	7	16	40
II				
Ymnus iste angel*ice* summeque sancte Br*i*g*ite*	e	6	16	38
fari non ualet om*nia* uirtutum mirab*i*l*ia*	a	6	16	34
que nostris nunquam aur*ibus* si sunt facta aud*i*u*i*m*us*	b	8	16	44
nisi per istam *uirginem* Marie sancte s*i*m*i*l*em*.	c	7	16	38

at 10–17; Howlett, 'Five experiments', 3–6. On Cummian, see also Walsh, this volume.
23 D. Howlett, 'The Brigitine hymn *Xpistus in nostra insula*', *Peritia*, 12 (1998), 79–86.

III

Zona sancte mil*i*t*ie* sanctos lumbos prec*in*ger*e*		e	6	16	40
hec consüeuit di*u*rn*o* nocturno quoque stud*io*		d	6	16	38
consummato certa*mine* sumpsit palmam uictor*ie*		e	5	16	40
refulgens magno splendor*e* ut sol in celi cul*mine*.	12	e	8	16	41
	12		78	192	471

'Christ in our island, which is called Ireland,
has been shown to men in the greatest miracles,
which he has performed through the happy virgin of celestial life,
precellent for great merit in the circle of the world
[or 'precellent for merit in the great circle (or "cycle") of the world'].

This hymn of the angelic and highest saint Brigit
is not proficient to utter all the miracles of virtues,
which, if they have been performed, we have never heard with our ears
except through this virgin similar to saint Mary.

With a girdle of holy soldiery this woman was accustomed to gird [her] holy loins
in daily, also nightly, zeal;
with the struggle consummated, she took up the palm of victory,
radiating brightly with great splendour as the sun in the height of heaven.'

The poet named himself in a central line of the central stanza. Although a hymn of only three stanzas could not record all Brigit's miracles, a complete Life could. As the letters of *fari non ualet* 'he is not proficient to utter' can be rearranged as *Ultán for in ea* [*sc.* *Vita*] 'I, Ultán, utter in that [*sc.* Life]', the anagram *Ultán for in ea omnia uirtutum mirabilia*, either with synizesis of *e* in *ea* or with elision of *a* in *ea*, is identical in rhyme and rhythm with the line in which it is embedded.

The hymn incorporates calendrical phenomena, in three stanzas of eight lines of end-rhymed octosyllabic verse for twenty-four hours of a day, or better in three stanzas of four lines of hexadecasyllabic verse rhyming at the caesura and the end of the line, the twelve lines representing twelve months of a year, to which the poet may allude in *magno in mundi circulo*.

Stanza 2 line 1 ends *Brigite*, the thirty-second word of the hymn, the letter *B* the thirty-second letter of the line, representing St Brigit's Day, 1 February, the thirty-second day of a year that begins on 1 January.

From | *sancte Brigite* to *istam uirginem Marie sancte similem* | there are 152 letters, coincident with the alphanumeric value in the Greek system of MAPIA, 40+1+100+10+1 or 152.

That each stanza contains sixty-four syllables is coincident with the alphanumeric value in the twenty-three-letter Latin system of HIBERNIA, 8+9+2+5+17+13+9+1 or 64, BRIGITA, 2+17+9+7+9+19+1 or 64, ULTAN, 20+11+19+1+13 or 64.

If these verses concluded the *Vita primitiua Sanctae Brigitae*, attributed to Ultán moccu Conchobuir bishop of Ard Breccáin, who died in AD657, they are the oldest extant Latin poetry in praise of an Irish woman. The connection of the poet's music with the stars is explicit in the last hemistich, *ut sol in celi culmine.*

CÚ CHUIMNE OF IONA

Cú Chuimne Sapiens, abbot of Iona, is with Ruben of Dairinis one of two compilers of the monumental *Collectio Canonum Hibernensis*, the first thematically organized collection of canon law, conceived and completed in Ireland during the second quarter of the eighth century and still in use in the University of Paris in the twelfth century. Cú Chuimne is also the author of the oldest extant Latin hymn to the Virgin Mary, published some time before his death in AD747. In the text that follows, edited from four extant manuscripts,[24] boldface marks alliteration; italics mark end rhymes in couplets; italics with underlining marks four- or six-syllable internal rhymes in the second lines of couplets.

Cantemus in **omni** die **concin**n*ante*s u*arie*
conclamantes **D**eo **d***ignum* *y***m***num* S*ancte* M*arie* .
Bis **per chorum hinc et in**de **conlaudem**us M*ariam*
ut uox pulset omn*e*m *aurem* **per** **laudem** uic*ariam* .
Maria de tribu Iuda Summi **m***a***ter** D*omini*
oport*unam* dedit c*uram* egrot*ant*i **h***omini* .
Gabriel aduexit Verbum s*inu* **prius** **P***aterno*
quod con*ceptum* **et** sus*ceptum* *in* *u*tero m*aterno* .
H*e***c** *e***st** **summa, hec est** sancta **u**irgo **u**ener*abilis*,
qu*e* *e***x** fide non r*ecessit* sed *exstetit* **st***abilis* .
Hüic matri **nec** *i*nuenta **ante** *ne***c** **p**ost s*imilis*
nec de **p**role fu*i*t **pl***ane* hum*ane* *o*rig*in*i*s* .
Per mulierem et lignum **mundus prius per***iit*;
per mulieris uirt*utem* ad sal*utem* red*iit*.
Maria **ma**ter **mir**anda **P**atrem s*uum* *edidit*

24 Howlett, 'Five experiments', 19–30.

per quem **a**qua **late l***otus* t*otus* **m***u*nd*u*s cr*edidit* .
Hec concepit margaritam — non **sunt uan***a* **s***omnia* —
pro **q**ua **s***ani* **C**hristi*ani* **uen**dunt **su***a omnia* .
Tunicam per **to**tum **te**xtam Christi m*a*t*er* f*ecerat*
que peracta Christi m*orte* **s***orte* **st***a*t*im* **st***eterat* .
Induamus **a**rma **lucis lo**ricam *et* g*aliam*
ut simus Deo **perf***ecti* susc*epti* **per** M*ariam* .
Amen, **Amen**, **a**diuramus merita pu*erpere*
ut **non p**ossit flamma **p***ire* **nos d***ire* **dec***erpere* .
Christi nomen **in**uocamus **an**gelis sub t*estibus*
ut fru*amur* **et** scrib*amur* litteris cel*estibus* .
Sancte **Marie meri***tum*
imploramus dign*issimum*
ut mereamur **sol***ium*
habitare alt*issimum* .
Amen .

'Let us sing on every day, putting together variously,
shouting together to God a hymn worthy of Saint Mary.
Twice in a chorus on this side and that let us together praise Mary,
so that the sound may strike every ear in successive praise.
Mary from the tribe of Judah, mother of the Highest Lord,
gave an advantageous cure to ailing man.
Gabriel conveyed first from the Father's bosom the Word,
Which [was] conceived and received in the mother's womb.
This is the highest, this is the holy venerable virgin,
who has not drawn back from the faith, but stood out stable.
To this mother a like has been found neither before nor since
nor after her offspring of fully human birth.
By a woman and a tree the world first perished;
by the virtue of a woman it came back to salvation.
Mary the wondrous mother brought forth her own Father,
through Whom the whole world far and wide washed by water [in baptism] has believed.
This woman conceived a pearl — these are not empty dreams —
for which sane Christians sell all their own possessions.
The mother of Christ had made a garment woven throughout,
which, when Christ's death was brought about, had remained constant in its order.
Let us put on the arms of light, breastplate, and helmet,
that we may be perfect for God, received through Mary.
Amen. Amen. We affirm on oath the merits of the child-bearer,

so that the flame of the dreadful pyre cannot snatch us away.
We invoke the name of Christ with angels as witnesses,
so that we may enjoy and be written in celestial letters.
We beseech the most worthy
merit of Saint Mary
that we may be worthy
to inhabit the loftiest throne.
Amen.'

A composer as well as a poet, Cú Chuimne has embedded within the text of the first quatrain instructions for performance, singing by a group (*cantemus*, *concinnantes*, *conclamantes*, *conlaudemus*), divided into an antiphonal pair (*per chorum hinc et inde*) to produce successive responsorial praise (*laudem uicariam*). Note that implying the idea of 'two' twice he began the second couplet with the word *bis*. He has stated within the last quatrain of long lines what the choir do and why, then restated this in the last quatrain of short lines. Enclosed within these four lines of introduction and eight lines of conclusion are eighteen lines in praise of the Virgin Mary.

The hymn's long lines echo the pentadecasyllabic verses of 'Saint Sechnall's Hymn' *Audite Omnes Amantes Deum* in praise of St Patrick.[25] Its short lines echo the rhyming and alliterating octosyllabic couplets and quatrains devised by Irish poets, from the fourth quarter of the sixth century onward.

The hymn relates music to the stars by exhibiting a wealth of calendrical phenomena. It consists of twenty-six pentadecasyllabic lines, four octosyllabic lines, and a concluding *Amen*, thirty or thirty-one lines, one for every day of the month. After the end of the twenty-sixth line manuscripts C and D read *Cantemus*, implying, as the second couplet states explicitly, that the hymn is to be sung *bis* 'twice'. Those who do that sing fifty-two pentadecasyllabic lines, one for every week of the year. The pentadecasyllabic lines sung and repeated and the octosyllabic quatrain and the concluding *Amen* sung once contain 365 words, one for every day of the year, appropriate for those who sing *Cantemus in omni die*.

The last, thirty-first, line ends with the word *Amen*, the alphanumeric value of which is 1+12+5+13 or 31. Line 23, which begins *Amen Amen*, contains thirty-one letters. After line 23 *Amen* 31 is the thirty-first word.

Note particularly the couplet *Gabriel aduexit Verbum sinu prius Paterno, quod conceptum et susceptum in utero materno*. In MS A this is preceded by the punctuation marks //., and followed by .., . In MS B the central word of the first line of the couplet is pointed *Verbum*. . The event referred to is the Annunciation, celebrated on 25 March, the eighty-fourth day, 23% of the way through a 365-

25 Howlett, *Celtic-Latin tradition*, pp 138–52.

day year. The seventh line is 23% of the way through a thirty-one-lined poem. The word *Verbum* is the forty-fifth of 188 words, 23% of the poem. The *m* of *Verbum* is 234th of 1,021 letters, 23% of the poem.[26]

From *Amen Amen adiuramus merita puerpere* 23 to the last mention of *Sancte Marie* in line 27 there are twenty-seven words, of which the central, fourteenth, word is *Christi* 25. From *Amen Amen* to the M of *Marie* inclusive there are 152 letters, coincident with the value in Greek alphanumeric notation of the name MAPIA, 40+1+100+10 +1 or 152. Eight-ninths of the way through this passage the poet refers to himself at *scribamur litteris* | *celestibus*. In the immediately preceding line he writes *Christi nomen inuocamus*, of which the last *s* is the 888th letter of the hymn, coincident with the value in Greek alphanumeric notation of the name IHCOYC, 10+8+200+70+400+200 or 888.

CORMAC'S POLYPHONIC COLOPHON

London, British Library, MS Additional 36929 is an Irish Gallican Psalter, written probably during the middle or second half of the twelfth century, its Psalms divided in the Irish way into three groups of fifty.[27] On folio 59r, after the canticles that follow the first group of fifty Psalms, the *Hymnus Trium Puerorum* and the *Canticum Isaie*, below the words *Domine saluum me fac: et psalmos nostros cantabimus cunctis* | *diebus uitae nostrae in domo domini*, the last of which echo Psalm 22:6, the scribe wrote a polyphonic colophon:

> Cormacus scripsit hoc ψalterium ora pro eo .
> Qui legis hec ora pro sese qualibet hora .

> 'Cormac wrote this Psalter; pray for him.
> You who read these things, pray for himself [perhaps meaning 'yourself'] in every hour.'

Here are two lines of dactylic hexameter verse, which both exhibit correct quantities, with two elisions in the first line in *psalterium ora* and *pro eo*, with nothing extraordinary in the second. Both verses exhibit internal vowel rhymes in the third and the sixth feet. In the first, compare the *o* in *hoc* with that in *eo*. In the second compare the *i*, *e* and *ora* in *legis hec ora* with those in *qualibet hora*.

26 The number 23 is half of the number 46, which represents the perfection of Creation, as above, n. 15. The triangular number of 23, 1+2+3 … 21+22+23, is 276. For repeated play on the number 23, see D. Howlett, *Muirchú Moccu Macthéni's 'Vita Sancti Patricii': Life of Saint Patrick* (Dublin, 2006), pp 140–2, 164, 168–70. 27 D. Howlett, 'The polyphonic colophon to Cormac's Psalter', *Peritia*, 9 (1995), 81–91, now recorded by the Schola Gregoriana Maynooth in *Saints & scholars, chant from Maynooth College Chapel*, track 7.

The second verse also exhibits internal alliteration on *qu* and *i* in *qui legis* and *qualibet*. Both verses exhibit chiastic disposition of sounds. In the first, around the central *o* in *hoc* in the third foot, compare the *ps-t* of *scripsit* with that of *psalterium*, the *ri* of *scripsit* with that of *psalterium*, the *u* of *Cormacus* with that of *psalterium*, and the *or-a* of *Cormacus* with that of *ora*. The first and last of these chiastically disposed sounds in the first verse are echoed at the centre of the second verse. Around the central *a* of *ora* in the third foot, compare the *o* of *ora* with that of *pro*, the *r* of *ora* with that of *pro*, the *e* of *hec* with that of *sese*, the *s* of *legis* with that of *sese*, the *i* of *legis* with that of *qualibet*, and the *e* of *legis* with that of *qualibet*. The first verse is linked to the second by parallelism of *Cormacus scripsit hoc* with *qui legis hec* and of *ora pro eo* with *ora pro sese*.

Cormac refers to himself and his work in places determined by calculation of sesquioctave ratio 9:8. The fifteen words of the colophon divide by sesquioctave ratio at 8 and 7, at the end of the first verse, *eo*. The twenty-nine syllables divide by sesquioctave ratio at 15 and 14, at *eo*. The seventy letters divide by sesquioctave ratio at 37 and 33, at *eo*. The fifteen words divide by one-ninth at 1.67, the twenty-nine syllables at 3, and the seventy letters at 8. The first word names the writer *Cormacus* in three syllables and eight letters. Let us note also the references to Cormac's work. From *hoc psalterium* to *qui legis hec* inclusive there are eight words, fourteen syllables, and thirty-two letters.

The reader is invited to pray for Cormac at *ora*, the fifth word from the beginning of the colophon, and for himself at *ora*, the fifth word from the end. Between *ora* and *ora* there are five words. Before the former *ora* there are ten syllables. From the latter *ora* inclusive to the end there are ten syllables. From the first syllable of the former *ora* to the first syllable of the latter *ora* inclusive there are ten syllables.

These verses are set to a polyphonic composition in three vertically arranged four-lined staves drawn in red ink, with red guilloche separating the stave of the upper voice from that of the middle voice and that of the middle voice from that of the lower voice. Bar lines in red ink make divisions of notation coincide with word divisions. Cormac the composer refers to himself at the symmetrical centre of his polyphonic music. In the upper voice the central twenty-eighth note of fifty-five is the last of *eo*. In the middle voice the central twenty-third note of forty-five is the last of *eo*. In the lower voice the central fifteenth note of twenty-nine is the last of *eo*.

The music in the first half of the first verse is identical with that in the first half of the second verse in all three voices. As the lower voice has one note for each syllable, the fifth note of the first verse, required for the second syllable of *scripsit*, though obscured by abrasion, is confirmed by parallelism with the fifth note of the second verse. Both verses of the lower voice are exactly parallel except for the antepenultimate note of the first, which has no pair in the second. Both verses of the middle voice are exactly parallel except for the tenth note of the first, which has no pair in the second. Both verses of the upper voice are exactly parallel except for the penultimate note of the first, which has no parallel in the second. The voices begin on DDG in both verses and end in unison on GGG in both verses. The only other points at which the voices sing in unison, AAA and AAA, coincide with the rhymes in the third foot of each of the hexameters, the latter sung on note A on the vowel *a* in *ora*.

We have considered five Hiberno-Latin texts: two computistic compositions, one in prose and another in verse with accompanying music, and three poems, the first the oldest extant Latin poetry in praise of an Irish woman, the second the oldest extant Latin hymn to the Virgin Mary, and the third the oldest extant composition in three-part polyphony. All are composed in Latin of a high artistic register, exhibiting faultless rhythms, informed by the ratios of cosmic and music theory. All five texts, illustrating complete mastery of techniques of varied liberal arts, not least gematria, were conceived and executed among the Irish, the first northern people drawn from outside the Roman Empire into the orbit of Latin Christian civilization.

As a postscript, for the most explicit relations between music and the stars one may consider four Carolingian attempts to represent the distances between planets (that made the music of the spheres) in terms of the intervals of music theory, the most arresting of these consistent with the series in a commentary on Martianus' *De nuptiis* by the ninth-century Irish scholar Johannes Scottus Eriugena.[28]

GEMATRIA

twenty-two-letter Hebrew alphabet, א = 1, ב = 2, ג = 3
  הֶבֶל 'Abel', 5+2+30 or 37
twenty-four-letter Greek alphabet, Α = 1, Β = 2, Γ = 3
  ΑΔΑΜ 'Adam' 1+4+1+40 or 46
  ΜΑΡΙΑ 'Maria' 40+1+100+10+1 or 152
  ΙΗϹΟΥϹ 'Jesus' 10+8+200+70+400+200 or 888.
twenty-three-letter Latin alphabet, A = 1, B = 2, C = 3
  HIBERNIA 'Ireland' 8+9+2+5+17+13+9+1 or 64
  BRIGITA 'Brigit' 2+17+9+7+9+19+1 or 64
  ULTAN 'Ultan' 20+11+19+1+13 or 64
Hebrew Book of Judges 3:7–11, כָּלֵב 'Caleb' 20+30+2 or 52 is the fifty-second word

28 S. Rankin, '*Naturalis concordia vocum cum planetis*: Conceptualizing the harmony of the spheres in the early Middle Ages' in S. Clark and E.M. Leach, *Citation and authority in Medieval and Renaissance musical culture* (Woodbridge, 2005), pp 3–19.

Greek New Testament Apocalypse 13:18, the number of the beast,
εξακοσιοι εξηκοντα εξ, 600+60+6 or 666
descending value of Roman numerals DCLXVI, 500+100+50+10+5+1 or 666
DICLVX 'he who says he is the light'
Latin name NERO CAESAR
spelled in Hebrew letters נרון קסר 50+200+6+50+100+60+200 or 666
Latin literary text Martianus Capella, *De nuptiis philologiae et mercurii* book II
Bede, *De temporum ratione* I, *De computo uel loquela digitorum*:

> Cum ergo dicis unum, minimum in laeua digitum inflectens, in medium palmae artum infiges. Cum dicis duo, secundum a minimo flexum, ibidem inpones. Cum dicis tria, tertium similiter adflectes. Cum dicis quattuor, itidem minimum leuabis [...] .
>
> 'When therefore you say 1, bending the smallest finger toward the left, you will fix the limb against the middle of the palm. When you say 2, you will put the second from the smallest bent in the same way. When you say 3, you will bend the third similarly. When you say 4, in exactly the same way you will lift the smallest [...] .'
>
> Si amicum inter insidiatores positum ut caute agat admonere desideras, iii et i et xx et xix et v et i et vii et v digitis ostende, huius namque ordinis litterae 'caute age' significant. Potest et ita scribi si causa secretior exigat. Sed haec Graecorum computo litterisque facilius disci simul atque agi possunt [...]
>
> 'If you desire to admonish a friend placed among enemies that he go cautiously, show with 3 and 1 and 20 and 19 and 5 and 1 and 7 and 5 fingers, for the letters of this order signify "go cautiously". It can also be written thus if a more secret case should compel. But these things can be learned and executed at the same time in the computus and letters of the Greeks [...]'

after which, Bede presents the numerical values of the letters of the Greek alphabet.

FOURTEENTH-CENTURY ANGLO-NORMAN ALGORISM

Dieux soit ou vous *quant* vous ali*etz*	a	7	8	30
Et de vos malx vous doynge par*donne*	b	7	8	29
Pensez de moy *quant* meux porr*etz*	a	6	8	27
Qe 78 [*i.e.* setante huit] hau*etz a* n*onne*.	ab	6	8	25
	2	26	32	111

'May God be along with you when you go,
and for your sins may He grant you pardon.
Think of me when best you can,
you who have for a name [the number] 78.'
E D V A R D I I I = 5+4+20+1+17+4+9+9+9 or 78

When art was informed by mathematics

ROBERT D. STEVICK

One of the aims set for the conference on the history of science in medieval Ireland was 'to change the way mathematics is perceived by the general public'. That is a tall order, and worthy. One promising way to do this was another of the conference aims: 'to explore ways of incorporating the skills of mathematics into subjects such as music, literature, art'. An ideal context for both is medieval Ireland, extended eastward into Northumbria and backward in time. That was the place and the time when some of the best art was informed by mathematics.

HOW MATHEMATICS INFORMED SOME OF THE FINEST ART

The notion of major creations of art being informed by mathematics may be radical, but I hope it is so in the root meaning of radical (!). Rather than referring to it as a notion, though, I would like to offer it as a proposition. One way to demonstrate that proposition is to lay out in mathematical notation the structures of various artefacts from that early Insular region. That entails using notation such as

$$\frac{2}{\sqrt{5}-1} \text{ or } \frac{\sqrt{3}+1}{2},$$

which does provide a rational, analytic account of the forms as we find them. This way tends to put some to sleep, cause others to become queasy. Another way should forestall slumbers and upsets. It is to demonstrate the proposition of art being informed by maths by showing how the primary forms of the finest Insular art objects must have been created. These forms had to be set out first, before any images could be inserted or any of the patterns and particulars of ornament laid out, whether on a manuscript leaf, a stone slab, a mould for casting a brooch. The forms have had little attention and less understanding beyond some subjective responses that tell as much about their viewers as they do about the forms. Understanding of the forms comes about when one grasps processes employed originally in creating them. My illustrations require only two hand-tools of tactile mathematics, compass and straight edge. No need for protractor or scale, for measuring angles and lengths: we don't need to read the measures of either angles or lengths when creating these forms. We don't record them; we manipulate them.

Here is how I believe some of these plans must have been devised.[1] The best demonstrations of the procedures of employing hands-on mathematics are found in pictorial art, especially the illuminated pages in the gospels' codices, and the most spectacular of these are in the Book of Kells. The St John portrait page (fo. 291v) in the Book of Kells has to be one of the finest and most fascinating portrayals of an author and his book that we will ever encounter (pl. 5). It is an icon in the etymological sense – a pictorial representation – but it is more than that. It also attracts a veneration of its artistic achievement. Thankfully, it was never actively venerated for its subject and used in preparing a cure for sick cattle, as was the Book of Durrow, in the eighteenth century, when at least one gathering from its St John's Gospel seems to have been soaked repeatedly to obtain its curative properties.

The fantastical decoration filling the cells that join to make the frame engages and re-engages one's attention, and so does the decoration inside the halo, as does the knotwork at the four corners of the frame. At midpoint of bottom, sides and top, the feet, hands and head outside the frame remain mysterious, most of the head having been lost to a binder's knife. The evangelist is not depicted as writing, as often is the case. He holds a pen in one hand, but not a knife in the other, and he does not have an open book in front of him. Instead, he is holding a quill in one hand (and there is an inkhorn by his foot) while in the other hand he is holding a closed book. The signification in all likelihood is that his book has now been completed.

In essence, here is a page in a book, depicting a writer (with pen and inkhorn) and his completed book (it even has a fine binding). And the author is nimbed as perhaps no other writer has *ever* been. What are we to say? Do we say: It is obvious that the artist was meditating so deeply on the meaning and the mysteries of the gospel according to John that he would not have distracted himself with elaborate mathematics? Try parsing *obvious* in that statement. And then try deconstructing (as it were) the composition – the form of this extraordinary portrait page. Or better, try constructing it by methods of medieval mathematics. The mathematics need not be a distraction, and in fact seem to have been the essential informing element for this page. The rectangle of the frame for John's portrait, without extensions or embellishments, is approximately 266 × 195mm. A synopsis of a procedure that will replicate – or create – the frame surrounding the evangelist is illustrated in figure 7.1.

1 Some examples in digital format can be viewed by going to http://content.lib.washington.edu/insdsgnweb/index.html and selecting 'Demonstrations'; the one for the 'Tara' Brooch is probably the best example to begin with.

THE FRAME

Begin (top, left) with concentric and commensurate square and circle, that is, the two having the same size and same centre. Then (top right) from it construct a frame in the ratio

$\frac{\sqrt{3}+1}{2}$, as shown.

(a) Sketch diagonals of this frame (lower left), and mark their intersections (1) with the underlying circle to be the primary set of corners of the inner panel area.

(b) Another set of corners of the inner panel area is located by (2) copying, onto the upper and lower sides, the vertical distance from the frame to the first set of corners and then sketching the paths of parallel lines joining them in pairs.

(c) A third set of corners of the inner panel area is located by (3) copying to the left and right sides the horizontal measures from the corners and then sketching the paths of parallel lines joining them in pairs.

(d) Midside 'notches' in the inner frame are plotted with separate diagonals, as shown (4, 5).

(e) The centre of the halo (lower right) is located along the centreline a distance from the upper border equal to half the width of the frame (see preceding step). Its outer circle is drawn tangent to 45° diagonals from midpoints of the sides. Its inner circle is drawn tangent to 45° diagonals from the centrepoint of the frame. The two circles are thus proportional to the frame and its underlying square. There are alternate constructions, some just as easy, all just as accurate.

It is a fair inference that the layout of the St John page (along with the Dunadd trial-piece, the Tara Brooch, and many, many more objects) was derived by steps like the ones just illustrated, using compass and straight edge and needing nothing more. Proof for any method of construction, of course, is hard to come by in the absence of documentation or in the absence of the tradition that produced it continuing intact. We have proof neither from a tradition still living, nor from documentation such as instructional manuals by the artists, or reports of observers of their craft. But we do have other kinds of evidence. One is comparative: shapes embodying the same geometrical terms and simplicity in other media and within the same era of Insular Christian art. Another kind of evidence is deductive: how else to explain designs whose measures *accurately* match simple geometrical constructions again and again throughout the entire form?

In the case of the John page there is a third kind, empirical evidence. It is not ample, and much of it is not easy to find. This is the evidence of construction marks, as they are called – small punctures and impressions, left by a sharp point,

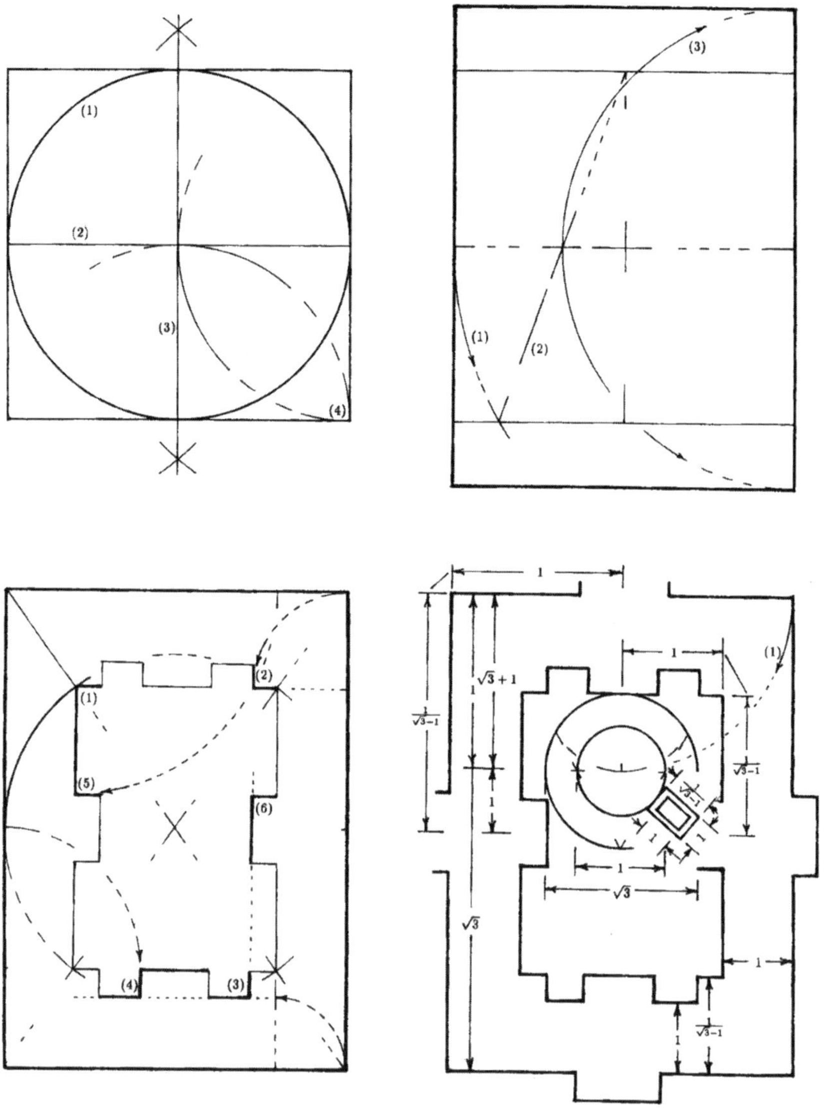

7.1 The layout procedure for the Portrait of St John, TCD MS 58, fo. 291v.

either a pricking knife used elsewhere to mark where the straight edge was to be used to rule the page for text, or the fixed foot of a compass (or dividers). They also include ruled lines, typically impressed in the surface of the leaf, without ink, or curved lines similarly drawn. In the upper portion of this page construction marks are to be seen everywhere: the centres of the eyes, the innermost circle of the halo where half-circle embellishments inside the circle have their centres marked, as well as their outer limits, a horizontal ruling level with the evangelist's eyes, on both left and right of his head, and so on. There are also prickings of the parchment leaf in corners of cells of decoration that make up the frame; in profusion they guide the interlace designs.

Still, any construction marks that correspond to the ones predicted by the construction method I have illustrated for frame and halo are difficult to detect. One reason is that on the portrait side of the leaf, many construction marks are covered by paint. Another is that some key parts of the derivation are located along the middle of the leaf. The parchment is extra thick across the middle of the page, the area of the membrane that lay along the back of the animal whose skin was turned into folio 291v; markings will not easily show through there. And another is that pages having elegant illuminations, in this manuscript and others, seem to have been laid out with light construction markings – part of the mastery of the artist. The main layout is lightly marked and probably copied directly from a model. The decorative filler, on the other hand, could not be copied with a few reference points but had to be constructed on the leaf so as to fit the specific area to be filled with the embellishing devices. Thus the profusion of construction marks for ornamental details and thus the elusiveness of marks for the overall plan.

There is, however, crucial empirical evidence in the pectoral area of the evangelist depiction. In the facsimile it cannot be made out with anything approaching certainty. On the manuscript leaf it is not easy to make out – in fact it may not be detectable by eye when the leaf lies flat. But when the leaf is held to the light, right there in John's chest, the exact middle of the frame, the point where diagonals of the frame intersect – there, exactly there, is a tiny hole letting light through from the other side. Here in the thick area of the parchment, the ridge of the animal's back, is a construction mark that does not guide interlace, or halo, or step-pattern, or boundaries of cells of the frame. It marks the point from which the mise-en-page of the illumination began, the central point of the framed image of St John, the point of origin of the icon. It is the beginning point of the formal plan of the St John page in the Book of Kells. Now, it *should* not come as a surprise to learn that the book in John's hand has the same ratio in its dimensions as the basic rectangle of the illumination itself.

The mathematics shaping early Irish art is wonderfully direct. It has been almost totally obscured by the mathematics of nowadays, especially as it is transmitted in schooling. At the same time, most visual art has gone its separate way, turning its back on disciplined deduction for its formal planar structures. The divergence of mathematics and visual arts has come about for any number of reasons, some practical, some necessary. There is the long history of arithmetic in commerce and trade; of the geometry of land surveying; in the computation of measures and quantities – in reporting the how much and the how many, not to mention all the modern forms and applications of mathematics. But mathematics in Ireland was well advanced before the adoption of arabic numerals and decimal notation that now appear everywhere in school curricula. The duodecimal system, with its dozens and its inches-per-foot and such, did not die even when decimal notation prevailed. Irish mathematics was advanced long before Euclidean geometry entered our educational systems; this kind of geometry, a system of postulates and proofs, has dominated in schools, but is hardly the stuff of art. Irish mathematics was advanced even longer before the teaching of algebra and calculus. There was a time when some important mathematical activity was visual and tactile, carried out in models, rather than in systems of abstract symbols or propositional proofs. It was from that time that mathematics was utilized in the creation of medieval Irish art.

THE ONTOGENY OF THE FORMS

When an overall form in Insular art is not determined directly by the utility of an object – a sword-hilt or a helmet cheek-guard, for example – the form nearly always is a circle, a rectangle, or a hybrid of the two. Commonly, circles are found in metalwork, rectangles in codex leaves and illuminations, hybrids in the standing stone crosses. The mathematics of these forms also underlie the segmentation of Old English verse narratives. Triangles, pentagons, hexagons and the like sometimes occur as overall form, but occur more often within other forms.

Art was informed by mathematics when the formal plan of an object *proceeds from* the enclosing configuration by geometric construction, as was best illustrated with the Tara Brooch and the John page. The mathematics is the formative principle for an art object. That is, the circle, or rectangle, or whatever is not simply an enclosure of the art within it. On the contrary, the enclosing configuration is itself the source of the primary plan of a piece. It is to express this feature that I have chosen the term 'ontogeny' for describing the generation of these forms in early Irish art.

To point up this feature, compare these formal schemes, say, to so-called magic squares. These are games with numerical patterns played within a grid; each magic square is a closed scheme, and the form is set out initially. The similarity to early Irish designing, though, goes no further. The square grid sets the rules for the relations among number sets that are put into it, yet the grid does not generate any of the numbers.

The tools needed to generate these forms, as already said, are no more than straight edge and compass. The creative method – the means of generation – is what I am calling the ontogeny of the formal schemes. This implies a method that produces the basic plans by continuous construction, their internal structures having a total coherence among the dimensions and relations within the forms. By this is meant that the resulting forms must have an evolutionary/ontogenic coherence eventuating from a continuous process of derivation, each step producing a structural or formal part linked to all others: the tactile process is totally efficient, without gaps, without redundancies, without patches or spurious elements. The process of designing, when successful, proceeds with ultimate efficiency. The objects shaped in this way, at their best, have a purity of something set forth with a simplicity that distinguishes searching art.

TRADITION AND THE INDIVIDUAL TALENT

It is not likely, I think, that loss of objects explains why the surviving art pieces are alike but never the same. A mould for casting the head of a brooch may be re-used, for multiple castings. But the next form to be created will re-use not the same measures and steps, but only the same principles of designing. Its creator devises form anew, *like* earlier ones, but not identical to any of them. The same can be said about the standing stone crosses and about cross-pages. The tradition of this art is to repeat the method of creating a form without copying an existing form. The tradition – as any other tradition – must have new development to stay alive. The forms devised for the visual and plastic arts of medieval Ireland and Northumbria are the works of individual creators, of course, yet there is no 'me', no self, no individual being expressed. They are the opposite of subjective works, for not expressing something personal or eccentric, not expressing anything untraditional.

That tradition of creating formal plans evolved, certainly, maintaining a vigorous life. Some of its evolution came with adaptation to new materials and object types. Incised and hammered and cast-metal objects had had their separate strands of evolution. And then, by saltation, it would seem, came the plotted and painted illuminations in codices of religious texts. This new medium, in use early in the medieval art of Ireland, was unlike the scroll: it had fixed rectangular shapes, for both text area and the surface for each text area; this provided a unique opportunity for development and linkage. And then came standing stone crosses – forms with an endoskeleton, as it were, quite different to exoskeletal forms of brooches, pictorial pages and other types of objects, yet with the same mathematical genealogy.

So while there is no lack of development over time, in materials, technology and topical matters in Irish art, the tradition of maths-based forms held steady, simply mutating here and there as needed to adapt. The tradition seems to reflect, in fact it seems to eventuate from, a fascination with visual forms that constitute closed systems. The signature forms have webs of interdependent elements forming a collective entity. They are interrelated and interacting elements within a maths-based system. Those systems, furthermore, are the opposite of patterns such as tilings and grilles, employing only repetition of shapes. Instead of aggregates of repeating shapes – sames occurring again and again, interlocked one way or another – the elements are ratios of measures – the sames being not configurations, but relations among measures of lines and radii.

This is getting a bit abstract, so let me return to perhaps the finest form in Insular art, the Tara Brooch. If possible, watch again the initial manœuvres that shape the brooch (website given above). Failing that, follow the steps in figure 7.2.

Figure 7.3 illustrates a very simple sequence in dividing a given measure – the radius on the right – into two parts, *a* and *b*. Then, by a further simple copying manœuvre, dividing that radius still further to set measure *c*; then another simple copying manœuvre divides the radius again to yield the measure *d*. The procedure has no in-built termination. No measure repeats. On the other hand, the relation of the members of the successive pairs of measures does repeat. And repeat. And repeat without end.

Look next at the same set of measures represented in modern alphanumeric notation. With the initial cut of 1 into *a* and *b*, we have these relations:

$$a+b=1$$

$$\frac{b}{a}=\frac{a}{b+a}=\frac{1}{1+a}$$

$$a/b=1/a=1+a$$

$$\varphi=1+a=\frac{1}{a}=\frac{a}{b}=\frac{b}{c}=\frac{c}{d}\ldots$$

With the further divisions into *c* and *d* measures, and any further divisions, measures that link together in pairs, in a chain of equivalent ratios, they form a continuous proportion – something like 'interlace':

$$\varphi=1+a=\frac{1}{a}=\frac{a}{b}=\frac{b}{c}=\frac{c}{d}\ldots$$

Another aspect of the linking appears in this display:

$$a=\frac{1}{\varphi}\quad b=\frac{1}{\varphi^2}\quad c=\frac{1}{\varphi^3}\quad d=\frac{1}{\varphi^4}\ldots$$

And yet further, the whole set can be reduced to combinations of 1 and *a*:

$b = 1-a$
$c = a-b$, thus $= a-(1-a)$
$d = b-c$, thus $= (1-a)-(a-(1-a))$
$e = c-d$, thus $= (a-(1-a))-((1-a)-(a-(1-a)))$

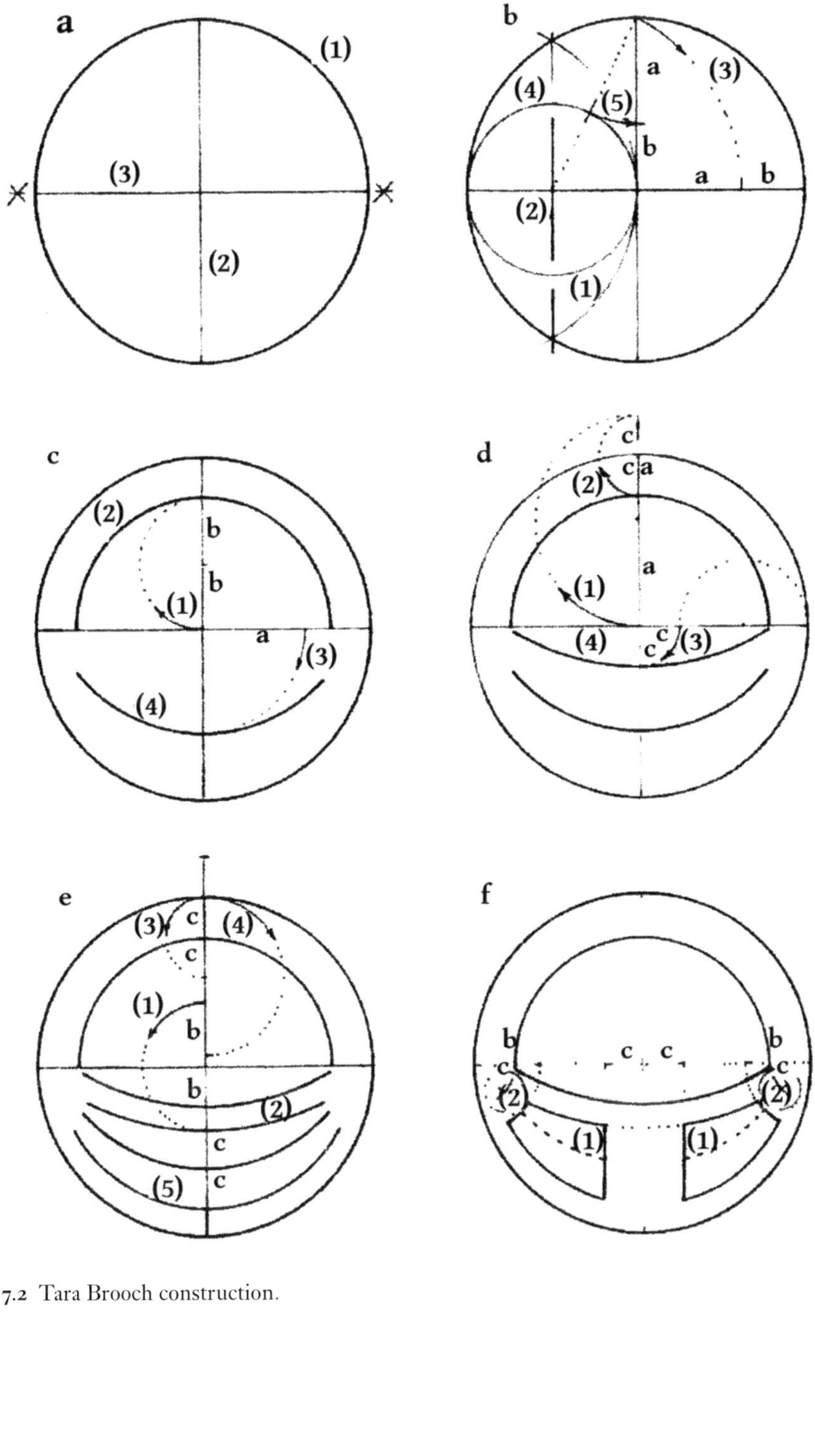

7.2 Tara Brooch construction.

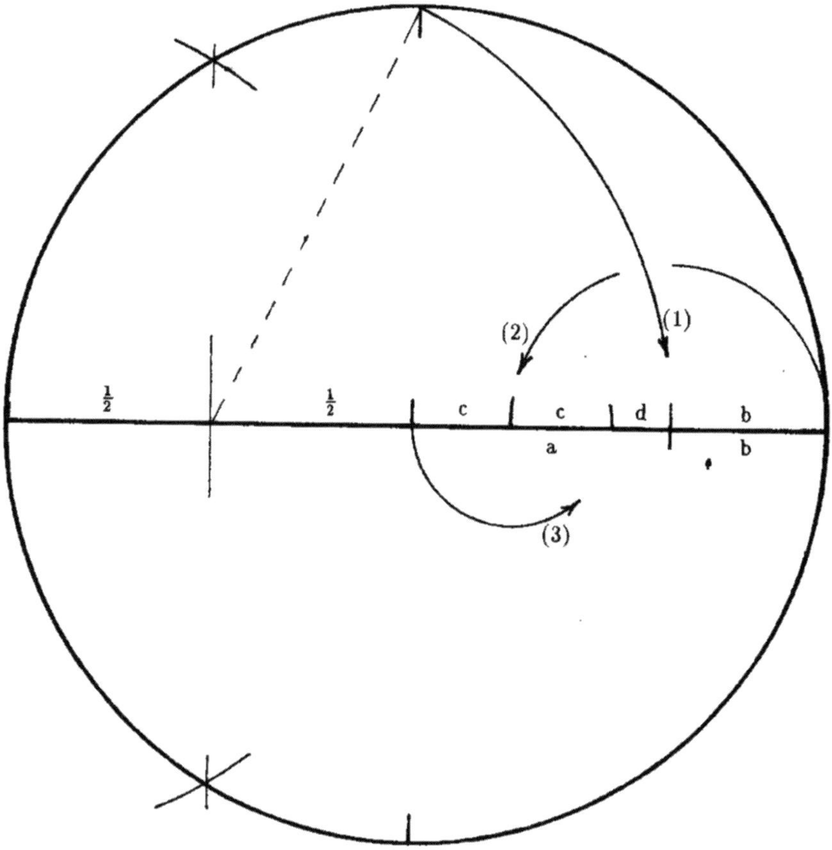

7.3 Sequence in dividing a given measure.

Again, this is getting a bit abstract. The way back to the everyday world, so to speak, is to retrace the graphic derivation of all this – the literally hands-on way of computation and designing with compass and straight edge.

WHAT THE INFORMING MATHEMATICS MEANS

These forms – the planar schemata constructed with dividers and straight edge embodied in sculptured crosses, page illuminations, metalwork – what may they mean? It is not unusual to hear pronouncements concerning medieval art to the effect that if something does not mean something, it does not matter. This too needs parsing.

Take the example of the rectangular forms enclosing evangelist portraits and evangelist symbols. Their being rectangular seems to be an unremarkable function of their being created on and informed by rectangular leaves of a codex. There is no lead to discovering a meaning in that. Go a step further and consider the frames individually. They all appear to have been constructed with compass and straight edge, construction marks in the parchment often confirm this; and constructed according to disciplined rules of devising – their proportional relations throughout demonstrating that.

Go yet another step and review the relations of the mathematics of the frames to the images they enclose (not to mention the structures of the frames and the filler ornament within them). There seem to be no correlations among the maths and the contents of any of the evangelist pages. In one codex all four evangelist frames have different shapes; in another, two shapes are used twice. In one codex, the frame around Matthew develops 1, 2, $\sqrt{2}$; in another it develops 1, 2, $\sqrt{3}$ and so on. With no correlation, gospel to gospel, codex to codex, of a frame's mathematical properties and the evangelist depicted or symbolized, there will be no reason to look for a specific meaning in any of the various shapes and their incorporated mathematical properties. In sum, it seems fairly sure that none of the shapes, none of the mathematical formulae of any of the shapes, has a meaning. None of them signifies, denotes, represents. The forms, as it were, are not predicative.

Or what about number symbolism? For many of the forms, the analyses given in earlier sections (if they are right) render dimensional relations in most full-form designs impervious to number symbolism as it is usually understood, and as it is widely documented in medieval Latin Christianity. This is the case with stone crosses, with illuminated Gospels pages, and with many other objects such as literary texts and fine metalwork. That tradition operates with numerals used to express multitude measures of quantity. The numerals are integers (one, three, seven, twelve, forty …); or they are expressible as integer ratios such as sesquitertial, sesquialter, sesquioctave and such. A few quantities in magnitude measure in any of the forms may be numerical – twos and threes, halves and thirds, mainly. But many are not. They are the 'irrational', which is to say the contrary of rational numbers, for being neither integers nor quotients of integers. So while seven and four and twelve, for example, have an extensive tradition of carrying symbolic meaning, such numbers as $\sqrt{2}$ and $\sqrt{5}$–1 do not seem to have. If they had been assigned symbolic meaning within the artistic tradition, one would expect, for instance, that portrait pages of a particular evangelist would have consistent shapes, or at least shapes sharing a governing ratio utilizing the same irrational number. The fact is that they do not.

That does not rule out maths-based form as it is found in early Irish art being somehow predicative. If it is consistently present, it is a fair guess that it mattered, that it was valued, and just maybe had an importance or a significa-

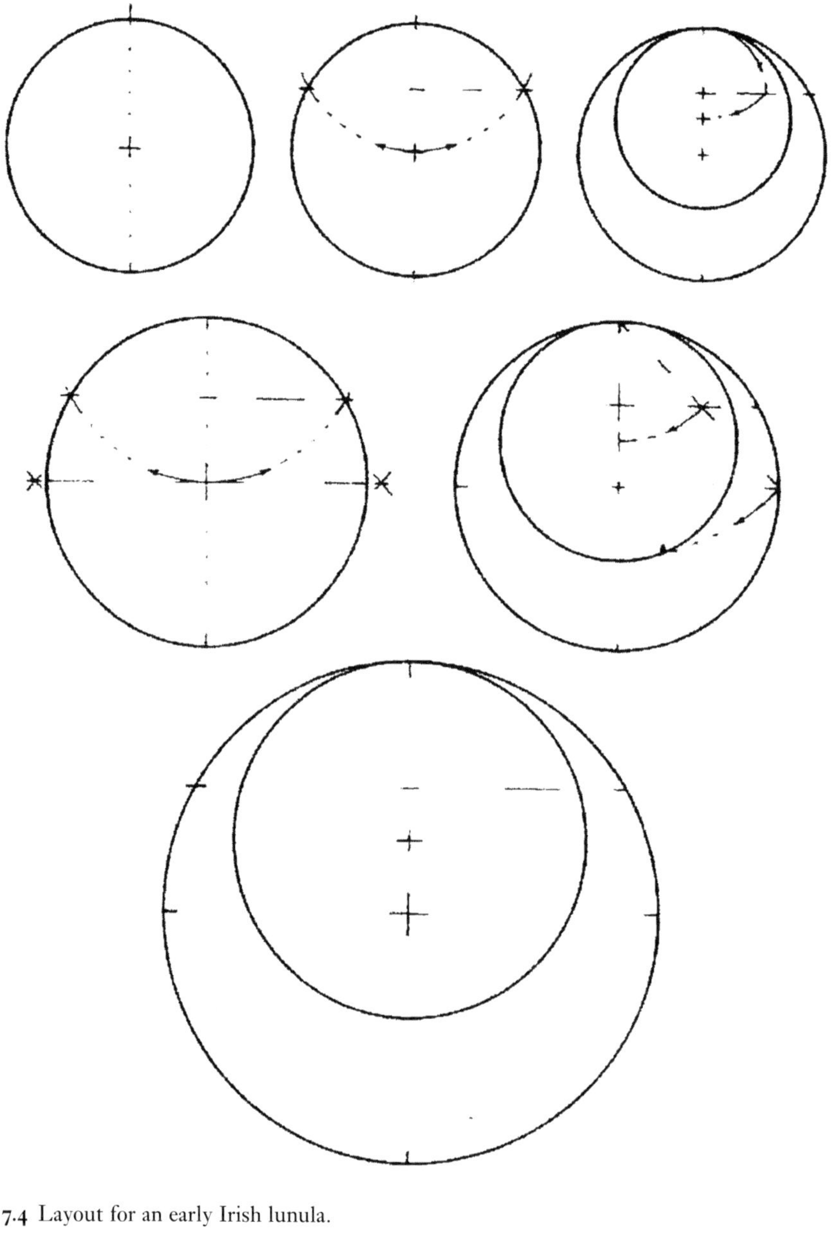

7.4 Layout for an early Irish lunula.

tion. The significance may have been somehow metaphysical – perhaps linked to a current cosmology. But who knows now about these things? It may have been essentially no more and no less than a cultural aesthetic imperative, untrammelled by the objects it informed, or their purposes.

ENVOI

As a parting shot I should like to propose translation of the designing methods into programmes for the new computer technology, to take its place alongside hand-tools for creating this maths-based art. Computer generation of interlace pattern, grilles, tilings is one thing, and that seems to be happening. These are fillers of the forms. Now we need to learn to develop the primary forms.

The challenge could be put into a school curriculum, as a project in algorithmic art (it was an algorithmic art originally, after all). And while we are at it, perhaps run a contest to find which schools produce the most successful 'apps' for replicating Insular designs, and if they are right, for creating new ones. Especially in schools' curricula, such a contest would be a natural way, I should think, for restoring understanding of the most admired forms of medieval Irish art, and leading (one can hope) to developing the skills to create new forms in the ancient mode.

One way or another – ancient tools or computer designing – whoever learns how to replicate the ancient and admired forms just may at the same time develop the knowledge and skills for creating new forms in the ancient way. This is distinct from generating, say, fractals, or any of the forms of computer doodling. The new 'app' approach just could be a way to regenerate the ancient art, to motivate a revival of the skills of early Ireland, to help restore and perpetuate the tradition.

POSTSCRIPT

Mathematics informing art in medieval Ireland and Northumbria – an *esthetische Mathematik* – may have found culturally new objects to design, new patterns and icons to enclose, new materials to inhabit, new functions to accompany, more elaborate expressions to exhibit. But it was not a new notion. It evolved within a tradition traceable through two millennia earlier. Watch the simple method of layout for an early Irish lunula (second millennium BC) at the website cited earlier in this essay or see figure 7.4. That leaves the question: When was excellent art in early Ireland *not* informed by mathematics?[2]

2 Note: when the enquiry came in December 2011, would I be willing to participate in this conference on the history of science and maths in medieval Ireland, I was just ready to hand off to a publisher a handbook on Insular designing, which represents my best efforts at understanding the primary forms of Insular objects I have been studying for thirty years. I now find myself in a brave new 'e-pub' world, where the handbook, *Insular art forms: their essence and construction*, is distributed in electronic bundlings, while I still try to reconstruct the ancient plans with the ancient tools.

Music and the stars in Cashel, Bolton Library, MS 1

CHARLES BURNETT

Three stages can be discerned in instructions in arithmetic in the early Western Middle Ages. The first stage is the instructions for working out the medieval church calendar (the 'computus' or 'compotus'), of which the earliest texts date from the fifth century AD. The second is the instructions for the use of the abacus, texts on which start to appear in the late tenth century. The third is those for calculating 'the Indian way', which came to be called 'algorisms'. The methods discussed in these instructions did not replace each other: we are not involved in stages in the development of mathematics. Nor were the earlier methods less sophisticated than the later. Rather, they enhanced each other. Both the use of the abacus and Indian methods of calculation contributed to the practice of computistics. But the earlier use of multiplication tables, and finger-reckoning continued in use. The Indian method (the algorism) was used alongside the abacus.

Irish scholars contributed greatly to computus literature from the seventh century onwards, as is shown in other articles in this book. An early example of abacus instructions survives in a fragment of a manuscript from Glendalough (now London, British Library, Egerton 3323, fos 18r–v).[1] Finally, a manuscript currently in Ireland, but not likely to have been the product of Irish scholars, exemplifies the third stage: namely, Cashel, Bolton Library, MS 1.

The first person to describe this manuscript in detail was Marvin Colker, in an unpublished typescript that he wrote when compiling his catalogue of the Medieval and Renaissance Latin manuscripts of Trinity College, Dublin, to which I am greatly indebted.[2] The manuscript consists of four major components. The first codex of the manuscript occupies the first 104 pages and to this I shall return. The second codex, pp 105–20, is an older manuscript (twelfth century), consisting of two works introducing Arabic knowledge – first, Qusta ibn Luqa's *De differentia spiritus et animae*, on the difference between the corporeal entity, spirit, which is responsible for various functions of the living human

1 See Mary Kelly, 'Twelfth-century ways of learning: from Worcester or Cologne to Glendalough', *JRSAI*, forthcoming. 2 I am grateful to Marvin Colker for allowing me to print his catalogue description (with additions and corrections) as an appendix to this essay. I am also indebted to Jennifer Moreton, who first brought this manuscript to my attention, and lent me her copy of Colker's description, and to Immo Warntjes and Philipp Nothaft for their comments on the description.

body, and soul, the incorporeal entity which rules the body: the second, the *Helcep Sarracenicum*, which by its very name (deriving from *al-hisab*, meaning 'calculation' in Arabic) reveals its Arabic origin, and is in fact one of the earliest accounts of how to do arithmetic in the Saracen – ultimately Indian – way. This is followed on pp 118–20 by a commentary on Porphyry's *Isagoge*, the first text in the logic curriculum, which occurs uniquely here.

The third codex, pp 121–83, has the function of an encyclopedia, reproducing, without acknowledgment, much of Honorius Augustodunensis' *Imago mundi*, including at the end four pages of diagrams: two schematic maps; one of the world, the other on the seven mountains in Thessaly (pl. 6).

The fourth codex, pp 185–240, is a book of instructive entertainment: Petrus Alfonsi's *Disciplina clericalis*, and two collections of fables, described as 'the three books of Aesop of Athens', and an anonymous collection, which corresponds in content, but not in language, with a collection associated with Marie de France.[3]

It is the first part of this manuscript that I am primarily concerned with. I call it a 'precentor's manuscript'. What were the precentor's duties? His name indicates that he directed the choir (literally 'infront-singer', 'fore-singer'). So, one would expect him to know the chants, and probably some theory of music too. But as director of the liturgy he would also have to know the times of the church festivals – especially of the moving festivals of Easter, Ascension Day, Whitsunday etc. He could, of course, just be following what had been ordained elsewhere by the bishop, or the archbishop or the pope. But it was helpful if he had the rudimentary astronomical knowledge to confirm the official calculations himself, or at least to justify them. But if he was 'predicting' future occurrences, such as full moons and new moons, and even solar and lunar eclipses, perhaps he would be expected to predict other things, too, such as the weather. All this material would be expected to be set out in clear and straightforward ways, even in verse, to make it easier to remember.

All these subjects appear in the first codex of the Cashel manuscript. On pp 59–70 there are texts on music. The first is a general introduction to music theory and the art of liturgical singing. This text corresponds (with omissions) to a text found in four further manuscripts, referred to as 'Augustinus Minor' or 'Pseudo-Thomas Aquinas II' (in reference to the false attributions of the manuscripts), and edited recently by Michael Bernhard. The theoretical parts are based largely on Boethius' *De institutione musica*, as can be seen from the very opening words:

> since it is that which deals with the complexion of the celestial bodies, the conjunction of the elements and the harmonic union of soul and body, we have thought it useful to consider what music is, what is taught in it, and how and to what purpose should it be taught.[4]

3 Léopold Hervieux (ed.), *Les fabulistes latins depuis le siècle d'Auguste jusqu'à la fin du moyen age*, 5 vols (Paris, 1893–99), ii, pp 483–586. 4 Cashel 1, p. 59: *Quoniam inter VII artes liberales*

Having described the divisions of pitches into low, high and very high, and of music into instrumental, mundane and human, the author focuses on instrumental music that concerns recognizing and understanding 'cantus' (the material of performed music), both in man-made instruments and in the human voice. Instrumental music is divided into the study of melodies (*melica*), that of metres (*metrica*) and that of rhythm (*rhythmica*).

Boethius himself, according to the author, was prevented from dealing with everything that he had hoped to deal with in music because he was sent into exile by the emperor Theodoric,[5] and so only wrote about instrumental music, and within that category, only about melodies. So his work has to be supplemented by that of his successors; in particular, Guido of Arezzo ('Wido'). In fact, the Cashel text, too, deals only with melodies and the voice as an instrument. The bulk of the treatise is devoted to explaining the Greater Perfect System: the musical scale, often depicted on a monochord, starting from Gammaut as the lowest note, the regular sequences of notes in a monophonic chant ('cantus planus'), and the possible resolutions of chords in a chant accompanied by a second voice ('organum').

The aim of this treatise is clear from phrases within it: 'These are the rules by which every cantor or every organist (*organizator*) should know – i.e. be able to understand or analyse – every chant (*cantus*) in music;'[6] and 'first the alphabet of the musical art should be demonstrated to the boy pupils'.[7]

The second aspect of the precentor's duties – that of the church calendar – is represented extensively in Cashel 1. Immediately following the musical items is a complete calendar, which Nigel Morgan has identified as being most likely drawn up for the use of the abbey of Tewkesbury, or one of its dependencies. Aside from the feastdays for Egwin, the founder of the abbey of Evesham and the third bishop of Worcester (d. 717), Birinus, the bishop of Dorchester (d. 649), and Machutus (Malo) the disciple of St Brendan, there is one for Cuthburga, who is distinctive of Tewkesbury, since her relics were at Cranborne priory, which was a cell of Tewkesbury.

The calendar is supported with a poem (pp 77–96) to enable one to remember the dates of the feasts. But it also includes some historical notes, such as: 'On the

primatum optinet musica, testante Boetio, utpote illa que de complexione celestium corporum et coniunctione elementorum et armonica unione anime et corporis continet, ideo quidem considerare utile duximus quid sit musica, quid in ea doceatur et ad quid et qualiter. Edited in Michael Bernhard, *Die Thomas von Aquin zugeschriebenen Musiktraktate* (Munich, 2006), p. 91. A summary of the three divisions of music described in Boethius, *De institutione musica*, I, 2, ed. G. Friedlein, *Anicii Manlii Torquati Severini Boetii De institutione arithmetica libri duo: De institutione musica libri quinque. Accedit geometria quae fertur Boetii* (Leipzig, 1867), pp 187–9. 5 Bernhard, *Die Thomas von Aquin*, p. 97: *Boetius itaque de musica tractaturus a Theodorico rege in exilium missus est et a proposito pro maxima parte inpeditus est.* 6 Cashel 1, p. 70: *Hee sunt regule quibus omnis cantor vel omnis organizator debet cognoscere omnem cantum in musica.* 7 Ibid.: *prius debet alphabetum musice artis ostendi pueris discipulis.*

arrival of the Normans into England', 'On the route followed by the Christians in their journey to conquer Jerusalem'. This is followed by questions and answers about the calendar – again a popular way of teaching, including questions that we would all like to know the answers to: 'Why does the Holy Church take into account the course of the sun and moon rather than that of the other planets?' But, on the other hand, 'When all the weekdays have their names from the planets, why do they not observe the order of the planets?' (In fact, there is an astrological explanation for this: when each of the twenty-four hours of the day is assigned to a planet according to the order of its sphere, starting with the sun for the first hour of Sunday, one will find that the first hour of each of the weekdays will be assigned to the planet after which that weekday is named.)[8]

At the beginning of the codex there is more calendrical information, mainly in verse, giving ways of calculating leap-years, and the date of Easter. Some of the verses coincide with those in Alexander de Villa Dei's *Massa compoti*.[9] Cycles and indictions are given between two sequences of dates: 1140–1642 and 1168–1223, in both cases suggesting a mid- or just after mid-twelfth-century date for the information. One text shows one how to work out the date of Easter using the hand as a mnemonic.[10] Again a verse recalls the defeat of the English in the Norman conquest, and its indication by a comet: *Sexagenus erat sextus millesimus annus/ Christi cum pereant Angli, monstrante cometa* ('It was the year 1066 when the "Angels of Christ" (a pun on *Angli*) perished, with a comet demonstrating this'). This reference marks the beginning of a change from computus to astronomy: lists of constellations (items 33 and 34), and the names of the signs of the zodiac (35) are included.

To use the computus information, it is handy to know some arithmetic: there are keys to the numerical value of the Greek letters, to Roman numerals, and to the Hindu-Arabic numerals. The latter two are included in three short texts on Indian arithmetic, and exercises in arithmetic (recreational arithmetic).

1) A number game *unus trium vocetur binarius [...]* (p. 14) is followed by a simple algorism, of which the most complex operation is finding the highest common factor (pp 16–17).[11] A more detailed text on the algorism follows on p. 41, discussing addition, subtraction, multiplication and division *Numerorum diverse sunt differentie. In prima differentia continentur omnes numeri ab unitate [...]* (unusual in starting from addition).

8 A. Bouché-Leclercq, *L'Astrologie grècque* (Paris, 1899), pp 479–80. 9 R. Steele (ed.), *Opera hactenus inedita Rogeri Baconi*, fasc. 6 (Oxford, 1926), pp 268–89. Other texts derive ultimately from Dionysius Exiguus' *Argumentum*, III (item 86) and VII (item 88), ed. B. Krusch, *Studien* I, pp 75–7. 10 Cashel 1, p. 11: *Si per iuncturas digitorum noscere curas/ Metas pascalem ...* 11 The text is edited in C. Burnett, 'Learning Indian arithmetic in the early thirteenth century', *Boletín de la Asociación Matemática Venezolana*, 9 (2002), 15–26, repr. in idem, *Numerals and arithmetic in the Middle Ages* (Farnham, 2010), article IX, see pp 20–3.

A third algorism, telling one the result of the multiplication of factors of 10 (p. 20), and the setting out of a multiplication (p. 21),[12] is followed by a very common form of divination based on numbers (add the number-equivalent of each letter of two clients' names, and divide the totals by 9), by which one can tell whether one client will die before the other (an alphabet with a numeral above every letter is also given on p. 97).[13]

Other divinatory tracts include weather prediction, unlucky and lucky days, and Egyptian days (the 24 days throughout the year on which no enterprise should be begun). There are two diagrams of the 'sphere of Pythagoras' for ascertaining the recovery or death of the patient, and finally, a charm for child-birth, and with this I shall end:

> Saint Quirina bore Remigius and Elizabeth bore John the Baptist. Saint Mary conceived and bore Christ without pain and labour [...] 'Come out child! The light is waiting for you!' Say this three times and put the charm under the right breast and destroy it immediately when the child is born.[14]

12 The table of powers of ten and the multiplication rules are edited in ibid., *Boletín*, 17–20; idem, *Numerals*, pp 20–3. 13 See C. Burnett, 'The Eadwine Psalter and the western tradition of the onomancy in Pseudo-Aristotle's *Secret of Secrets*', *Archives d'histoire doctrinale et littéraire du moyen âge*, 55 (1988), 143–67, reprinted in idem, *Magic and divination in the Middle Ages* (Aldershot, 1996), article XI, see p. 166 (the Cashel MS has not been used in this article). 14 Cashel 1, p. 22: *Sancta Quirina peperit Remigium et Elizabet Iohannem Baptistam. Sancta Maria concepit et peperit Christum sine dolore et sine labore ... Infans exi foras! Lux te expectat .... Ter debet hoc dici cum pater noster et sub dextra papilla poni et cito deleri dum infans nascetur.*

APPENDIX 8.1

Cashel, Bolton Library, MS 1
Described by Marvin L. Colker, revised by Charles Burnett

The items in this manuscript are numerated within brackets.

A)
pp 1–4 (1–3) Calendrical tables:
p. 1 (1) indicates Septuagesima, Lent and Easter.
p. 2 (2) concerned with leap-years. At the foot of the table: A prima creacione terre et aque ad primum annum huius tabule sunt anni .v̄cccxl. (5,340).
pp 3–4 (3) indicates indictions, epacts, Easter, lunations for the years 1168–1223.
pp 6–7 (4–10) Short pieces on time-reckoning, with (4–8) possibly intended as one work:
p. 5 (4) Verses about Lent:
Ebdomada prima Martis ieiunia prima
non hinc mutantur nisi cum cineres ibi dantur
[…] par sibi cessat ordo quam ratio. Also in MS Roman 353, fo. 88r.
A circular diagram about leap-years, of which the inner rim is labelled: Regula de bissexto uera.
(5) Prognostica from or about the weather: Si dies dominica fuerit kl. Ianuarii, hiems bona et uentosa erit […] oues peribunt senes morientur. Cf. L.S. Chardonnens, *Anglo-Saxon prognostics, 900–1100* (Leiden, 2007), pp 497–8.
pp 5–6 (6) *De solsticiis et equinoctiis*. Quatuor sunt distinctiones anni, scilicet duo solsticia et duo equinoctia […] retrocedit solarem.
Incipit as in the chapter *De .iiii.*[or] *anni distinctionibus* in *Summa Magistri Williami de compoto*, edited by J. Moreton, *Computus ecclesiasticus*, p. 22 (typescript).
pp 6–7 (7) *Quare preceptum sit ut pascha celebretur luna xiiii*[a]. Idcirco preceptum est ut celebretur pascha luna xiiii[a] […] hos duos terminos.
p. 7 (8) *De saltu lune*. Saltus lune dicitur subtractio unius diei a decemnouenali ciclo […] in quolibet ciclo fiet dies unus.
(9) Luna est turgide nature et argentei coloris […] Saturnus est frigide nature et ferrei coloris (= *Liber Alchandrei*, 1, edited by D. Juste in idem, *Les* Alchandreana *primitifs: étude sur les plus anciens traités astrologiques latins d'origine arabe* (Xe siècle) (Leiden, 2007), p. 433).
(10) Mnemonic: Post S. I. M. S. V. M. (glossed 'Saturnus Iupiter Mars Sol Venus Mercurius) sequitur ultima luna subest.
pp 8–10 (11)–(13) Calendrical tables:
p. 8 (11) indicates Lent, Easter and Pentecost.
p. 9 (12) claims to show 28 year cycles and indictions, applied to particular years from 1140 to 1642 (the years are reported in Hindu-Arabic numerals)

p. 10 (13) lunations.

pp 11–15 (14–67) Calendrical verses. Many of these pieces, but not (14), are to be found in Alexander de Villa Dei, *Massa compoti*, Steele, *Opera hactenus inedita Rogeri Baconi*, pp 268–89.

p. 11 (14) *De aureo numero.*

Aureus hac arte numerus formatur aperte
[…]
linea sola tenet nemo sic esse refutet.

Cf. Steele, *Opera Rogeri Baconi*, pp 257–8 (= Grosseteste's *Compotus*)

(15) A (*sic!*) Augustus Iano triginta diebus et uno
[…]
Octo tenent idus menses communiter omnes.

(15) follows on from (14) without a break.

(16) Si per iuncturas digitorum noscere curas
Metam pascalem normam sumas tibi talem
[…] metam postrema notabit.

(17) A G preposito quota sit mete data scito (metre?)
[…] solum te fallere nescit.

(17) is possibly part of (16)

(18) Aduentus domini non est celebrare Decembris
[…] quinas ante kalendas.

Steele, *Opera Rogeri Baconi*, p. 276.

(19) Andree festo uicinior ordine quouis
[…] prima colit feria.

Steele, *Opera Rogeri Baconi*, p. 276.

(20) Quisquis in f(eria) Martis ueniat numerus ferialis (one foot too many)
[…] concurrenti socialis.

(21) *Concurrentes et dominicales litere.*

Fons Eue deitate cluens ab gente furenti
[…] clipeum bellantibus astans.

A roman numeral above each word.

(22) Fons educit agrum fons escas bos agit edus
[…] fronde cibos (cibor?) ars.

Steele, *Opera Rogeri Baconi*, p. 277.

(23) *Ad epactas inueniendas.* Nex (?) labor […] obit euge fidelis
[…] fulgor gloria sint.

A roman numeral above each word.

(24) Addens cum numero .d. bis .t. non superabis
quo pariter iuncto tu .i. .k. et .ı. uiduabis.

Repeated in (54)

(25) Addicies eta sed non superes iotha theta
differ maiorem lateratim pone minorem.

Repeated in (55)

(26) (later addition) Que tenet undenas Aprilis iiii (luna?) kalenda
[...] per quemlibet annum.
Steele, *Opera Rogeri Baconi*, p. 282.

p. 12 (27) Immo (glossed 'Iulius') tex (?) captant reliqui sibi quatuor aptant
[...] idus omnibus octo.

(28) A quinta decima donec uicesima prima
[...] festum pasche celebratur.

(29) Est anni factus mensis quasi ianua Ianus
[...] orbe refulsit.

(30) Circulus hic magis paschalis dicitur annus
[...] continet annos.

(31) Sexagenus erat sextus millesimus annus Christi
cum pereant Angli monstrante cometa.

(32) Anno millesimo sexagesimoque seno
Anglorum mete f<1 >ammas sensere comete.
Hoc anno seno milleno bisque trigeno
Anglorum gentes bello periere cade<n>tes.

(33) Exoriens Aries Libram demerget in ima
[...] muictoria uicte.
H. Walther, *Initia carminum ac versuum medii aeui posterioris latinorum* (Göttingen, 1959), No. 6089.

(34) A boree partes arctos uertuntur et anguis
[...] Eridanique fluenta.

(35) Est Li. Ari. Scor. Tau. Sa. Gemi. Capri. Can. A. Le. Pis. Vir.
The names of the signs of the zodiac (in pairs of opposite signs). Steele, *Opera Rogeri Baconi*, p. 273.

(36) Solari ciclo bessexti noscitur ordo
[...] bissextum clarificabit.

(37) Normales lune tali lugubramine sume
[...] tot normales tibi tolle.

(38) Christi duc annos per denos atque nouenos
[...] tollens sumis epactas.

(39) Normales lune pariter necnon et epacte
[...] fulge luna kalendis.

(40) Sume dies mensis quot sunt. His adde kalendis
[...] semper tibi pandet.

(41) Horis bissenis annus sibi uendicat omnis
[...] dies solet uno.

(42) Quo Februi decem (decimam?) cernes consistere lunam
[...] retro tardius uno.
Cf. Steele, *Opera Rogeri Baconi*, p. 275, Walther, *Initia carminum*, no. 16214.

(43) A festo sancti patris nostri Benedicti
[...] semper tibi pasche.

(44) Tercius et quintus necnon et septimus annus
[…] Maiique Iuliique kalendis.
p. 13 (45) Aprilis none ter<m>inant hic quinque repone
[…] Maii sumpsere kalende.
(46) Primus in orbe dies lucis primordia sumpsit
[…] requies his rite peractis.
P.F. Alberto (ed.), *Eugenius Toletanus opera omnia*, CCSL, 114 (Turnhout, 2005), p. 252, *Carmina* 37).
(47) Alfa boni cuncti dux fax genitricis amator
[…] decet edere fructum.
(48) Inspice quot uicibus paschalis sit modo ciclus
[…] cernes quod gliscis habere.
(49) Primalis merito numerus succrescit ab octo
[…] prope portet epactam.
(50) Etatem lune quarter e duo bis fore quinque
[…] tot lune signa notabis.
(51) Quisquis normales dinoscere uult feriales
[…] pascales uel normales feriales.
(52) Si peruenture feriam uis scire kalende
[…] feriam quam restat habeto.
(53) Digna gerens et amanda deus bonus emoderator
[…] altus doce beata.
p. 14 (54) appeared previously as (24).
(55) appeared previously as (25). Followed by a table illustrating Greek alphanumerical notation, giving (a) the roman numerals from .i. to .m.; (b) the names of the letters of the Greek alphabet, together with the extra letters used for numerals; (c) the Greek letters themselves; (d) the equivalent Latin letters. The Hebrew alphabet is added twice in the margins.
(56) Quatuor et penta duo monos tres mias unus
hinc dias ambo trias unus dias et duo monas.
Walther, *Initia carminum*, no. 15298.
(57) Prima dies mensis et septima truncat ut ensis
[…] undenus ut anguis.
Unlucky days. Walther, *Initia carminum*, no. 14563.
(58) *De epactis.*
Ter tria pre triplo ter sex cum monade triplo
[…] unum deponere debes.
(59) *De regulis lunaribus.*
Post .e. bis .g. bis post .i. k. et .1. manet ordo
En ex grege iokus it klamannus ovem (?)
(60) *De regulis ferialibus.*
Est. ars. crux. fas. ac. dos. grus. bos. est. gula. cur (?)

(61) Altitonans dominus diuina gerens bonus extat
[...] invenies cum religiosis.
About the months, which are named in a gloss above the poem. The first two verses correspond to the first two verses in (92). The final verse reappears in (81) as item c.
Cf. Steele, *Opera Rogeri Baconi*, p. 272.
p. 15 (62) Bissena mensis uertigine uoluitur annus
[...] trigenis quinque diebus.
Cf. Walther, *Initia carminum*, no. 2187.
(63) Est aries taurus gemini cancer leo uirgo
[...] urnaque piscis.
Cf. Steele, *Opera Rogeri Baconi*, p. 273 and Walther, *Initia carminum*, no. 5599.
(64) Lustra tenent annos quinos indictio ter tot
[...] date secula dena. Also in Alexander Neckam, *Sacerdos ad altare*, ed. C. McDonough, CCCM, 227 (Leiden, 2010), p. 31.
(65) Tres obolos siliqua siliquas .iii. dramma sed October[...]
perygenesis hec bene complet.
(66) Circumcisus Adam facit ephiphania feritur
[...] festa sequentur.
Walther, *Initia carminum*, no. 2805.
(67) Ordine continua nathalia fest uocato
[...] Thome siluestri [...] anni.

(68–9) Short pieces.
p. 11 (68) Yris nic<h>il aliud est quam quedam repercussio solaris impetus resultans in nube aquosa composita soli.
p. 14 (69) Unus trium uocetur binarius alius ternarius quarter quaternarius
[...] tercium scire non erit difficile.
pp 16–17 (70) Cum numero numerum addere uolueris [...] tunc operandum ut supradiximus.
An algorism edited in C. Burnett, 'Learning Indian arithmetic', 15–26.

(71–5) Prognostica
p. 17 (71) *Abute (?) Pytagore seu gratois (?)*. Discc lune diem in quo (?) decubuit
[...] hoc modo cetera inspicienda et colligenda sunt.
The *Tetragonus* subiectus, edited by E. Wickersheimer, 'Figures médico-astrologiques des IXe, X^{e} et XIe siècles', *Ianus* (1914), 157–77 at 171–4.
p. 18 (72) Hoc est organum Pitagore de unaquaque re inuenienda iuxta hominum uitam [...]
p. 19 (73) [...]siris Necepso regi salutem. De his que a me ad humane uite cautelam [...] invenies zoemegale que positio duorum (numerorum?) victorem Ach<illem> et Hec<torem> uictum significat.

Epistola Petosiris, edited by Wickersheimer, 'Figures médico-astrologiques', 166–7.

p. 20 (74) Collige per numerum quicquid cupis esse probandum [...]morietur et infra.

Explanatory verses preceding diagram for prognosticating life and death.

p. 21 (75) *Artis prima distinctio*. Unus et unus, minor uincet. Unus et duo, duo uincet. Unus et tres, unus uincet [...] Nona distinctio. Nouem et nouem, minor uincet.

Followed by the onomantic alphabet with a Roman numeral over each letter, and by a note: Si par est tunc per medium semel adde pro toto (?), si non per partem maiorem collige totum.

The victorious and the vanquished, edited (without the last sentence) in C. Burnett, 'The Eadwine Psalter', 143–67.

p. 20 (76) Cencies m m, decies m m m m cm xm m c x i.

pp 20–21 (77) Si uis ducere digitum in se scribe eius decuplum et aufer eum a suo decuplo [...] prouenient ex multiplicacione unius digiti in alium tot .c. etc.

(76–7) edited in Burnett, 'Learning arithmetic', pp 17–20.

p. 22 (78) Sancta Quirina peperit Remigium et Elizabet Iohannem Baptistam. Sancta Maria concepit et peperit Christum sine dolore et sine labore [...] Infans exi foras! Lux te expectat [...] Maria peperit dominum nostrum Ihesum Christum saluatorem mundi. Panditur interea domus omnipotentis Olimpi. Ter debet hoc dici cum pater noster et sub dextra papilla poni et cito deleri dum infans nascetur.

Charm for childbirth. In the margin there is the palindrome 'Sator arepo tenet opera rotas'.

p. 22 (79) *Ratio de terminis mobilibus*. Post .xvi. kl. Februr (*sic!*) ubi luna fuerit .x. ibi erit terminus septuage(sime) [...] ibi terminus pent(ecoste).

About time-reckoning for Septuagesima, Easter, Rogations, Pentecost, also about epacts and concurrents.

p. 23 (80) Sexaginta dies tricentum quinque dat annus [...] grauat Adam.

Calendrical verses added in a later hand. p. 24 is blank.

p. 25 (81) Table of common and embolismic lunations. The table includes mnemonic verses: (a) Esse grauem nobis bello karnem kaueamus; (b) Bellum sepe gerens etenim puto deicit hostem; (c) Mox anime lucrum inuenies cum religiosis. (c) already appeared as the final verse of (61). The first three and a half verses appear in Steele, *Opera Rogeri Baconi*, p. 275. All three verses appear in the *Computus ecclesiasticus*, ed. J. Moreton, p. 44 (typescript).

pp 26–7 (82) added by a later hand. Decree concerning church ornaments, to settle disputes arising among churches in the province of Canterbury: Ad doctrinam presentium et memoriam futurorum. Cum plerumque altercacio inter rectores ecclesiarum in prouinciis Cantuar. constitutos eorumque parroch. super uariis ornamentis rebusque ecclesiasticis que et quanta quomodo ad rectores pertineant infrascripta [...] Scient et intelligant et obseruent taliter uniuersi quod calix missale uestimentum ipsius ecclesie principale [...] approbatas a locorum rectoribus seu uicariis reparari. p. 28 is blank.

pp 29–35 (83) Qui facturi essent et cum ita ibi congregabantur nundine magne fiebant. Ipsi uero Romani quoniam nescierunt principium cuiuslibet mensis constituerunt inter se ut preco ciuitatis in principio cuiuslibet mensis ascenderet portam ciuitatis et clamaret sic Calo calo pluries [...] remanebit erit clauis in secundo. An acephalous fragment about the solar and lunar years, with diagrams.

The same information in Moreton, *Computus ecclesiasticus*, p. 8 (typescript).

pp 36–7 (84)–(97) Short calendrical pieces.

p. 36 (84) An<n>o .i. .g. et sic deinceps formantur usque ad xix annos [...]

Claues terminorum erarum with diagram. On 36 a note is added in a later hand about the claues: Ut clauis docet a Iani G[...]

(85) *Ad inueniendam epactam in quolibet anno talis datur noticia.* Consideretur quotus sit annus [...] epacta in illo anno.

(86) *Ad epactas.* Diuide presentes annos ab incarnacione domini per xix [...] epactarum sufficiant.

(87) (Ad cognoscendum quis concurrens sit in quolibet anno [...] *cancelled*)

(88) Ad cognicionem bisexti. Diuide presentes annos ab incarnacione domini per quatuor [...] seques annus bis erit.

(89) Si uis scire infra quot annos quota feria aliquis mensis incipiat [...] est operandus.

pp 36–7 (90) Si uis scire quota erit luna in aliquo sequentis anni feria [...] feria sequentis anni.

p. 37 (91) Verses:

Hec duo solsticia faciunt cancer capricornus
[...] septem decimoque kl.

Walther, *Initia carminum*, no. 7502.

(92) Verses (as in 61 above):

Altitonans dominus diuina gerens bonus extat
[...] celestibus estibus ardens.

The first verse corresponds to Steele, *Opera Rogeri Baconi*, p. 224 (= Grosseteste's *Compotus*)

(92) Si uis scire quota sit luna in aliqua feria preteriti anni [...] prius facta est.

(94) Verses:

Embolismales compos sic inspice sedes
[...] et tercio Martis. Cf. Hrabanus Maurus, *De computo*, 10 (CCCM, 44, p. 227).

pp 37–9 (95) Atomus ostentum momentum pars [...] Athomus interpretatus indiuisibilis pars temporis [...] omnes cicli in unum conueniunt.
About divisions of time. Cf. *Bobbio Computus*, *PL* 129, col. 1279D.

pp 39–40 (96) Si uis scire cotidie quota sit feria computa dies a principio mensis cuiuscumque uolueris et unum semper subtrahe [...] studenti manifestabitur.
On 39 an alphabet with a numeral above each letter.

p. 40 (97) *De indictionibus*. Consideretis illis que ad hanc artem utiliora uidentur [...] hoc tam multi obseruabant <quoniam?> nescierant.
Ends probably imperfectly.

pp 41–53 (98) Numerorum diuerse sunt differentie. In prima differentia continentur omnes numeri ab unitate usque ad .10.[...] et etiam de cubicis dicemus numeris.
A text on arithmetic. No section on cubic numbers follows.

pp 54–8 (99) Si proposito numero preponatur cifra uel figura medietatem habens [...] Amplius sciendum est quod numerorum alius primus alius compositus [...] deinceps secundum ordinem usque ad milies.
A text on arithmetic. Not published.

pp 59–70 (100) Quoniam inter .vii. artes liberales primatum optinet musica testante Bo(ethio) utpote illa que de complexione celestium corporum et coniunctione elementorum armonica unione anime et corporis continet[...] ex supposita diuisione artis musice. Musica alia instrumentalis alia mundana [...] ut diximus et hoc scimus esse in acutis. Hec de hoc sufficient.

pp 69–70 Hoc est principium artis musice. Gammaut. Gamma Grece ut Latine ut prima uox artis ut gubernat quinque uoces. Hoc est dispente ut re mi fa sol [...] Hee sunt regule quibus omnis cantor uel omnis organizator debet cognoscere omnem cantum in musica.
'Augustinus Minor' or 'Pseudo-Thomas Aquinas II', *De musica*, edited by Michael Bernhard in idem, *Die Thomas von Aquin*, pp 89–147.

pp 71–6 (101) Calendar for a dependency of Tewkesbury Abbey.[15] Each month occupies a column of text. Among the entries in red are: Wulstan, Oswald (p. 71); Benedict, 'Adam creatus' (p. 72); Dunstan, Augustine of Canterbury (p. 73); Swithin, translation of Thomas martyr, translation of Benedict, translation of Swithin, Kenelm, octave of Benedict (p. 74); translation of Egwin, 'dedicacio ecclesie S. Mich. Cap.' (29 September), Michael in Monte Tumba (p. 75); Edmund archbishop, Edmund king, Birinus, Thomas martyr (p. 76). Among the entries in black: octave of Thomas martyr, Werburg, Edward king, Milburg (p. 71); Edmund king, Cuthbert, Alphege, Ethelbert martyr (p. 72); Aldhelm, translation of Wulstan, Alban, Etheldred (p. 73); Wilfrid (p. 75); Machutus (p. 76), with numerous calendrical verses (e.g. p. 76: Scorpius

15 I owe this judgment to Nigel Morgan; this is based on the fact that the calendar commemorates Cuthburga, who is distinctive of Tewkesbury, since her relics were at Cranborne priory, which was a cell of Tewkesbury.

hibernam preceps iubet ire novellam). Includes prognostica, e.g. for January (p. 71): optima dies nona, nocet hora septima quinta.

pp 77–96 (102) Calendrical poem:

Aduentum Christi lux preuia quinta Decembri
[...]
Maii et epta et quarta kalenda.

Includes (pp 78–9): *De aduentu Normannorum in Angliam.*

Iste quidem geminos cyclus compleuerat annos
quando Normanni pugnasse feruntur et Angli.

It also includes (p. 79):

De itinere Christianorum ad expugnandam Ierusalem.

Et quater octonos idem transegerat annos
Christicole sacrum quando peciere sepulcrum.

Some of the verses of (102) may be found elsewhere: e.g. Iunius Aprilis Septemberque Nouember [...] additur unus ei (p. 86) cf. Steele, *Opera Rogeri Baconi*, p. 271 (= Walther, *Initia carminum*, no. 9945).

pp 96–104 (103) Queritur quare pocius non ponitur unitas primo die Ianuarii quam ternarius [...] Solutio. Ad hoc dicimus quod non debet poni aliquis numerus nisi ternarius. [...] inpare gaudet eo quod numerus inpar perfectus est. Among the questions are: Queritur cum in kalendario in quolibet mense assignati sint duo dies Egypciaci quare dicantur Egypciaci (p. 99); Queritur quare sancta ecclesia pocius consideret cursum solis et lune quam ceterorum planetarum (p. 100); queritur cum omnes dies nomina habeant a planetis quare ordinem planetarum non obseruant (p. 103); queritur quare Fabruarius pauciores habeat dies quam aliquis alius mensis (p. 104).

Questions and answers about the calendar, ends imperfectly. Cites Bede (p. 103) and Ovid's *Fasti* (p. 104).

B)

pp 105–11 (104) *Incipit liber differentie inter animam et spiritum.* Constaben Luce cuidam amico suo scriptori cuiusdam regis eddit (*sic!*) et Ioh(ann)es Ispolensis (*sic!*) episcopus ex Arabico in Latinum Raimundo Toletano archiepiscopo transtulit. Interrogasti me honoret te deus de differentia [...] cuius spiritus prolongatus est et hoc tibi sufficit in hoc quod interrogasti. Explicit.

Costa Ben Luca, *De differentia spiritus et animae*, edited by J. Wilcox (in preparation).

pp 111–17 (105) Virtus amicicie inter eos qui eius habitu [...] [...] uel trigies quinque bis.

H. Ocreatus, *Helcep Sarracenicum*, edited by C. Burnett in '*Algorismi vel helcep decentior est diligentia*: the arithmetic of Adelard of Bath and his Circle' in M. Folkerts (ed.), *Mathematische Probleme im Mittelalter: Der lateinische und arabische Sprachbereich* (Wiesbaden, 1996), pp 221–31, reprinted in Burnett, *Numerals*, article III at pp 261–97.

pp 118–20 (106) *Cum sit necessarium* etc. Cum in reliquis dial(ecticis) libris [...] Dicunt enim moueri.
An unidentified commentary on the *Isagoge* of Porphyry.

C)

pp 121–83 (107) Ad instructionem multorum quibus deest copia librorum hic libellus edatur, nomenque ei ymago mundi detur, eo quod disposicio tocius orbis in eo quasi in speculo conspiciatur[...] nisi quod maiorum commendat tradicio. Mundus dicitur quasi undique motus [...] Nabogodonosor annis xliiii. Sibilla Cumana claruit et Ezechiel finit.

Honorius, *Imago mundi*, ends incomplete after the start of the *quinta aetas* (= *PL* 172, 119–75), ed. V. Flint, *Archives d'histoire doctrinale et littéraire du moyen âge*, 48 (1982), 7–153 at 49–137.

pp 183–4 (108)–(11) Diagrams, probably meant as an appendix to (107).

p. 83 (108) Circular diagram correlating the zodiacal signs and the months.

(109) Incomplete diagram consisting of an unlabelled circle.

p. 184 (110) circular diagram indicating wind-directions, the different divisions of the world and the rivers Danais (Don) and Nilus (Nile).

(111) Circular diagram of Thessaly and her five mountains.

D)

pp 185–240 (112) *Libellus magistri Petri Anf.* Dixit Petrus Anfulsus serus Christi Ihesu compositor huius libri Gratias ago deo qui primus est [...] sue displiceat uoluntati. Enoch philosophus qui lingua Arabica [...] perfrui mereamur in aula celesti prestante domino nostro Ihesu Christo cui est honor et gloria per infinita seculorum secula amen. See pl. 7.

Petrus Alfonsi, *Disciplina clericalis*, Cristiano Leone (ed.), Pietro Alfonsi: *Disciplina clericalis. Sapienza orientale e scuola delle novelle* (Rome, 2010).

pp 240–349 (113) Fables, in two parts:

pp 240–86 (a) *Incipit prephatio fabularum primi libri Aesopi Atheniensis.*

pp 240–1 Romulus urbis Rome imperator Liberino filio salutem. Esopus quidam Grecus sapiens et ingeniosus [...] acuunt sagax ingenium.

pp 241–2 List of contents for book 1.

pp 242–86 *Incipit liber primus Aesopi Athen<i>ensis.* Hesopus hanc primam fabulam dixit de his qui despiciunt sapientiam [...] In sterquilinio quidam gallinacius dum quereret escam [...] cunctos uolentes legere ostendens mores et uitam hominum. *Explicit. iii. fabularum liber.*

Romulus Nilantii, edited by Léopold Hervieux, *Les fabulistes latins*, ii, pp 329–64, i, pp 419–31.

pp 286–349 (b) Fables, usually beginning 'Ferunt fabule iam dudum quendam hominem[...]', many of which correspond with tales that Hervieux assigns to the Latin Romulus collections associated with Marie de France (see the texts edited in Hervieux, *Les fabulistes latins*, ii, pp 483–586). These correspondences are in theme but not in language, which is often markedly different between this manuscript and Hervieux' edition. References in parentheses below are to his texts in vol. ii:

pp 286–8 (pp 569–70, nos cxiii–cxiv)
pp 288–307 (pp 524–37, nos xxxv–lvii)
p. 308 Wolf, tonsured as a monk, is revealed when his hair grows back (cf. pp 610–11, no. xxv)
pp 308–11 (pp 538–40, nos lviii–lix)
pp 311–13 (pp 541–2, no. lxi)
pp 313–14 (pp 540–1, no. lx)
pp 314–16 (pp 542–3, no. lxii)
pp 316–19 (pp 570–1, no. cxv)
pp 319–22 (pp 646–7, no. lxxv)
pp 322–3 (pp 485–6, no. vii)
pp 323–31 (pp 573–7, nos cxviii–cxxvi)
pp 331–6 (p. 586, appendix).
pp 336–7 (p. 493, no. xviii)
pp 337–8 (p. 492, no. xvi)
p. 338 (p. 547, no. lxxi)
pp 338–42 (pp 544–7, nos lxv–lxx)
pp 342–6 (pp 547–9, nos lxxii–lxxiv)
pp 346–7 (pp 346–7, nos cxxvii–cxxviii)
pp 347–9 (p. 580, nos cxxxii–cxxiii)
pp 350–2. Originally blank.

Parchment i + 352 pp (1rv is a front flyleaf, pp 351–2 is an end flyleaf). 118 x 88mm (A and B 80–106 x 51–83mm, C and D 100 x 72mm). 2 cols on pp 11–15, 71–6. A and B number of lines irreg. C 22/23 lines, D 24 lines. 2 flyleaves, i^{6} ii^{10} (wants 7, 8) iii^{8} (wants 2, 6) iv^{8} v^{10} vi^{8} (wants 1) + 1, vii^{6} viii^{10} (wants 4, 8) ix^{14} (wants 3, 5, 11, 12) x–xi^{8} xii^{8} (wants 5, 6) xiii^{8} (wants 1) xiv–xvi^{8} xvii–xviii^{12} xix–xx^{10} xxi^{8} flyleaf with conjoint stub. D quires numbered. A several hands and the additions as indicated. B, C, and D each in a single hand. In heading on p. 185, after the first letter (which is green) each letter is red but filled with green. Initials are red or green throughout. A and D red initials sometimes with green flourishes; C red initials sometimes with red flourishes; D green initials sometimes with red flourishes. Initial on p. 30 is blue with red and blue flourishes; initial on p. 37 is blue. A capitals filled with red or blue or green; D capitals filled red. Considerable use of green in A for lines in diagrams and for filling words or numbers in diagrams. C and D thirteenth century corrections. Medieval binding of white leather over wooden boards, traces of a thong. Second folio: A anni domini/mclxviii; B ad pulmonem; C mensuratur; D que fecerat.

p. I (flyleaf) indicates, by a modern hand, that the classing of the codex was changed from C.R.B. MS 14 to MS 1. Iv has Archbishop King's shelfmark BX 194/N 2823 (eighteenth century). P. 350 (divided into grids by a drypoint) has the verse: Post barbam collum, post collum rade coronam (fourteenth cent.).

The heavens, earth and imagined islands: an introduction to the medieval medical and astronomical resources of the Royal Irish Academy Library

SIOBHÁN FITZPATRICK

Almost one hundred medical manuscripts created in Ireland from the fifteenth to the seventeenth century survive. They are held in a number of repositories, chief among these being the Royal Irish Academy, which holds over thirty. If some nineteenth-century transcripts are included, the total rises to forty; of which, twenty-two are on paper and eighteen were written on vellum. Some illustrations are to be found in the more important copies and some of the manuscripts contain decorative elements, as shown by the first two examples. Plate 8 shows p. 52 of RIA MS 23 A 4, a manuscript compiled in the sixteenth and seventeenth centuries, at least in part in Co. Donegal. It contains texts by Aegidius and Hippocrates *inter alia*. Plate 9 is from RIA MS 23 H 19, fragments of medical treatises, compiled in the fifteenth century, including texts by Ibn de Mesue, Hippocrates and others. The detail shown (fo. 7r) is from a segment of a tract ascribed to Galen.

Seventeen of the Academy's most important medical manuscripts have been digitized and are accessible on the Irish Script on Screen (ISOS) resource.[1] The second largest collection of these manuscripts is at Trinity College Dublin and the third is at the National Library of Scotland. Smaller collections are held by the National Library of Ireland, NUI Maynooth, King's Inns Library, the British Library, the Bodleian Library and other repositories.

These books were compiled in each of the four provinces. They consist mainly of translations or adaptations of Latin treatises that in turn contain the medical matter taught in the European medical schools from the twelfth to the seventeenth century. This learning had Graeco-Arabic roots. As Aoibheann Nic Dhonnchadha explains,[2] the bulk of the treatises are compendia of general medicine which conform to the standard Arabic arrangement of diseases of the

1 Irish Script on Screen is a collection of digitized Irish-language manuscripts held in public and private collections in Ireland and internationally. The resource is managed by the Dublin Institute for Advanced Studies, School of Celtic Studies, and is accessible at www.isos.dias.ie.
2 Aoibheann Nic Dhonnchadha, DIAS, School of Celtic Studies, has carried out extensive research on the Gaelic medical texts. This paper has drawn heavily on Nic Dhonnchadha's writings and the author acknowledges her generous support.

body, *a capite usque ad pedes* ('from head to toe'). The texts deal with a whole range of topics – anatomy, diagnosis and prognosis, diet and regimen, gynaecology, obstetrics and pharmacology – and with common medical problems and diseases – coughs, colds, skin complaints, problems associated with the blood, pain and so on. The texts sometimes contain more local compilations, notes, glosses and commentaries compiled by the Irish doctors; only occasionally do they provide details of patients treated.

Before considering the texts and examining pages from some of them, it is important to note the following points:

1. The texts are almost all written in Early Modern Irish. This is one of the more unusual aspects of them. Irish doctors were unique in translating texts, copied throughout the Continent through the medium of Latin;
2. Latin was frequently retained for introductory sections and was used throughout for citations;
3. Texts are often abridged, or excerpts are translated;
4. This was not an academic exercise – the compilations were made for practical purposes – for the use of Irish doctors in the course of their working lives. However, the translators sometimes used several versions of a text to compile their own copy of a tract; thus there is often an element of selectivity or editing;
5. Several key texts occur in a number of manuscripts compiled by different scribes in different places.

The manuscripts to which I shall refer here are all in the Royal Irish Academy collection. Much of the information about the doctors who translated the texts and practised in Ireland, and about the original texts on which the translations were based, is based in turn on the Academy's *Printed catalogue of Irish manuscripts*[3] and on the writings of Aoibheann Nic Dhonnchadha, who is the recognized authority on these texts with an unrivalled knowledge of the content of these works and of the medieval Irish medical schools.

THE SOURCES

Who were the authors copied by the Irish doctors?

Hippocrates, born in 460BC on the island of Cos in Greece, where he founded a medical school. Regarded as the greatest physician of his time, he based his

3 T.E. O'Rahilly, Kathleen Mulchrone et al., *Catalogue of Irish manuscripts in the Royal Irish Academy*, fasc. 1–28 (Dublin, 1926–70).

medical practice on observation and on the study of the human body. Believing that illness had a physical and a rational explanation, he rejected the prevailing view that it was caused by possession by evil spirits, or resulted from having displeased the gods.

Hippocrates considered that the body should be treated holistically. He accurately described the symptoms of diseases, being the first physician to describe correctly the symptoms of pneumonia, as well as epilepsy in children. He believed in the natural healing power of rest, good diet, fresh air and hygiene. He noted that illnesses presented in different degrees of severity from one individual to another and that some were better able to cope with their disease than others. He was the first physician to connect thought, ideas and feelings with the brain rather than the heart.

Hippocrates developed the eponymous oath of medical ethics. He died in 377BC and is considered to be the 'Father of Medicine'. Hippocrates' *Aphorisms*, *Diagnostics* and *Prognostics* feature hugely in our manuscripts. The *Aphorisms*, consisting of four hundred statements, were, from the time of Galen, divided into seven books. They were central to the *Articella*, which formed the basis of advanced teaching in Europe for four centuries from the twelfth.

Claudius Galen was born in Pergamum *c.*AD130, of Greek parents. He studied in Greece, in Alexandria and in other parts of Asia Minor and returned home to become chief physician to the gladiator school in Pergamum, where, not surprisingly, he gained experience in the treatment of wounds. In the early 160s, Galen moved to Rome and, with the exception of a brief return to Pergamum, spent the remainder of his life there. He became physician to the emperor Marcus Aurelius and several of his successors. Galen was the originator of the experimental method in medical investigation, dissecting animals in order to gain an understanding of the functions of the body. Some of his anatomical and physiological observations were accurate – for example, he proved that urine was formed in the kidney, as opposed to the bladder, which was the commonly held theory up to then. He discovered that arteries carry blood but he did not discover circulation.

Galen was a prolific writer and was responsible for compiling all significant Greek and Roman medical thought up to his own time, adding his own discoveries and theories. One of his central tenets was the concept of innate heat, which maintains life, aids nutriment and so on. This was based on the writings of Aristotle. According to Galen's theories, women were colder than men. His influence dominated medicine for fifteen centuries, until the Renaissance. He is cited on numerous occasions throughout the Irish medical texts.

Bernard of Gordon (*fl.* 1283–1308) was born *c.*1258 in France; he taught at the medical faculty of the University of Montpellier from 1283. He was a teacher, writer and practising physician. His most influential work was the *Practica, seu Lilium medicinae*, compiled in 1305 'to provide physicians with a comprehensive guide to the theory and practice of pathology and therapeutics'.[4] The *Lilium* features frequently in the Irish medical books.

Gerard de Solo (*fl.* fourteenth century) was connected with the school at Montpellier and wrote an influential *Practica*.

Lastly, **John of Gaddesden** (Johannes de Gaddesden), born *c.*1280, is thought to have hailed from Gaddesden on the borders of Buckinghamshire and Hertfordshire. He studied at Merton College, Oxford, obtaining degrees in arts, theology and later, medicine. According to Winifred Wulff's biography in the introduction to her edition of the Irish translation of his *Rosa Anglica*,[5] John was the first eminent English physician to have completed his medical training in England. He used Bernard of Gordon's *Lilium* in compiling the *Rosa*. Furthermore, the Irish translator of the *Rosa* borrowed extensively from the *Lilium* in compiling his translation of the *Rosa*. John refers constantly to his practice at Oxford. He eventually became court physician to Edward III, whom he is supposed to have cured of smallpox. Not known for his humility, John explained the title of his text – 'just as the rose excels among flowers, this book excels among textbooks on practical medicine'.[6] He died in 1361. Wulff points out that while there are many reasons to criticise Gaddesden's text (for example, his poor efforts at interpreting Greek terms – he was widely regarded later as a charlatan), the text has a lasting value for the insights it provides on daily life during the Middle Ages, the remedies used, the contents of the larders, charms etc. – 'a hot-potch of medical teaching, genuine or fabulous results of the application of remedies, oriental leechcraft and superstition, native English cures and charms, prayers and religious practices, interwoven with the native beliefs of the country'.[7] There were many others whose works were known and transmitted through the medical schools and were translated in whole or in part by the Irish medical confraternity.

Medicine in Ireland, like genealogy and law, was a hereditary affair – professional medical families worked on a hereditary basis. Prominent families in Leinster included the Ó Bolgaidhe and Ó Conchubhair; in Connaught, the Ó Fearghusa and Mac an Leagha, physicians to the O'Flahertys; in Ulster, the Ó Caiside, physicians to the Maguires of Fermanagh, Ó Duinshleibhe, who looked after the O'Donnells, Ó Siadhail, physicians to the MacMahons of Oriel; in Munster, the Ó hIceadha (note all the Hickey chemists still to be found in

4 Aoibheann Nic Dhonnchadha, 'The "Book of the O'Lees" and other medical manuscripts and astronomical tracts' in B. Cunningham and S. Fitzpatrick (eds), *Treasures of the Royal Irish Academy Library* (Dublin, 2009), pp 81–91 at p. 86. 5 Winifred Wulff (ed.), *Rosa Anglica; seu, Rosa medicinae Johannis Anglici: an early modern Irish translation of a section of the mediaeval medical text-book of John of Gaddesden* (London, [1923] 1929). The *Rosa* was written 1304–17 and first published in print at Pavia in 1492. 6 Citation from British Library website featuring a fourteenth-century copy of the *Rosa*, BL Sloane MS 280; *http://*www.bl.uk/onlinegallery/onlineex/illmanus/slomanucoll/r/011sl000000280u00009000.html. 7 Wulff, *Rosa*, p. XVII.

Ireland), physicians to the O'Briens of Thomond, Ó Leighin and Ó Nialláin. There were many others.[8] They were responsible for the organization and regulation of medical schools in Ireland, training students and compiling and translating texts. They were accorded prominence in Irish society and were provided with hereditary tenure of land in exchange for their services. One of the schools about which we have a certain amount of information was at Aghmacart in Co. Laois, run by the Ó Conchubhair in the Mac Giolla Phádraig lordship of Upper Ossory. Two of the main scribes at this school were Risteard Ó Conchubhair (1561–1625) and Donnchadh Albanach Ó Conchubhair (1571–1647). Another important member of the clan was Donnchadh Óg Ó Conchubhair (*fl.* 1581–1611) who was *ollamh leighis* to Finghin Mac Giolla Phádraig. This Donnchadh had studied medicine in Ireland exclusively. The evidence for this is a comment in RIA MS 3 C 19 – *Donnchadh Óg Ua Conchubhair .i. ullamh Osraighi (re leghes ⁊ rogha legh Erenn ina aimsir fen tuig riot gan dul a hÉirinn do dhenam foghluma)* ('the best of the doctors of Ireland in his own time – and that without leaving Ireland to study').[9] Nic Dhonnchadh points out that this shows that by this time the medical education being offered in Ireland was regarded as on a par with that on offer on the Continent, but she notes that comment might suggest that 'for many of Donnchadh Óg's contemporaries, and perhaps especially for those aspiring to the higher echelons of the profession, travelling abroad for study was a desirable, and feasible, option'.[10] Let us now turn to the manuscripts, in which can be found some of these medical men, their signatures and glosses.

HIPPOCRATES' *PROGNOSTICA*: AN EXAMPLE

RIA MS 24 P 22 is a seventeenth-century compilation, on paper, of medical treatises. Unusually, we do not know the name of the scribe of this particular book. It opens with Hippocrates' *Prognostica* with accompanying commentary by Gulielmus de Copo of Basel. In this manuscript each change in subject is noticed by a heading in capitals. Note the retention of the Latin text for the heading, common to practically all of the translated texts, on p. 1 (fig. 9.1).

The Hippocratic text breaks off at p. 20. The following tract is taken up with texts on the blood, humours, fevers etc. The text shown here lists the colours of the urine from a section dealing with stones, from Aegidius, *De urinis*, another much-cited work (fig. 9.2).

8 A very late manuscript RIA MS 24 B 2 (nineteenth century), compiled for the Cork antiquarian scholar and collector, John Windele, includes information on learned medical families on the flyleaves. 9 Aoibheann Nic Dhonnchadha, 'The medical school of Aghmacart, Queen's County', *Ossory, Laois and Leinster*, 2 (2006), 11–43 at 16 and n. 17. 10 Ibid., 17.

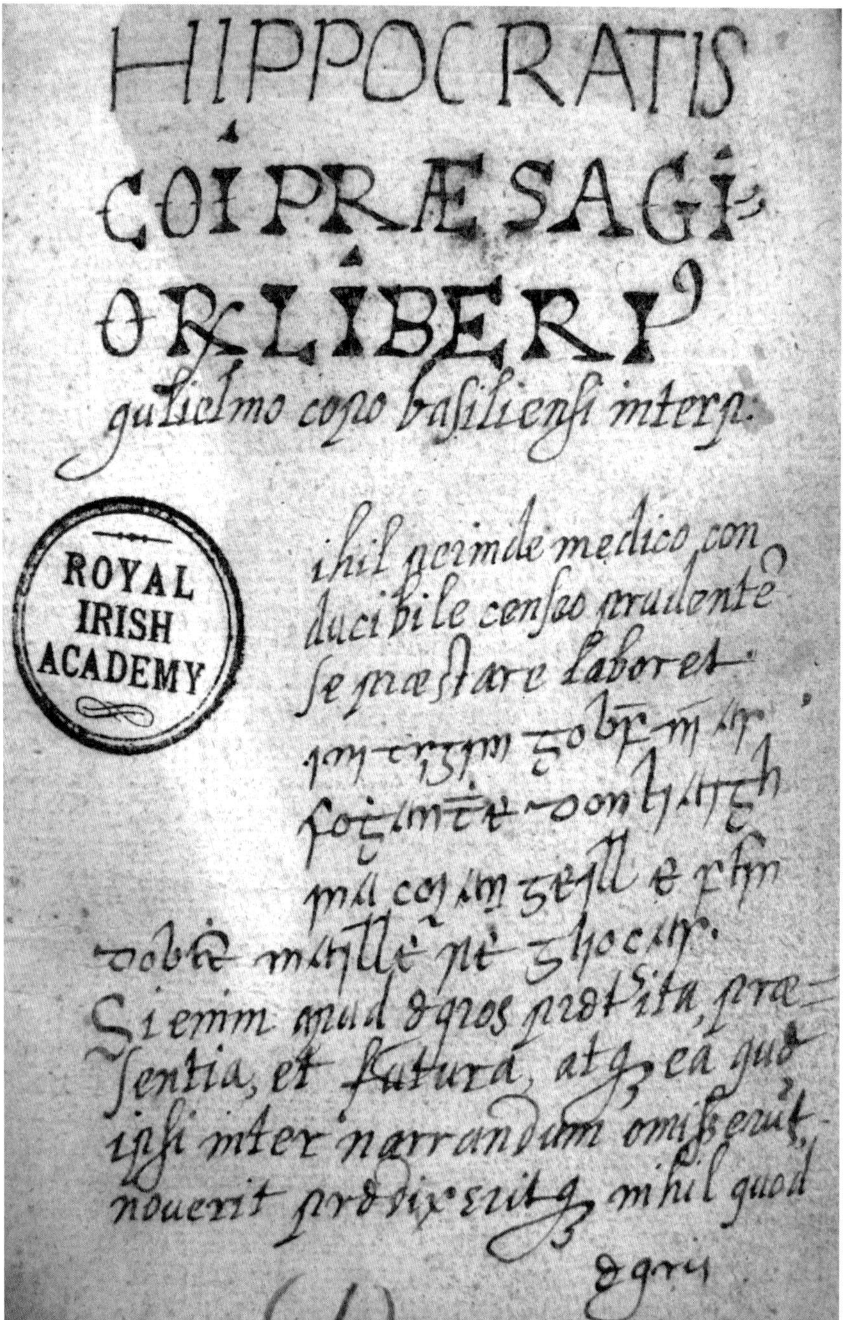
HIPPOCRATIS
COI PRÆSAGI-
ORUM LIBER I
guilielmo copo basiliensi interp.

[N]ihil perinde medico con-
ducibile censeo prudentem
se præstare laboret

Si enim apud ægros præsentia, et futura, atque ea quæ
ipsi inter narrandum omiserint,
nouerit prædixeritque, nihil quod
ægri

(1)

9.1 RIA MS 24 P 22, p. 1 (image courtesy of the Royal Irish Academy © RIA).

Finally, on p. 196 (fig. 9.3) we find a note 'This book is the property of John McDonald of the parish of Lead and County Antrim. February 12 1782'. Lead (Layde) is in the barony of Lower Glenarm. Such notes are a common occurrence in these manuscripts and they provide a context for their use. A later owner of the book was Revd William Reeves (1815–92), bishop of Down, Connor and Dromore, librarian at Armagh and president of the Royal Irish Academy at the time of his death. Reeves was a wonderful Irish scholar who translated and edited many important Irish texts. He owned several medical manuscripts, a number of which he bought from the collector, Robert MacAdam, in November 1889; MacAdam was a Belfast engineer and scholar about whom Annraoi de Paor has written in the Royal Society of Antiquaries of Ireland's publication, *Pathfinders to the past.*[11] On Reeves' death, the Academy raised the money to purchase his Irish manuscript collection, thus ensuring that a major collection was secured for continued scholarly use.

BERNARD OF GORDON'S *LILIUM MEDICINAE*

The *Lilium medicinae* occurs in several Academy manuscripts. The aforementioned MS 3 C 19, sixteenth and eighteenth centuries, was written by Risdeard mac Muircertaigh Ó Conchubhair (1561–1625) with some replacement leaves written by Tadhg Ó Neachtain and another scribe a couple of centuries later. Ó Conchubhair was a member of the medical school at Aghmacart referred to above. Risdeard made this copy of the *Lilium* from an exemplar written by Donnchadh Óg.[12] The first page shown from MS 3 C 19 is folio 2r – *Febris est calor innaturalis* – a tract on fevers (fig. 9.4).

On folio 116r we find personal scribal information: *Tairrnic ann sin in treas partical don Lili maille cabhair an tslanuightheora. An 18 la do November [...]* ('There finished, with the help of the Saviour, the third part of the Lily. The 18th of November [...]'). Ó Conchubhair was at Courtstown, Co. Kilkenny, at this time, having begun the *Lilium* at Cullahill Castle, Co. Laois, on 7 May that year. There follows a long note recording the hospitality of the couples, in five counties, with whom he had stayed during this time. He particularly mentions the role of Gráinne, daughter of Brian Mac Giolla Phádraig and wife of Edmund Butler, second Viscount Mountgarret, in promoting his education from the age of twelve onwards, that is after the death of his father (fig. 9.5, see col. 2, lines 7ff). A further note occurs on folio 145v: *Misi Risderd mac*

11 Annraoi de Paor, 'The life and work of Robert MacAdam of Belfast, 1808–95' in Próinséas Ní Chatháin et al., *Pathfinders to the past: the antiquarian road to Irish historical writing, 1640–1960* (Dublin, 2012), pp 72–85. See also A.J. Hughes, *Robert Shipboy MacAdam: his life and Gaelic proverb collection* (Belfast, 1998). 12 Based on Nic Dhonnchadha, 'The "Book of the O'Lees"', p. 87.

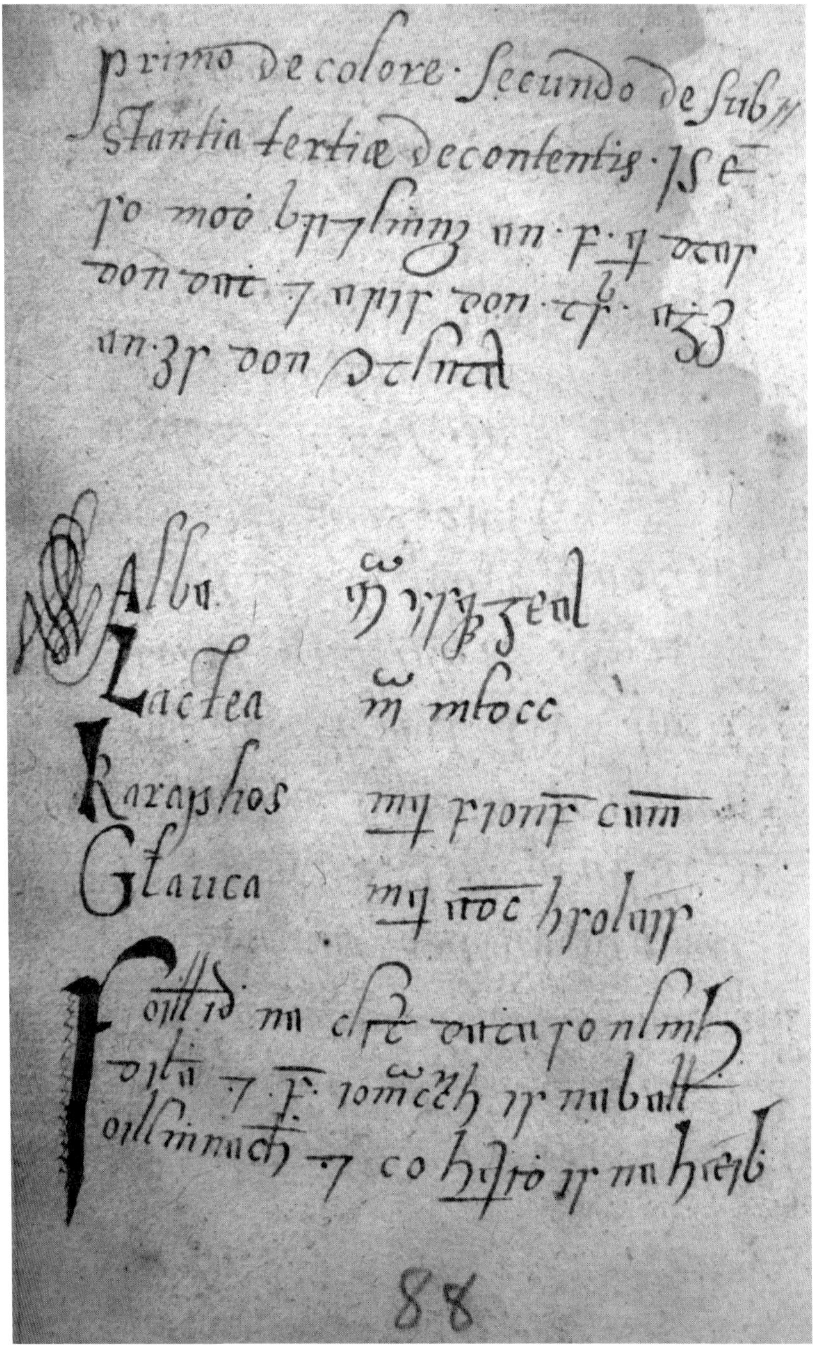

9.2 RIA MS 24 P 22, p. 88 (image courtesy of the Royal Irish Academy © RIA).

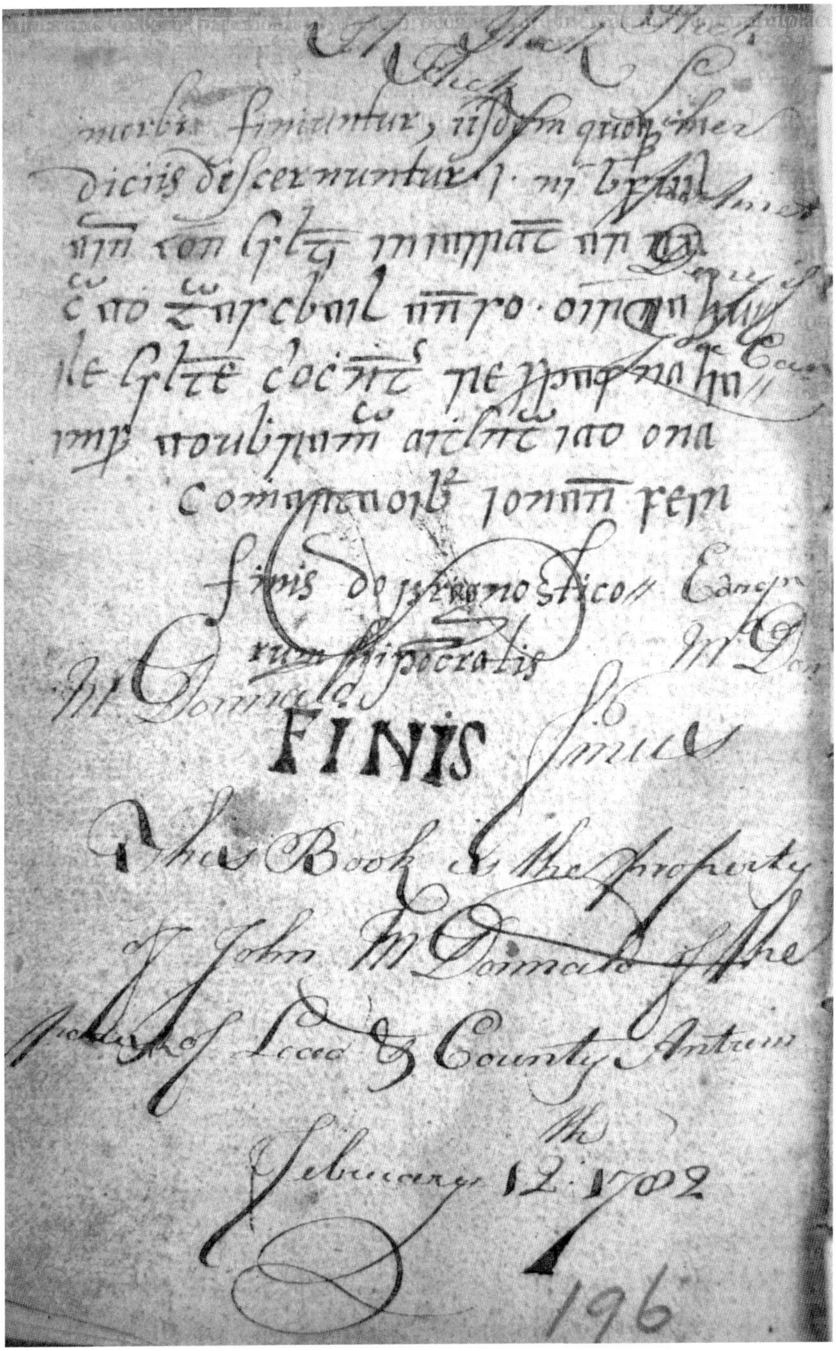

9.3 RIA MS 24 P 22, p. 196 (image courtesy of the Royal Irish Academy © RIA).

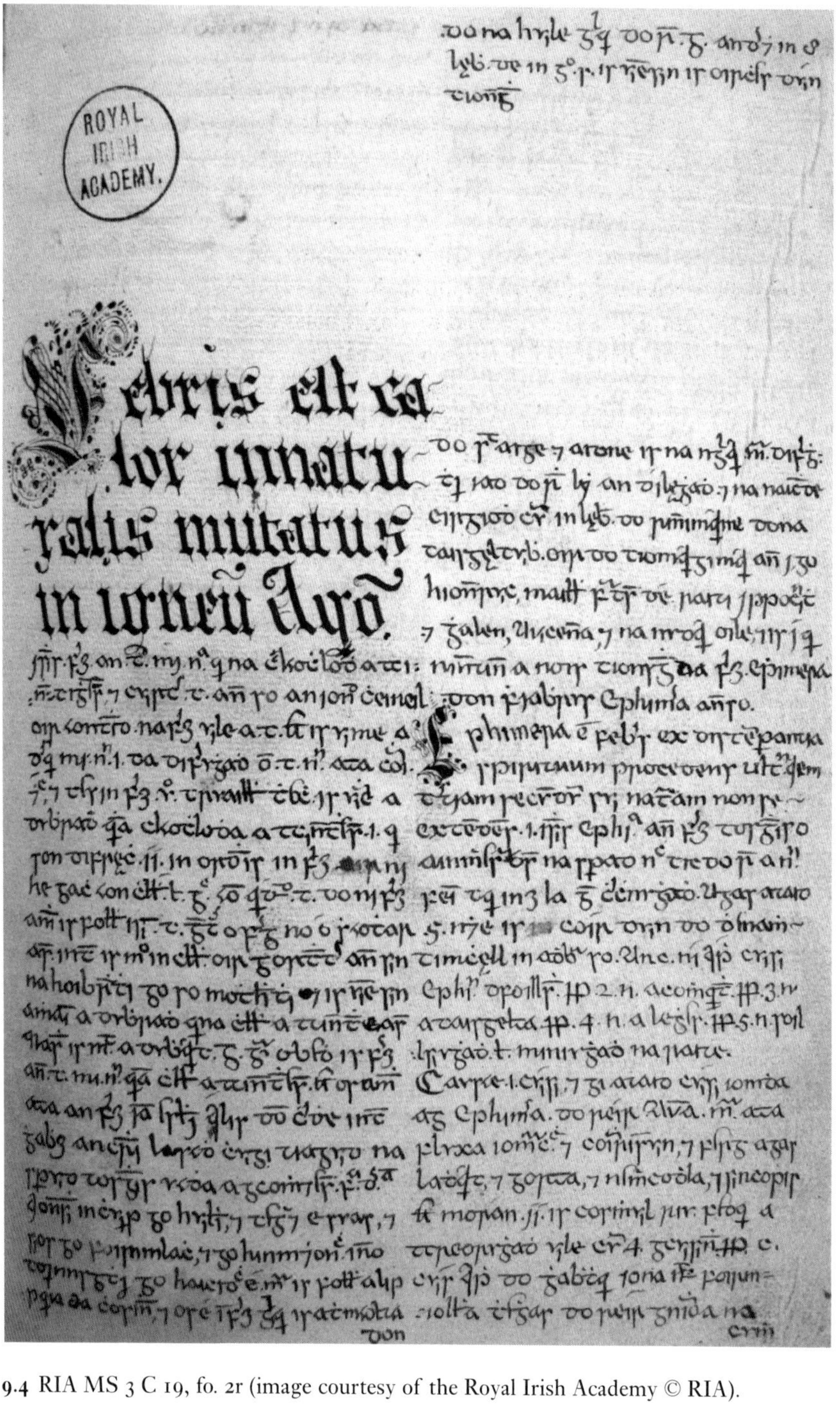

9.4 RIA MS 3 C 19, fo. 2r (image courtesy of the Royal Irish Academy © RIA).

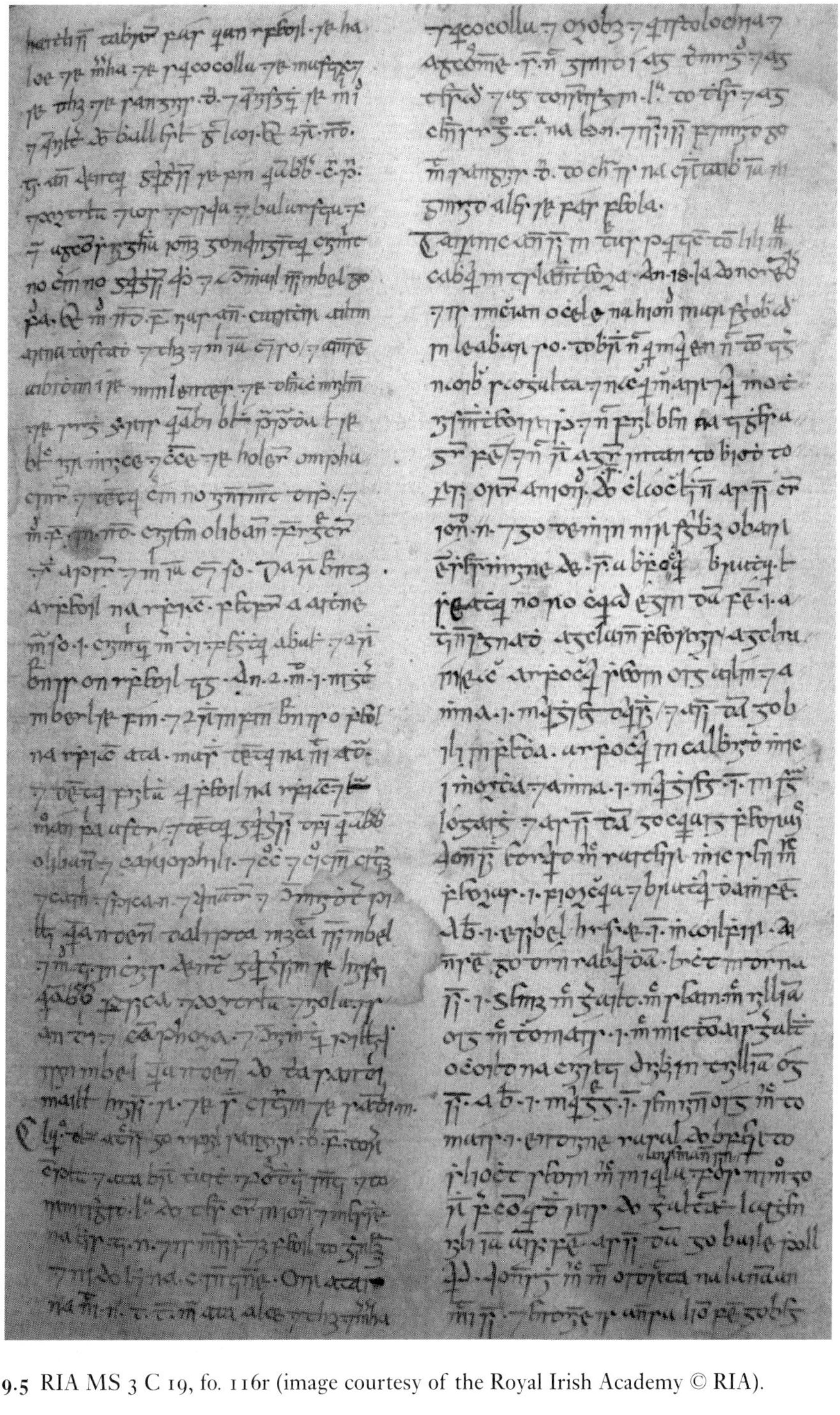

9.5 RIA MS 3 C 19, fo. 116r (image courtesy of the Royal Irish Academy © RIA).

Muircertaigh rosgriobh agus Baile in Gronta mo log a nOsraighi a bfochuir mic Gilla Patruic i Finghin mac Briain mic Briain. 1590 ('It is I, Risderd son of Muircheartach, who wrote [this] and Grantstown [bar. Clarmallagh, Co. Laois] is my place [of writing] in Ossory, in the company of Mac Giolla Phádraig, i.e. Finghin son of Brian, son of Brian. 1590') (fig. 9.6, see lines 9ff).

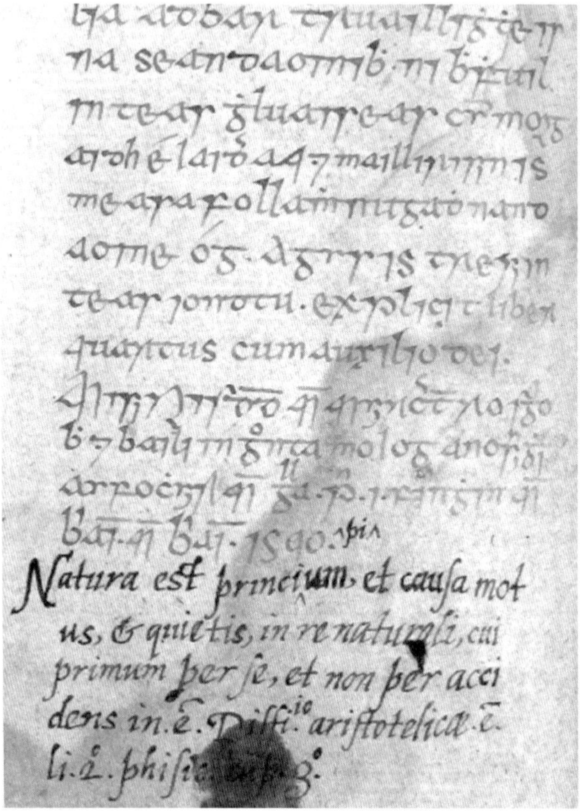

9.6 RIA MS 3 C 19, fo. 145v (image courtesy of the Royal Irish Academy © RIA).

RIA MS 23 E 24 is another fifteenth-century copy of the *Lilium*, scribe unknown. It is part of the Betham Collection purchased by the Academy in 1850 from Sir William Betham (1779–1853) who was Ulster King of Arms and a member of the Royal Irish Academy. The text on p. 102 is about food and it deals with weak appetite and indigestion (p. 104). Plate 10 shows an illustration on p. 100.

RIA MS 24 P 14 is a very important copy of the *Lilium*, whose scribe Corc Óg Ó Cadhla (*fl.* 1577–84), was a Leinster-based physician. It was written between 1577 and 1578, at Teagh na Ronna (Cournellan, Co. Carlow) and Graiguenamanagh, Co. Kilkenny. According to Nic Dhonnchadha, it is the only complete copy of the Irish *Lilium* extant.[13] This manuscript was owned by Edward O'Reilly, then by Robert MacAdam, who purchased it for 11*s*. at the O'Reilly library sale in 1830, and later by Bishop Reeves. The O'Reilly sale catalogue refers to it as follows: 'the subject medicine, written in the School of Montpellier in the year 1305; the book is perfect [...]'.

On p. 76a, we can see that Corc is starting to copy the *Lilium* at Tinnaranny (fig. 9.7) and on p. 170a (fig. 9.8), he is completing part 3: *22 la do mhi mharta agus a nGrainsigh na Manach a bfochuir Briain mic Cathaoir mic Airt Caomhanaigh do crichniges e [...]* ('In Graiguenamanagh, in the company of Brian, son of Cahir, son of Art Kavanagh, I completed it on 22 March [...]'). In his reference to Graiguenamanagh, Corc mentions that he treated the two daughters of Brian Caomhánach, chief of his sept, for menstrual irregularity. This is one of the few allusions made to the active treatment of female patients in these manuscripts – *⁊ me ag leiges a deise ingen o hsechran fola mista* ('while I was treating his two daughters for a menstrual irregularity').

Another manuscript written by Corc, and by Cairbre Ó Cennamhain, is RIA MS 24 P 15. Corc was writing at Cell Clogain, Templetown parish, Co. Wexford, in 1583–4 and Cairbre, at Ros Broin, parish of Schull, Co. Cork, in 1478. The contents include a tract on medicine, fragments of a translation of *Decem ingenia curandorum morborum* and Bernard of Gordon's *Prognostica*, as well as *De chirurgia* by Petrus de Argelata.

The manuscript was owned by a Paul Peter Layburn and a Louras Cuminn (pp 150 and 160, figs 9.9–9.10) and includes a note on the murder of the earl of Desmond on p. 193 (fig. 9.11) at the end of the *Prognostica*:

> Finis. Anno salutis 1583. Tarnuig annsin Prognostica Bernardus Gordonius, an 3 la do mi Februarius maille re cabuir an tslanaigtheora Isa. ⁊ a gCill Clogain do crichuighedh gin gurob ann do tinnscnadh é. Ecda aibhle ar ndenamh an bliagain so .i. Gearoid mac Semuis mic Seaain iarla Deasmuman do dicennadh lesna sighdiuireadhaibh ⁊ Seaan [...]
>
> Finis. The year of our Lord 1583. There ended, on 3 February, with the help of the Saviour Jesus, the Prognostica of Bernard of Gordon. And at Kilcloggan [Co. Wexford] it was finished though it is not there it was begun. Great slaughters have been made this year, i.e. Gearóid son of Séamus, son of Seán, the Earl of Desmond, was beheaded by the soldiers ⁊ Seán [...].

A very important medical book is the Book of the O'Shiels (RIA MS 23 K 42), a compendium of medical treatises, written by Pádruic gruamdha Ó

13 Nic Dhonnchadha, 'The "Book of the O'Lees"', p. 86.

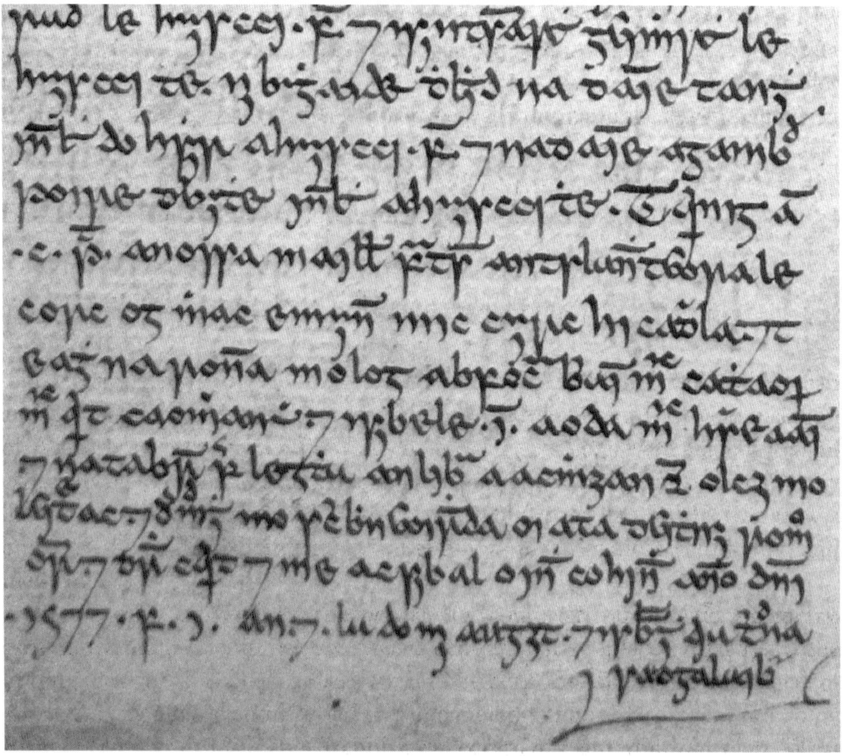

9.7 RIA MS 24 P 14, p. 76a (image courtesy of the Royal Irish Academy © RIA).

Siadhail, at Coill Lagach near Cruachan Gaileang and several other locations, 1657–8. It contains the Aphorisms and *Prognostica* of Hippocrates, Aegidius' *De urinis*, Bernard of Gordon's *Lilium*, *particula prima*, on fevers, with extracts from Valescus de Taranta, interspersed by extracts from Galen and Avicenna.

As in many of the other medical manuscripts, we see the writer close up, as on p. 211 (fig. 9.12) *ataim tuirseach gan chodladh – I nDruim Scuabach dham anocht – Novembr. 26, 1657. Patr: Shyell* ('I am tired for lack of sleep – I am in Drumscoba tonight – 26 November 1657') – Drumscoba is in the barony of Attymas, Co. Mayo – and on p. 396 (fig. 9.13): *I gCúil Iorra dhamh anocht Oidhche Fheile Martain, 1658. Misi Padruic gruamdha Ó Siaghail: agus tabhradh gach áon leighfeas so a bheannacht ar mh'anmuin* ('I am in Killaspugbrone tonight, St Martin's Eve, 1658. I am grim Pádraig O'Sheil: and let everyone who will read this bestow a blessing on my soul'). Killaspugbrone is in the barony of Carbury, Co. Sligo.

The *Lilium* also takes up RIA MS 23 G 14, a seventeenth-century compilation of which the scribe is not known to us. Former owner Terence O'Cassidy's signature occurs throughout (fig. 9.14). This manuscript is imperfect, having lost the opening section as well as text at the end. Yet another copy forms RIA MS 3

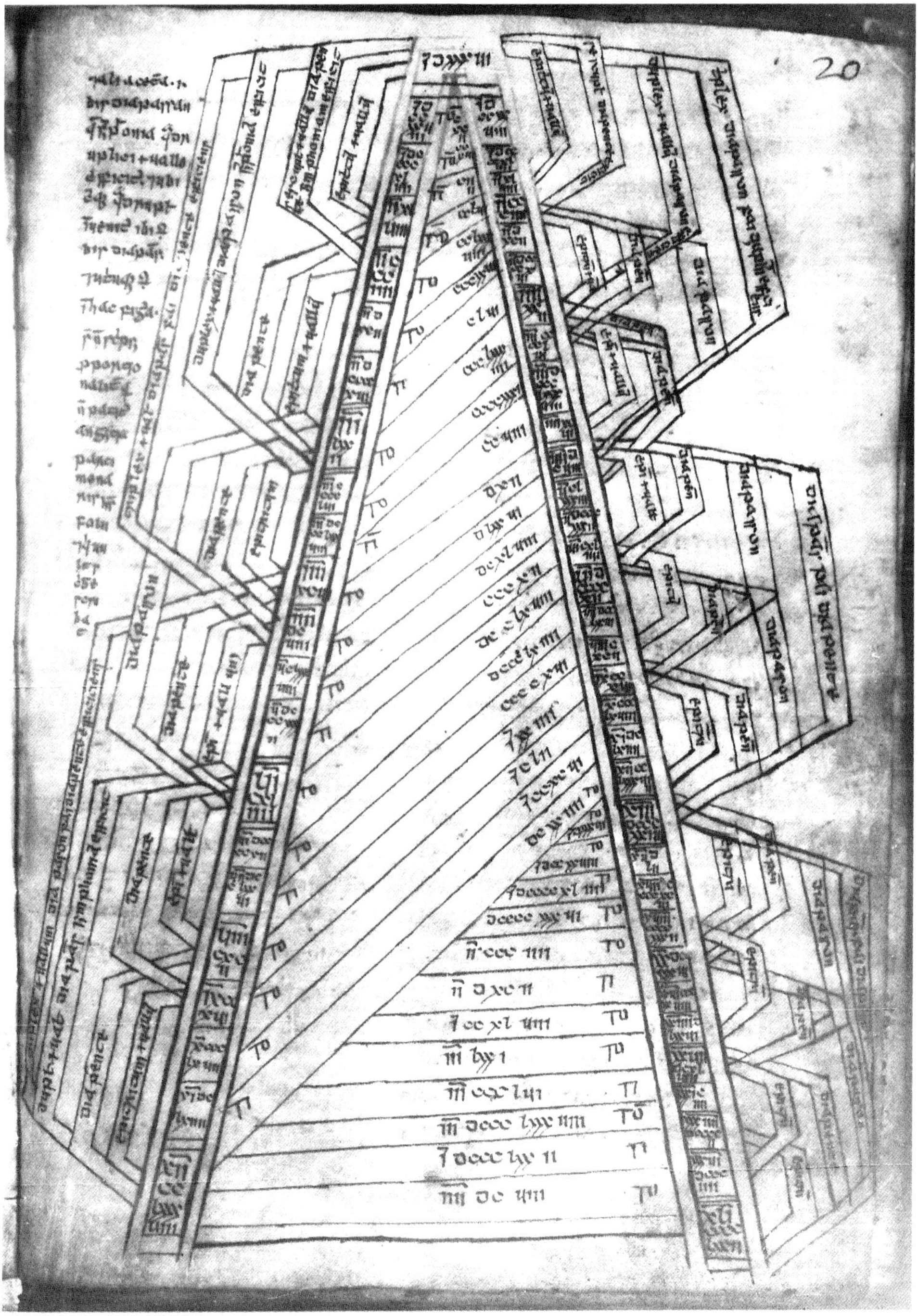

1 Lambda diagram of musical tones and intervals expressed in Boethian/Chalcidian mathematical terms: Oxford, Bodleian Library, MS Auct. F.3.15, fo. 20r (reproduced by kind permission of the Bodleian Libraries, University of Oxford).

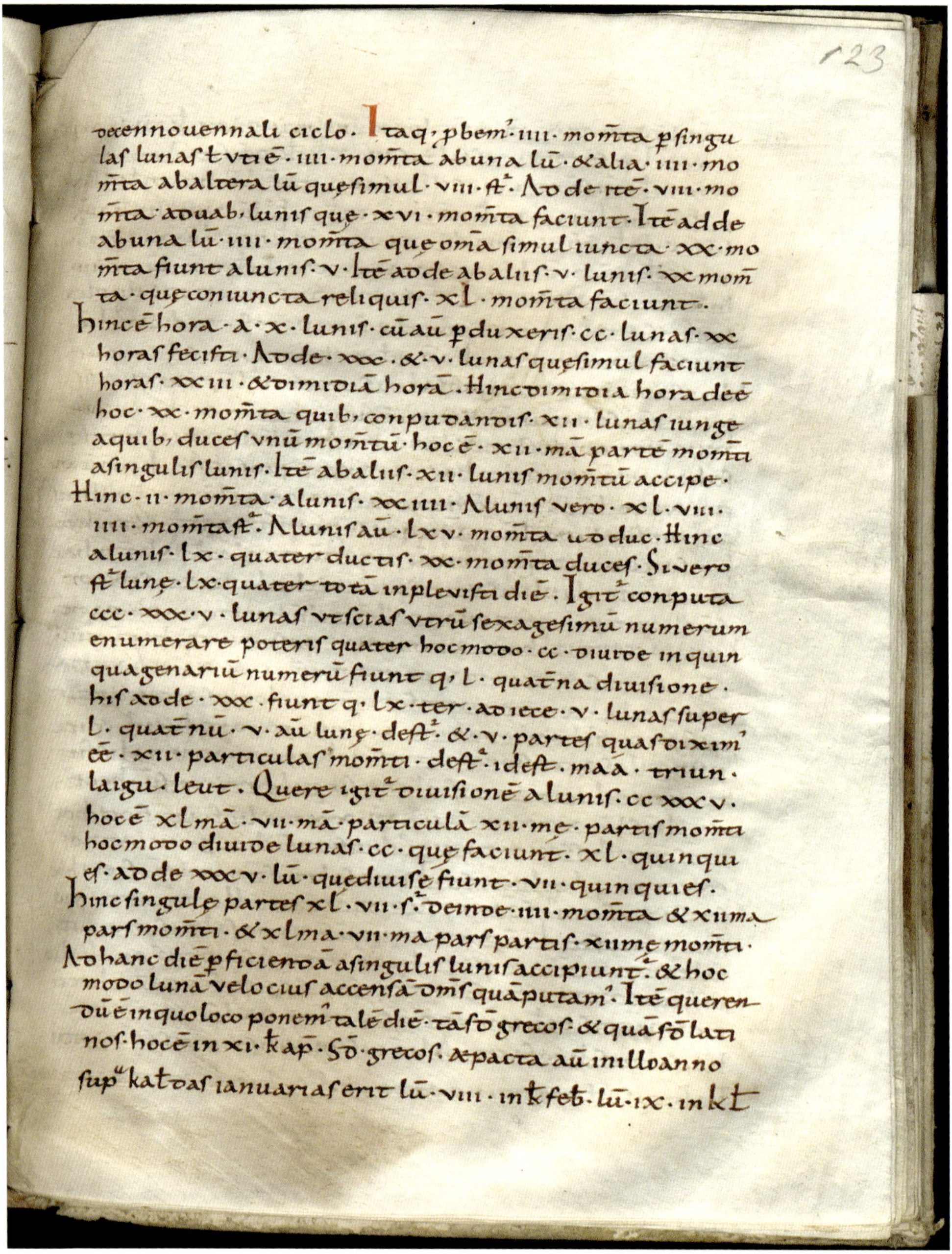

2 Einsiedeln, Stiftsbibliothek, 321 (647), p. 123 (© Einsiedeln Stiftsbibliothek).

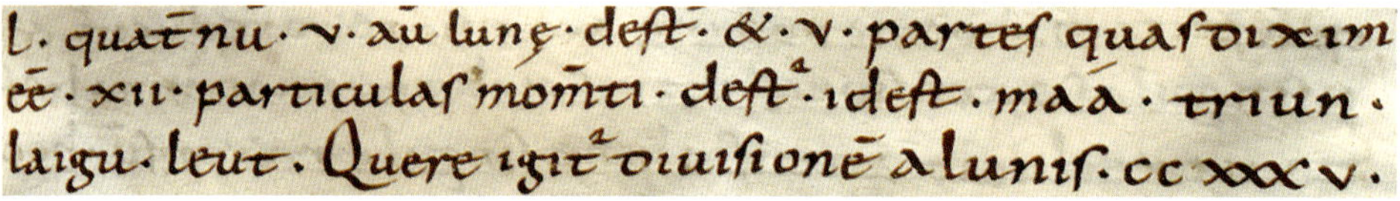

2.1 Detail from pl. 2: the Old Irish explanation of the fraction 5/12 in the *Computus Einsidlensis, maá triun laigu leut* ('more than a third, less than a half')

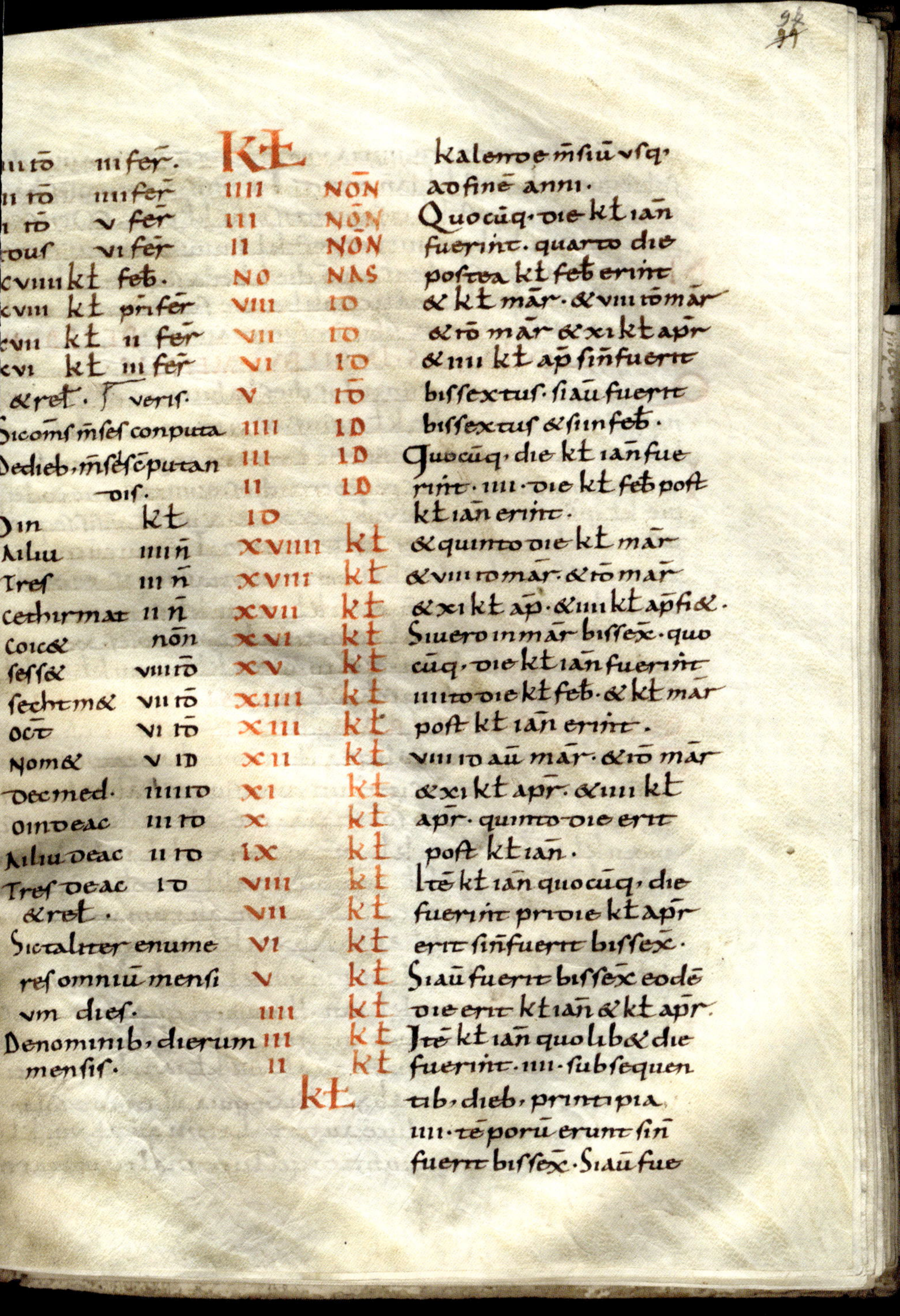

iii id	iii fer.	KL		kalendae mensium usq
ii id	iiii fer	IIII	NON	ad fine anni.
i id	v fer	III	NON	Quocumq die kl ian
idus	vi fer	II	NON	fuerint. quarto die
xviiii kl	feb.	NO	NAS	postea kl feb erint
xviii kl	pri fer	VIII	ID	& kl mar. & viii id mar
xvii kl	ii fer	VII	ID	& id mar & xi kl apr
xvi kl	iii fer	VI	ID	& iiii kl ap sin fuerit
& ret.	veris.	V	ID	bissextus. si autem fuerit
Si omnes menses computa		IIII	ID	bissextus & sin feb.
De dieb; menses computan		III	ID	Quocumq die kl ian fue
dis.		II	ID	rint. iii. die kl feb post
Oin	kl	ID		kl ian erint.
Ailiu	iiii n	XVIIII	KL	& quinto die kl mar
Tres	iii n	XVIII	KL	& viii id mar. & id mar
cethirmat	ii n	XVII	KL	& xi kl ap. & iiii kl ap fi&.
Coic&	non	XVI	KL	Si vero in mar bissex. quo
sess&	viii id	XV	KL	cumq die kl ian fuerint
sechtm&	vii id	XIIII	KL	iiii to die kl feb. & kl mar
oct	vi id	XIII	KL	post kl ian erint.
Nom&	v id	XII	KL	viii id aut mar. & id mar
Decmed.	iiii id	XI	KL	& xi kl apr. & iiii kl
Oindeac	iii id	X	KL	apr. quinto die erit
Ailiudeac	ii id	IX	KL	post kl ian.
Tres deac	id	VIII	KL	Item kl ian quocumq die
& ret.		VII	KL	fuerint pridie kl apr
Sic taliter enume		VI	KL	erit sin fuerit bissex.
res omnium mensi		V	KL	Si autem fuerit bissex eodem
um dies.		IIII	KL	die erit kl ian & kl apr.
De nominib; dierum		III	KL	Item kl ian quolib& die
mensis.		II	KL	fuerint. iii. subsequen
		KL		tib; dieb; principia
				iiii. temporum erunt sin
				fuerit bissex. Si autem fue

3 Einsiedeln, Stiftsbibliothek, 321 (647), p. 97 (© Einsiedeln Stiftsbibliothek).

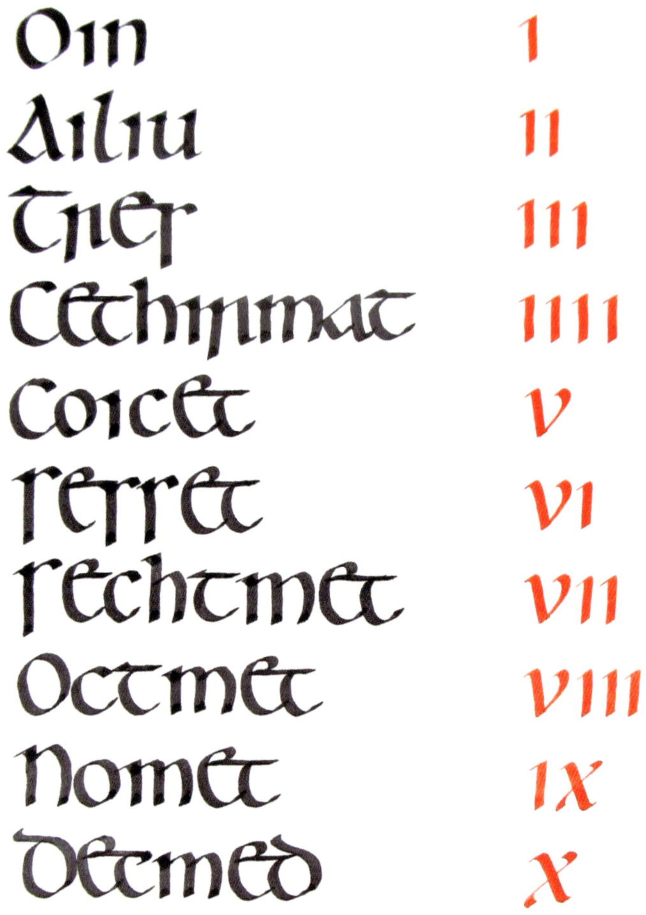

4 Conjectural representation of the earliest Irish numerals (Tim O'Neill).

5 (*opposite*) Portrait of St John, *Book of Kells*, TCD MS 58, fo. 291v (reproduced by kind permission of the Board of Trinity College Dublin).

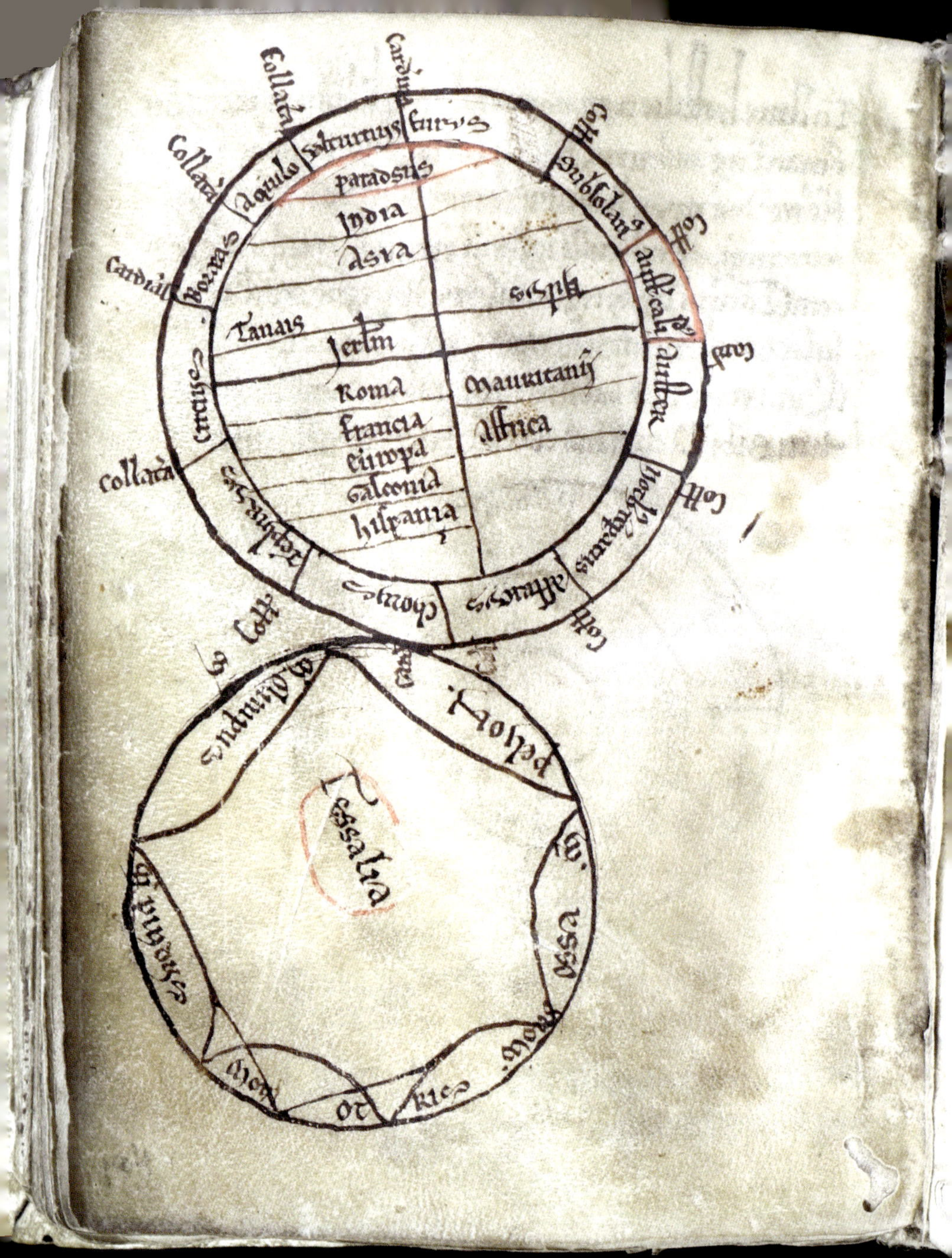

6 Map of the world and the mountains of Thessaly: map of Cashel, Bolton Library, MS 1, p. 184 (courtesy of the Very Revd Dr Philip Knowles, dean of Cashel)

7 Cashel, Bolton Library, MS 1, p. 185: the beginning of Petrus Alfonsi's *Disciplina clericalis* ('The education of the clerk'), moralizing tales translated from Arabic (courtesy of the Very Revd Dr Philip Knowles, dean of Cashel).

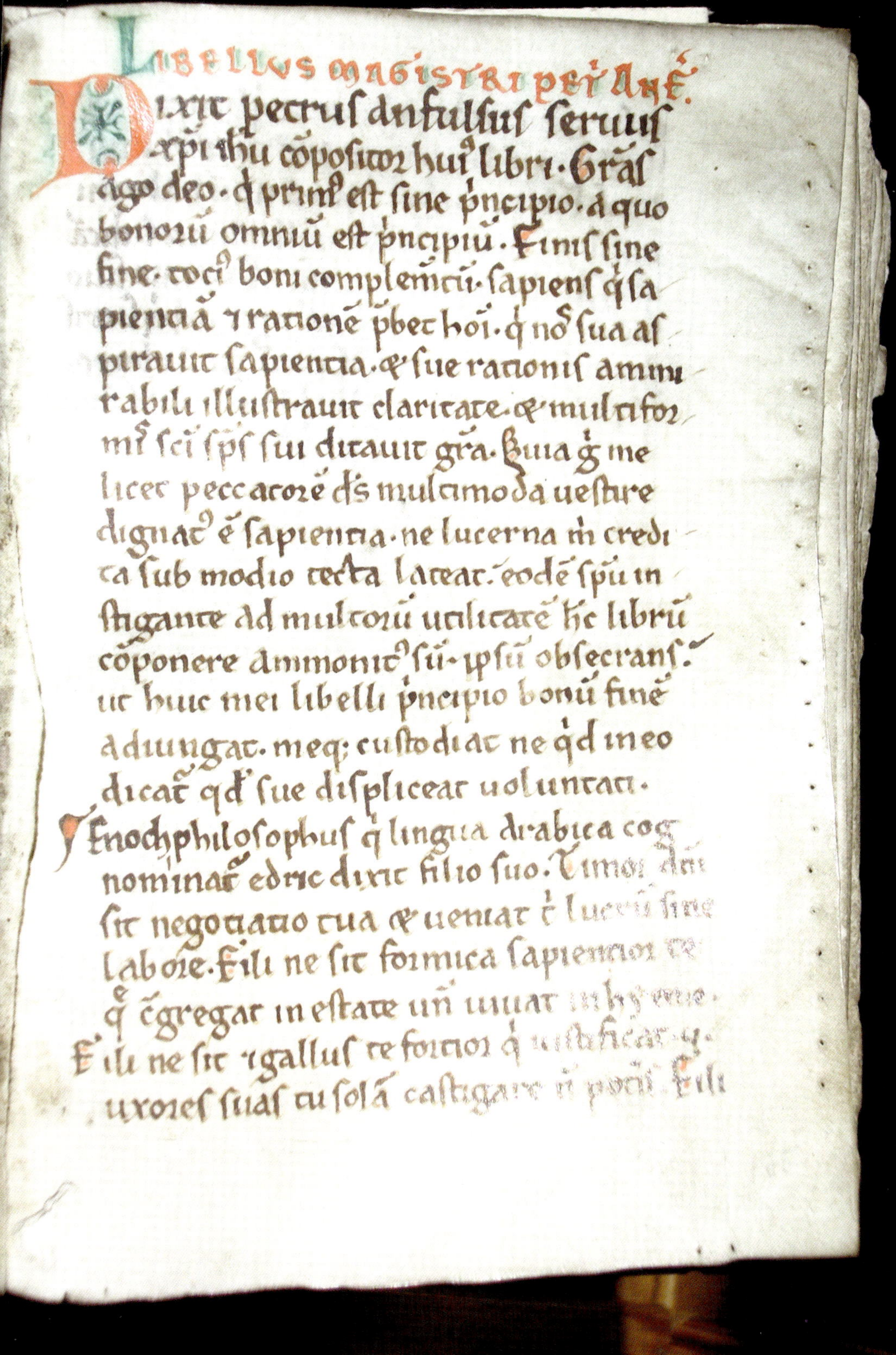

8 RIA MS 23 A 4, p. 52 (detail) (image courtesy of the Royal Irish Academy © RIA).

9 RIA MS 23 H 19, fo. 7r (detail) (image courtesy of the Royal Irish Academy © RIA).

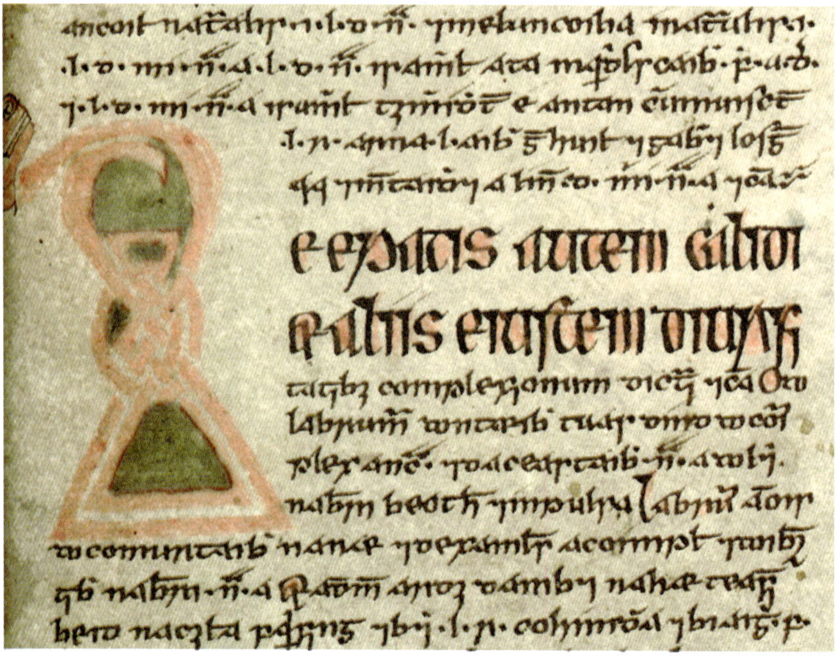

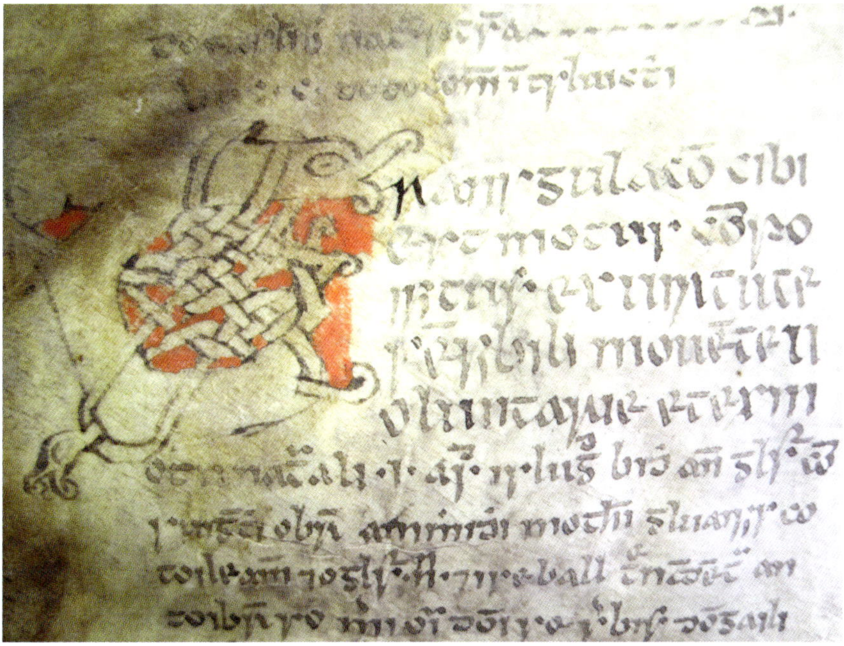

10 RIA MS 23 E 24, p. 100 (detail) (image courtesy of the Royal Irish Academy © RIA).

11 RIA MS 23 P 20, p. 1 (detail) (image courtesy of the Royal Irish Academy © RIA).

12 RIA MS 23 P 10 (i), p. 1 (detail) (image courtesy of the Royal Irish Academy © RIA).

13 RIA MS 23 P 10 (ii), p. 10 (image courtesy of the Royal Irish Academy © RIA).

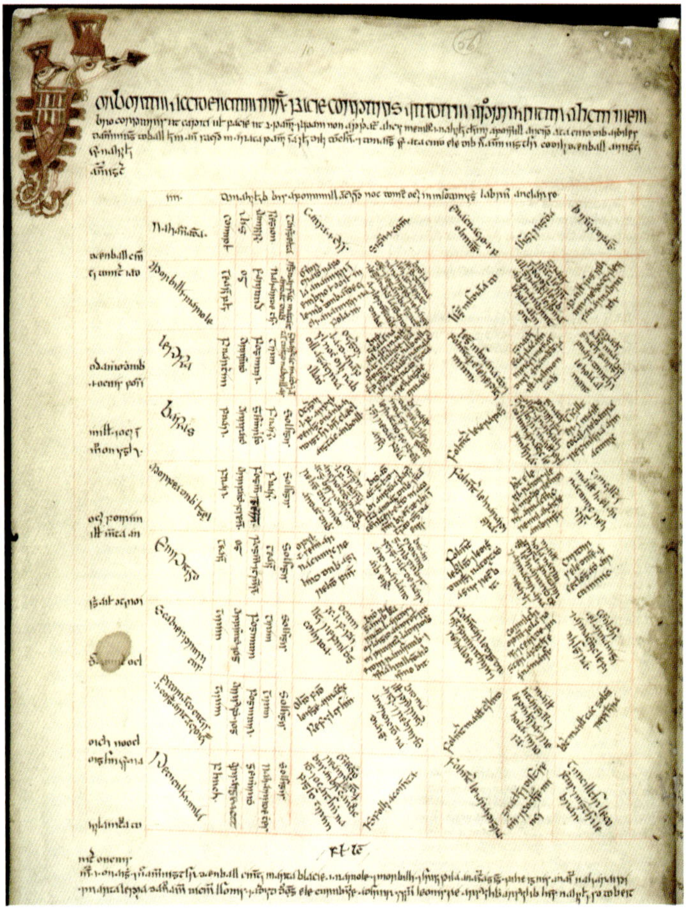

14 RIA MS 23 P 10 (ii), p. 11 (image courtesy of the Royal Irish Academy © RIA).

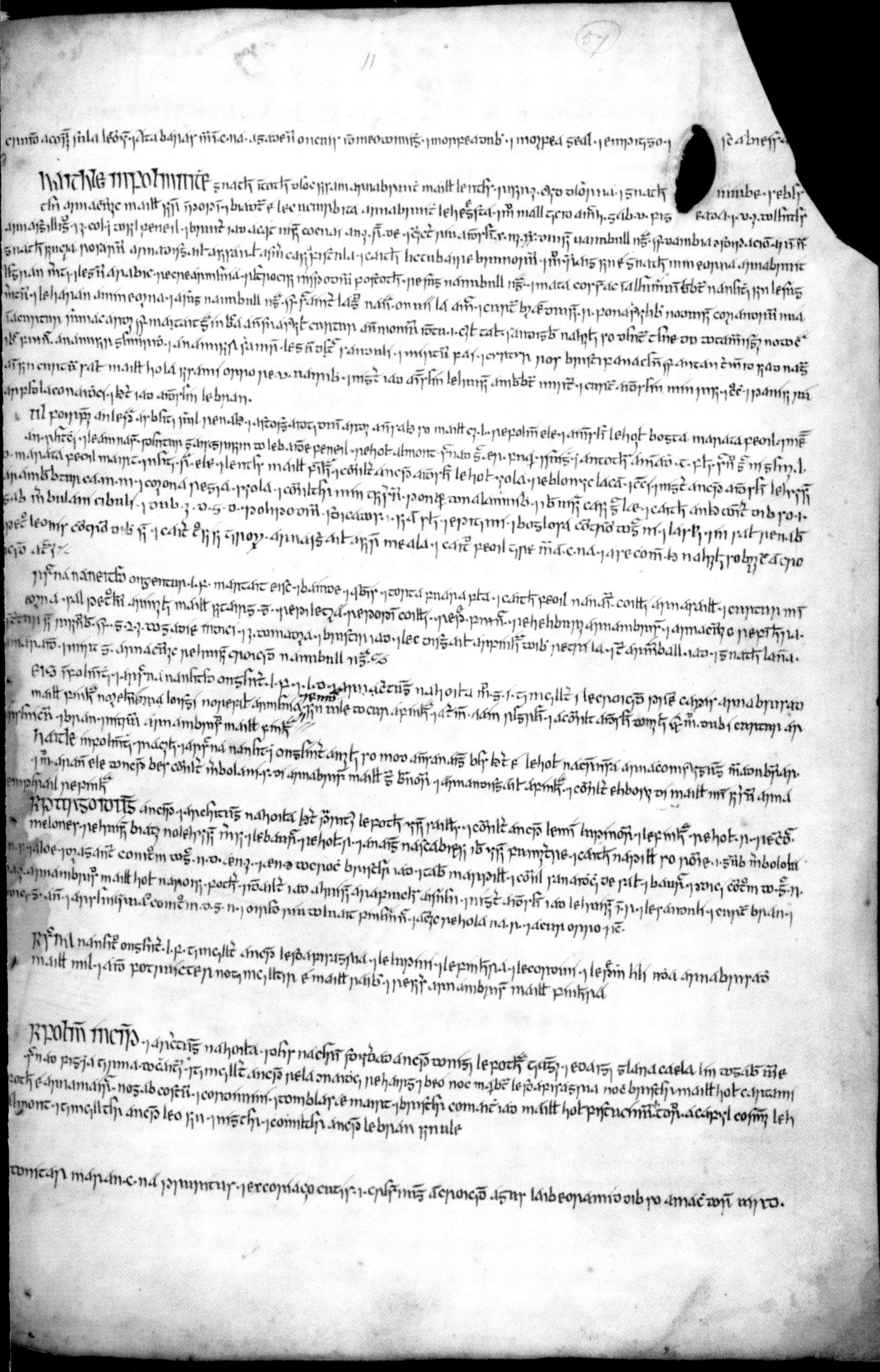

15 RIA MS 23 P 10 (ii), p. 48 (detail) (image courtesy of the Royal Irish Academy © RIA).

16 RIA MS 23 N 8, fo. 18v (detail) (image courtesy of the Royal Irish Academy © RIA).

17 RIA MS 23 P 10 (iii), p. 1 (image courtesy of the Royal Irish Academy © RIA).

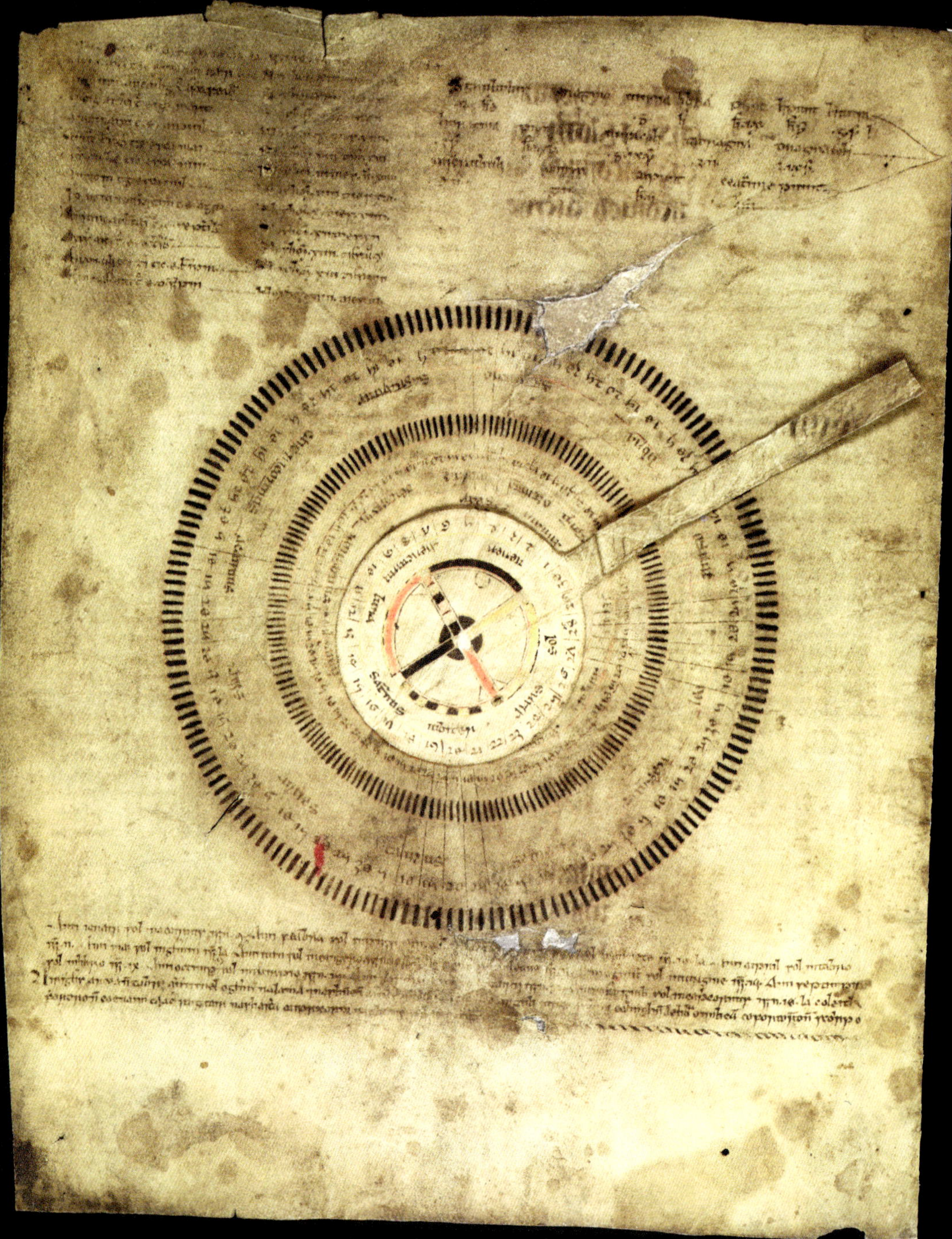

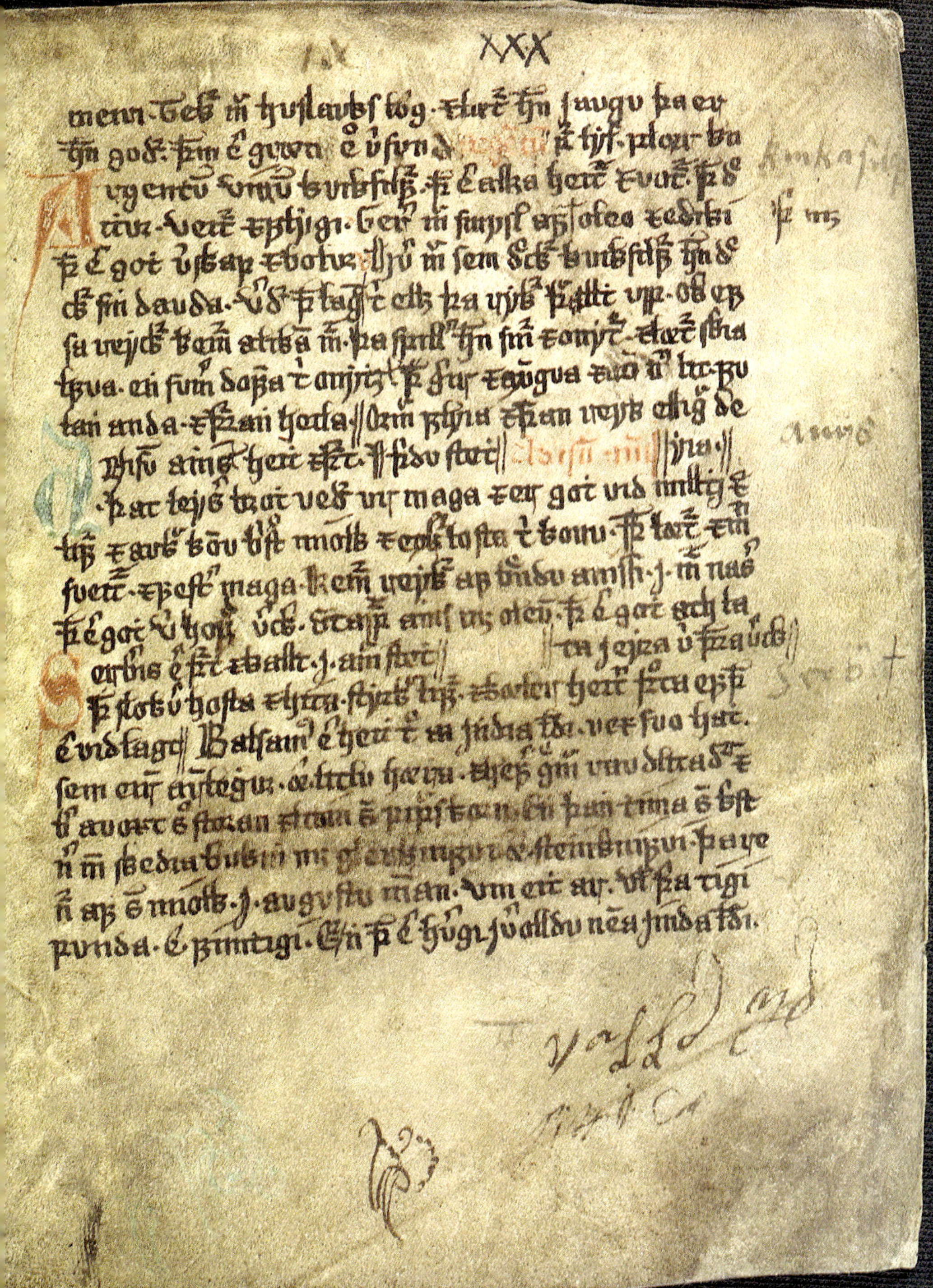

19 RIA MS 23 D 43, fo. ixr (image courtesy of the Royal Irish Academy © RIA).

20 Modern reconstruction of Giovanni di Dondi's astrarium completed in 1364 (image courtesy of the Science Museum, Smithsonian Institute).

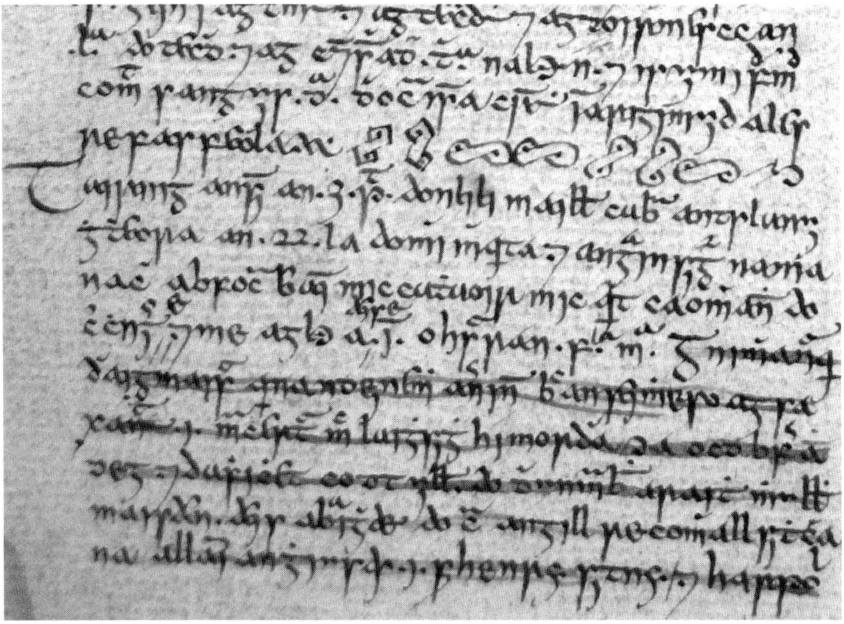

9.8 RIA MS 24 P 14, p. 170a (image courtesy of the Royal Irish Academy © RIA).

9.9 RIA MS 24 P 15, p. 150 (image courtesy of the Royal Irish Academy © RIA).

C 22, the translation of the *Lilium* being the same as that in MS 3 C 19, above. This copy, dating from the eighteenth century, probably belonged to Bishop Reeves. It contains an enormous number of marginal data, often IOUs and other notes. See, for example, folio 218v, 'Balimena, 15 Dec. 1759' (fig. 9.15); folio 63r: a section on Alopecia (fig. 9.16) and folio 107v, referencing Galen: 'Sicut scribit Galenus' (fig. 9.17).

RIA MS 24 B 3 is a fifteenth- to sixteenth-century compilation of medical

9.10 RIA MS 24 P 15, p. 160 (image courtesy of the Royal Irish Academy © RIA).

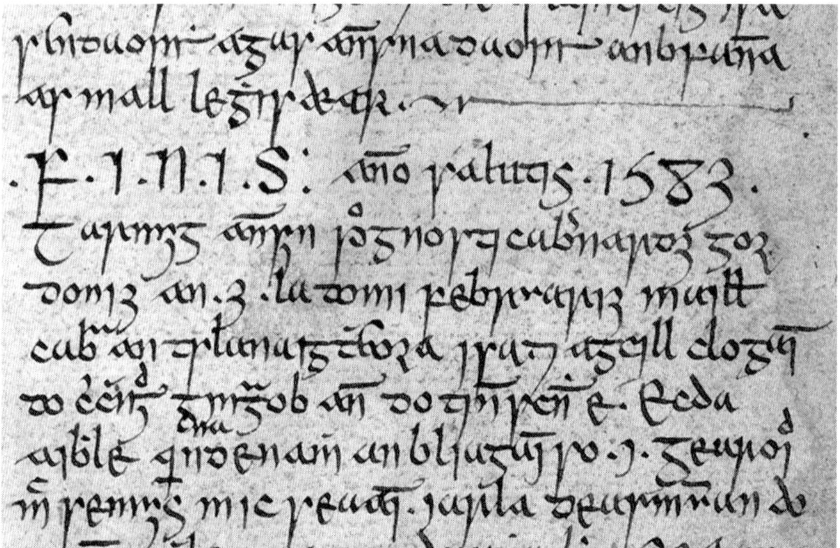

9.11 RIA MS 24 P 15, p. 193 (image courtesy of the Royal Irish Academy © RIA).

treatises, written by Conla mac an Lega at Granseach Muintere Fitcheallaigh in 1496; other scribes added to it later. It includes part of the *Lilium*, some of Gualterus (Walter de Agelon), *De dosibus*, extracts from Johannes Damascenes and others. The virtues of *aqua vitae* are elaborated on p. 94 (fig. 9.18)!

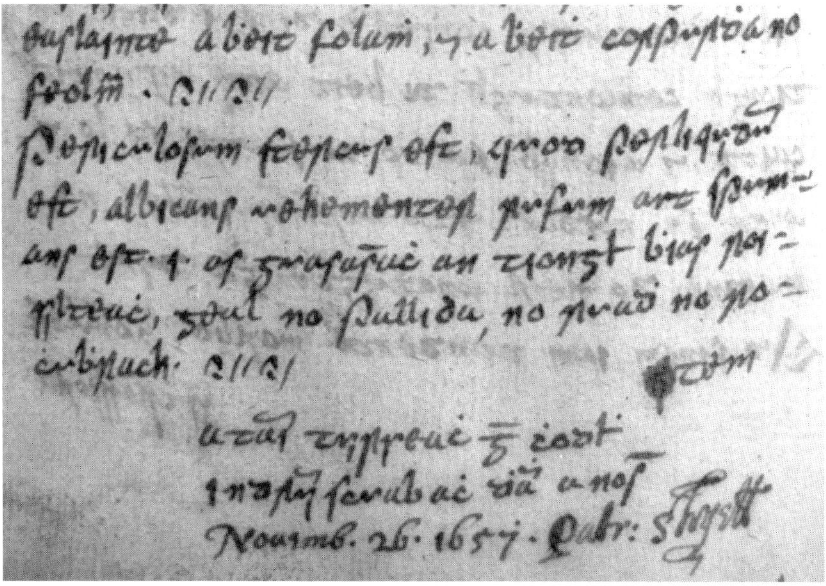

9.12 RIA MS 23 K 42, p. 211 (image courtesy of the Royal Irish Academy © RIA).

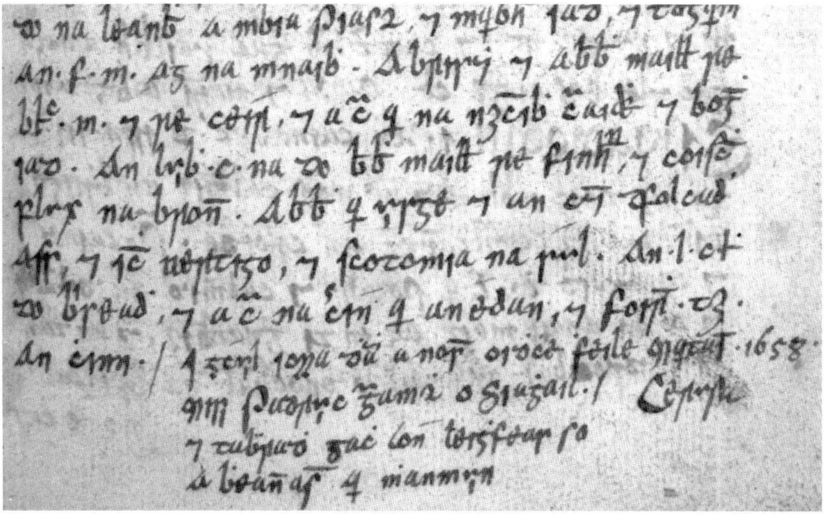

9.13 RIA MS 23 K 42, p. 396 (image courtesy of the Royal Irish Academy © RIA).

Manuscript 23 N 16, another of Betham provenance, contains part of the *Collectorium* of Niccolò Bertruccio (d. 1347), physician, anatomist and professor of medicine at Bologna. The *Collectorium* was first published at Lyon in 1509. This is the only extant Irish version of the section of Bertruccio's treatise on

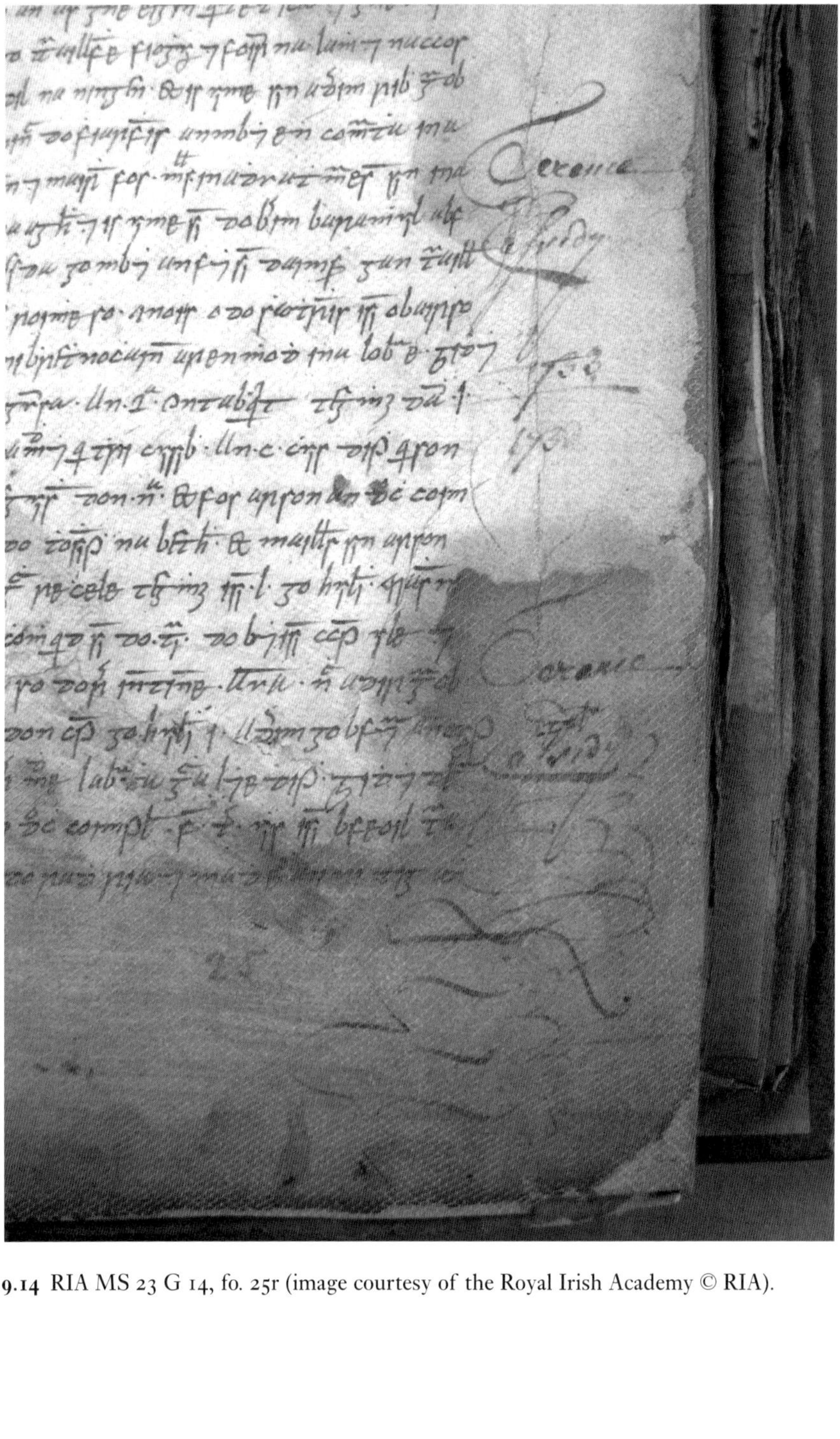

9.14 RIA MS 23 G 14, fo. 25r (image courtesy of the Royal Irish Academy © RIA).

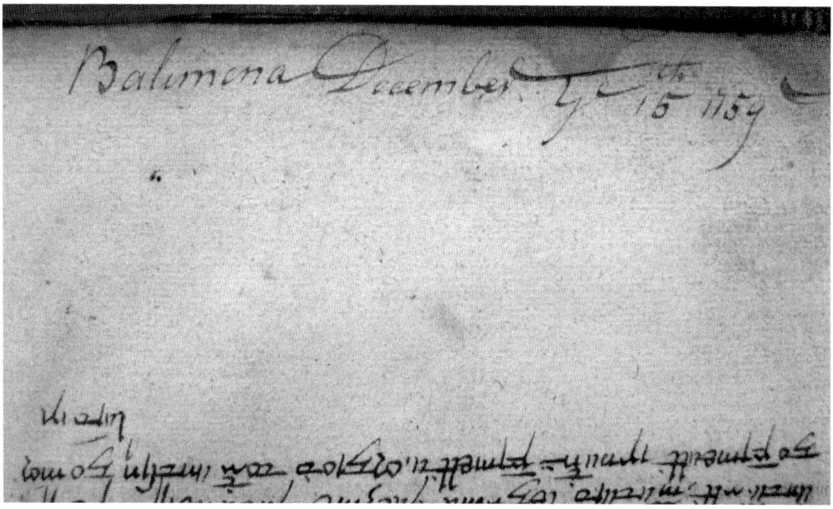

9.15 RIA MS 3 C 22, fo. 218v (image courtesy of the Royal Irish Academy © RIA).

stretching and yawning – *De alitatione et oscitatione* – translated by Donnchadh Óg Ó Conchubhair (*fl.* 1581–1611), Finghin Mac Giolla Phádraig's chief physician. It contains the scribal note: *Trocaire don ti do chuir so a nGháoidilg .i. Donatus Iuvenis an .20. la do Iunius. 1611.* ('Mercy on him who translated this into Irish i.e. Donatus Iuvenis, on 20 June 1611').[14]

The overall manuscript is a miscellany and includes works by Hippocrates, Aegidius – treatises on fevers, humours – Avicenna on paralysis, Galen on dropsy, Lanfrancus de Mediolano, and others. It is dated throughout – 1596 (fo. 61r); 1597 (fo. 23v); 1606 (fo. 133v); 1610 (fo. 122r) – and contains details of a number of locations, as for example this reference to Aghmacart:

> anno domini 1596 an 13 la do Maius. a nAchaidh mhic Airt dam an tansa a bfhochair Dhonnchadha óig í Chonchubhair agus is urusa dhamh anois bheith dobrónach oir adime mo charaid agus mo chumpanach uaimh = Mairghrég inghen Donnchadh agus dar an leabhar ni fes dam créd do dhén ina hégmuis fesda.
>
> On 13 May in the year of our Lord 1596, I am in Aghmacart now in the company of Donnchadh Óg Ó Chonchubhair. And it is easy for me now to be sad for my friend and my companion has left me, i.e., Mairghrég, daughter of Donnchadh, and upon my word I don't know what I shall do without her from now on.

14 See Nic Dhonnchadha, 'The medical school of Aghmacart', 20.

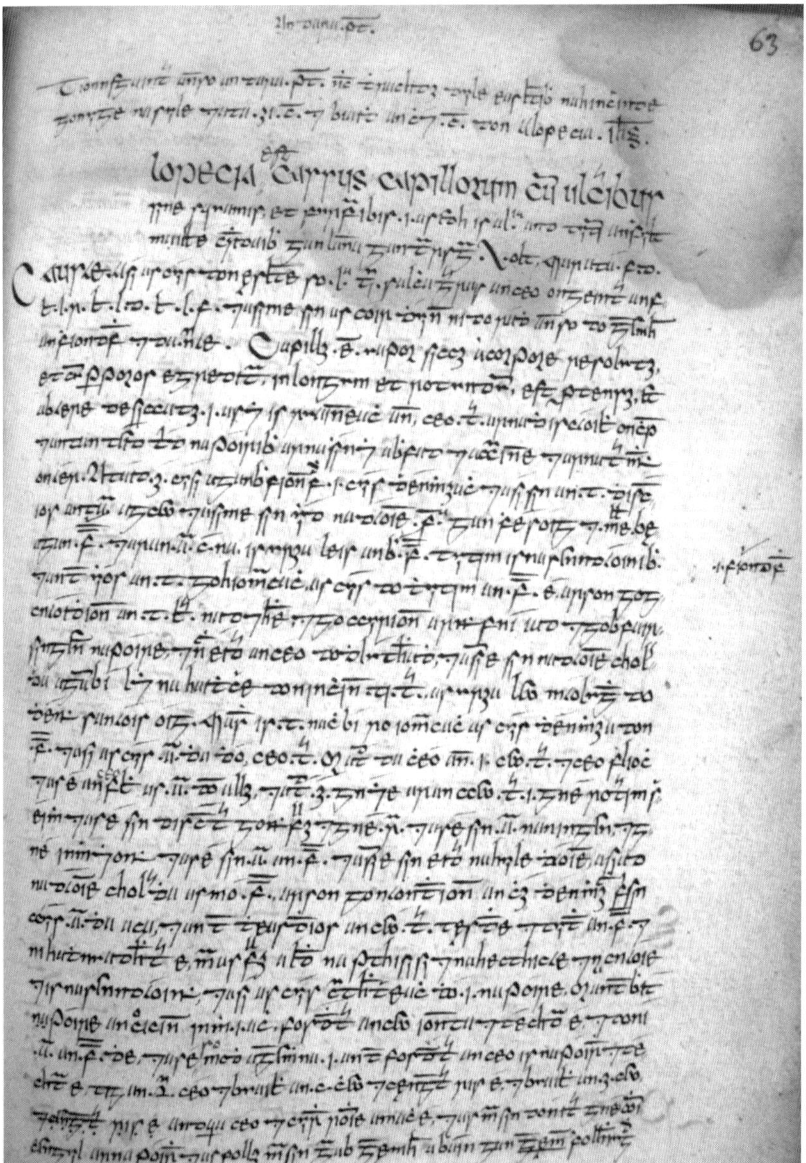

9.16 RIA MS 3 C 22, fo. 63r (image courtesy of the Royal Irish Academy © RIA).

RIA MS 24 P 3 is an interesting miscellany containing material compiled at various times during the seventeenth and eighteenth centuries. The record for this manuscript in the *Catalogue of Irish manuscripts in the Royal Irish Academy* notes the description in the Reeves sale catalogue of 1892:

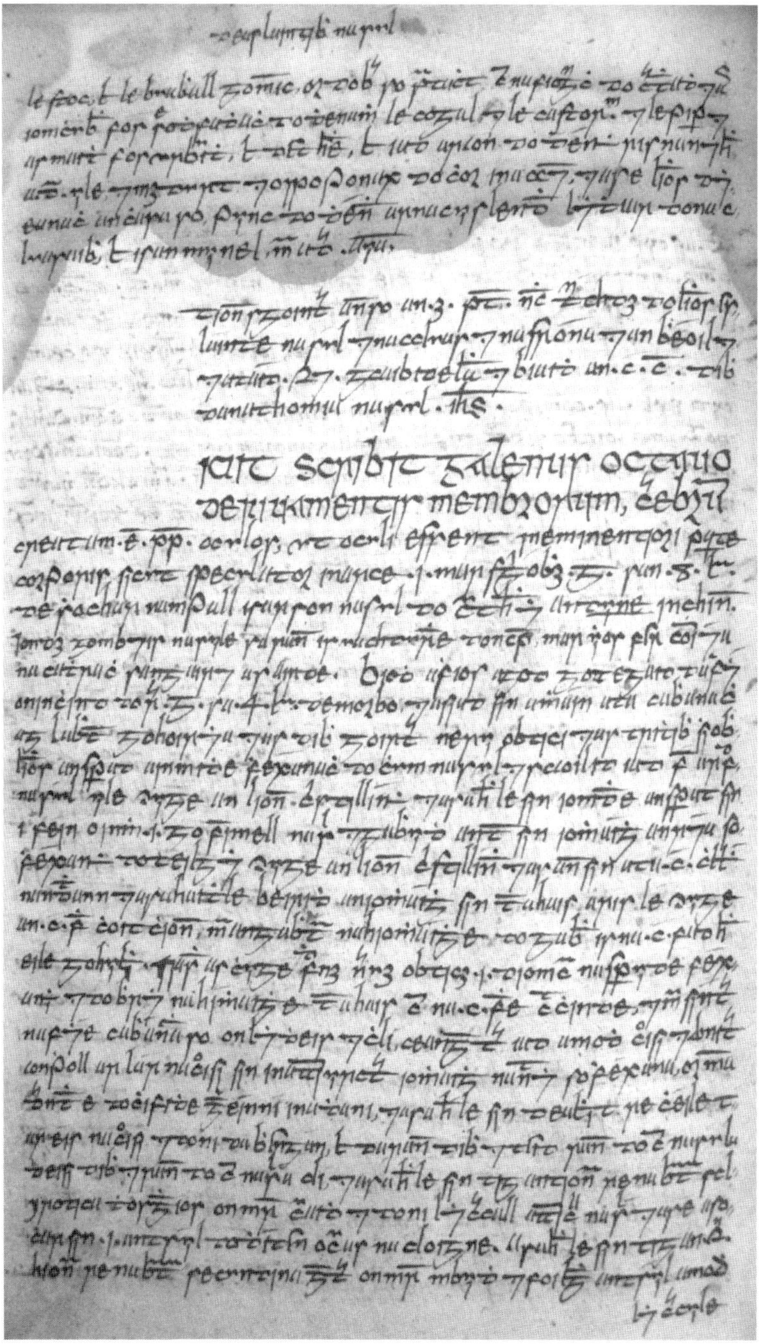

9.17 RIA MS 3 C 22, fo. 107v (image courtesy of the Royal Irish Academy © RIA).

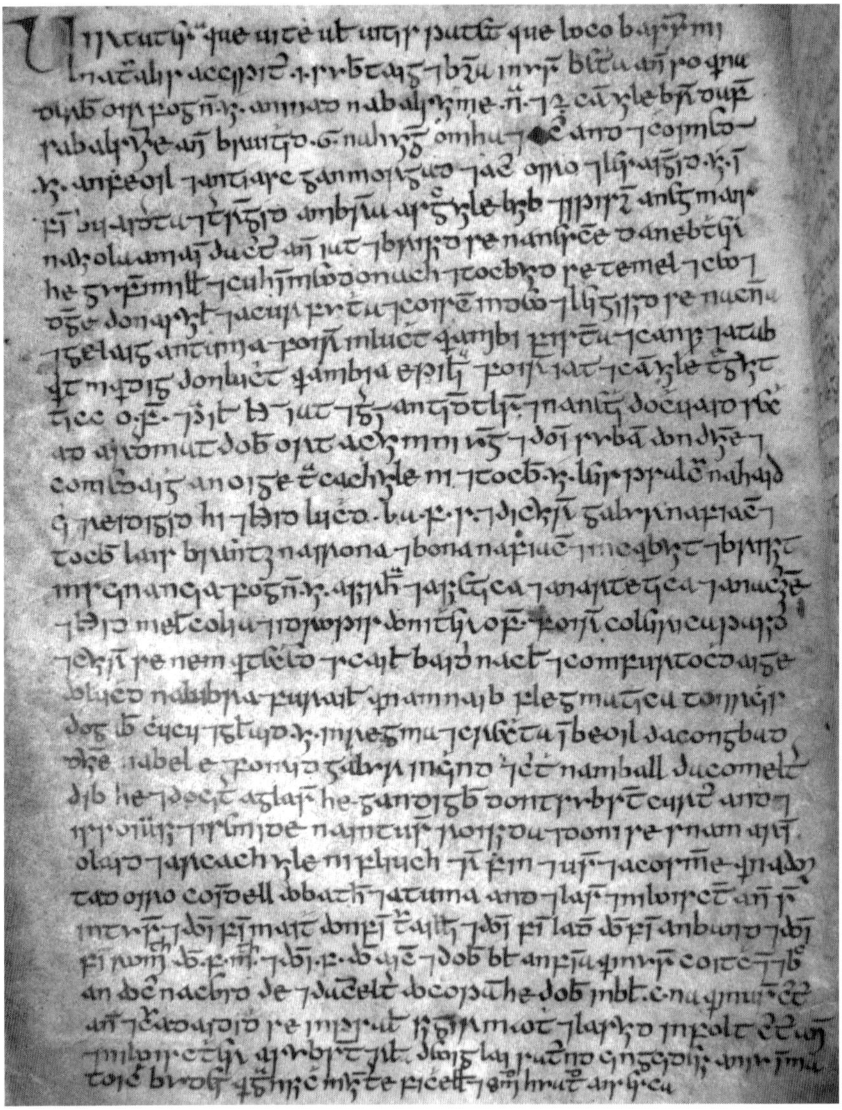

9.18 RIA MS 24 B 3, p. 94 (image courtesy of the Royal Irish Academy © RIA).

816. Medical MSS of various dates, 1616, 1624, 1718 etc., containing the Aphorisms of Hippocrates and Aurelius Celsus in Latin, with explanations in Irish by Tomas Ua Cuinn (Quinn), Giollagaire Ua Callanain, Seamas Ua Niallain, Risteard Tuibear and other Leeches and Herbalists, Curious Prescriptions etc. On Paper, quarto, hf. Vel.

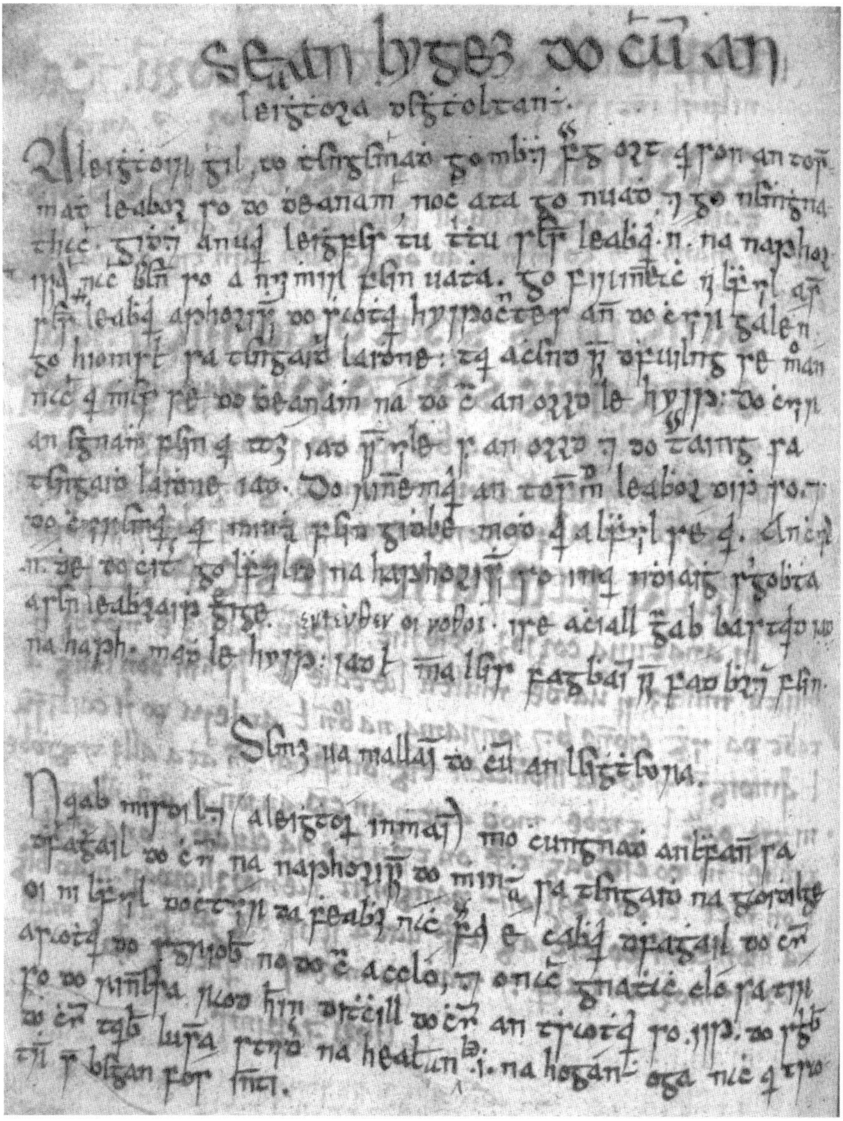

9.19 RIA MS 24 P 3, p. 130 (image courtesy of the Royal Irish Academy © RIA).

It appears to be a compilation of six different texts, including Gordon's *Lilium*, by various hands, the known scribes being Séamus Ó Nialláin, Risteard Tuibear (Richard Tipper) and Éamonn Ó Caiside. It is a Munster compilation, written partly at Baile na mBeuc, possibly Beakstown in the barony of Holycross, Co. Tipperary, or Mabbotstown, Co. Kilkenny, Baile in Fedha (1614) and Baile

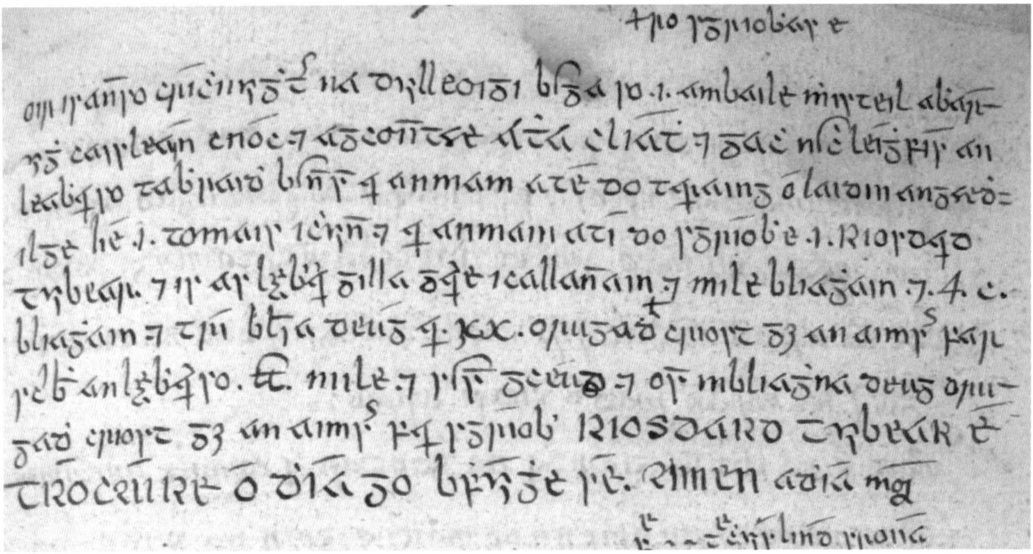

9.20 RIA MS 24 P 3, p. 195 (image courtesy of the Royal Irish Academy © RIA).

Mhisteil (Baile Mhistéala/Mitchelstown), Co. Cork. In a section containing seven books of Hippocrates' Aphorisms there is an extra book by Seaan Lygeus with a long explanation to the reader (fig. 9.19).

On p. 195 (fig. 9.20) we find the later, Dublin scribe, Risteard Tuibear, writing at Castleknock paying tribute to Tomás Ua Cuinn, who translated the work from Latin into Irish:

> Oir is annso críchnuightear na duilleoigi bega so .i. a mBaile Mhisteil abháruigh Caisleán Cnóc ₇ a gconntae Atha Clíath ₇ gach nech léighfís an leabhar so tabhraidh bennacht ar anmain até do tharaing ó Laidin a nGaedhilge hé .i. Tomais i Chuinn ₇ ar anmain atí do sgríobh e .i. Riosdard Tuibear ₇ is as leabhair Gilla gaire i Callannain ro sgriobhas e ₇ mile bliaghain ₇ 4 céad bliaghain ₇ trí bliaghna deúg ar XX o rugadh Criost gus an aimsir far scribhadh an leabhar so. Et mile ₇ secht gcéud ₇ ocht mbliaghna deug o rugadh Criost gus an aimsir far sgríobh RIOSDARD TUIBEAR e. Trócaire ó Dhía go bfhúighe sé. Amen a Dhia moir.

> For it is here these small leaves are completed i.e. in Mitchelstown close to Castleknock and in the county of Dublin. And everyone who will read this book, let him bestow a blessing on the soul of him who translated it from Latin into Irish, i.e. Thomas Ó Cuinn, and on the soul of him who

> transcribed it, i.e. Richard Tipper; and it was from the book of Giolla Gaire Ó Callanáin I copied it. And 1433 years from the birth of Christ to the time in which this book was written. And 1718 years from the birth of Christ to the time in which Richard Tipper copied it. God's mercy upon him. Amen, O great God.

The third part (pp 199–230) deals with remedies for different complaints, including ones that purge the stomach if cold humours are present.

RIA MS 23 O 6, dating from the very early part of the fifteenth century, is a miscellany, the owners and scribes of which are not known. It details the ingredients of poison and antidotes; colours of the urine; and properties of beans and barley. A list of plant names occurs on p. 34. Botany was an important element in medieval medicine (fig. 9.21).

RIA MS 23 N 20, an eighteenth-century Munster manuscript, compiled by Micheál óg Ó Longáin, owned by Sir William Betham, includes cures for all kinds of minor ailments including warts, sprain and colic, as well as more worrying illnesses – jaundice and the like; hiccups and nosebleeds are dealt with on p. 21 (fig. 9.22). We find Ó Longáin's colophon, written at Coppan, Co. Cork, on p. 134 (fig. 9.23):

> Air na ghraphadh liomsa Micheál óg O Longáin cuim mh'uasaide fein a ttigh chSeaghain shoilbhir ui Dhuill a cCuil ui Murchadha laimh le Curracha Chiopain a Musgraide san mbliaghain 1794, December 2d.
>
> Compiled by me, Michael Longan, for my own use, in the company of Seán Soilbhir Doyle, at Cuil Ui Murchadha by Coppan in Muskerry, 2 December 1794.

RIA MS C iv 2 is a sixteenth-century compilation by several scribes, including Niall Ua Cuinn and a member of the Mac Maol Tuile family, under the signature 'in Tillech'. The book includes a calendar, a fragment of a medical treatise, a copy of a *materia medica*, and several poems. The provenance of this manuscript is important. The name, 'Rd. Sheffield Grace' is incised on the inside of the wooden boards. The manuscript was later part of the Stowe Collection, the library of Richard Temple-Nugent-Bridges-Chandos-Greville, first duke of Buckingham and Chandos (1776–1839). Buckingham's wife was a member of the Grace family. One hundred and fifty Stowe manuscripts relating to Ireland were transferred to the Academy in 1883, on the dispersal of the collection by the British parliament, which had purchased the collection (by then it was the property of the fifth earl of Ashburnham, having been purchased by the fourth earl on the sale of the Buckingham Collection in 1849). The remainder of the collection was sent to the British Museum; the Stowe

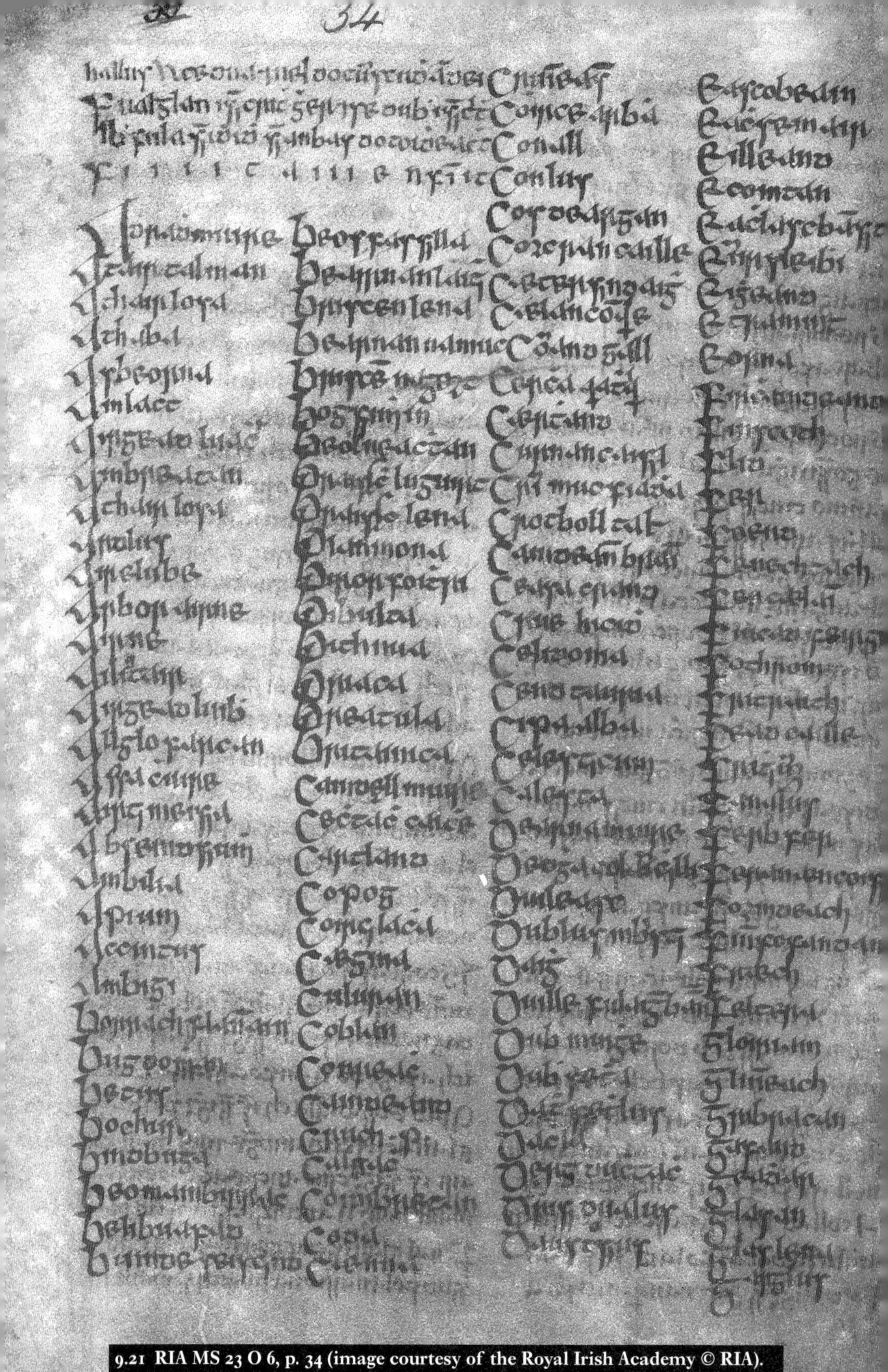

9.21 RIA MS 23 O 6, p. 34 (image courtesy of the Royal Irish Academy © RIA).

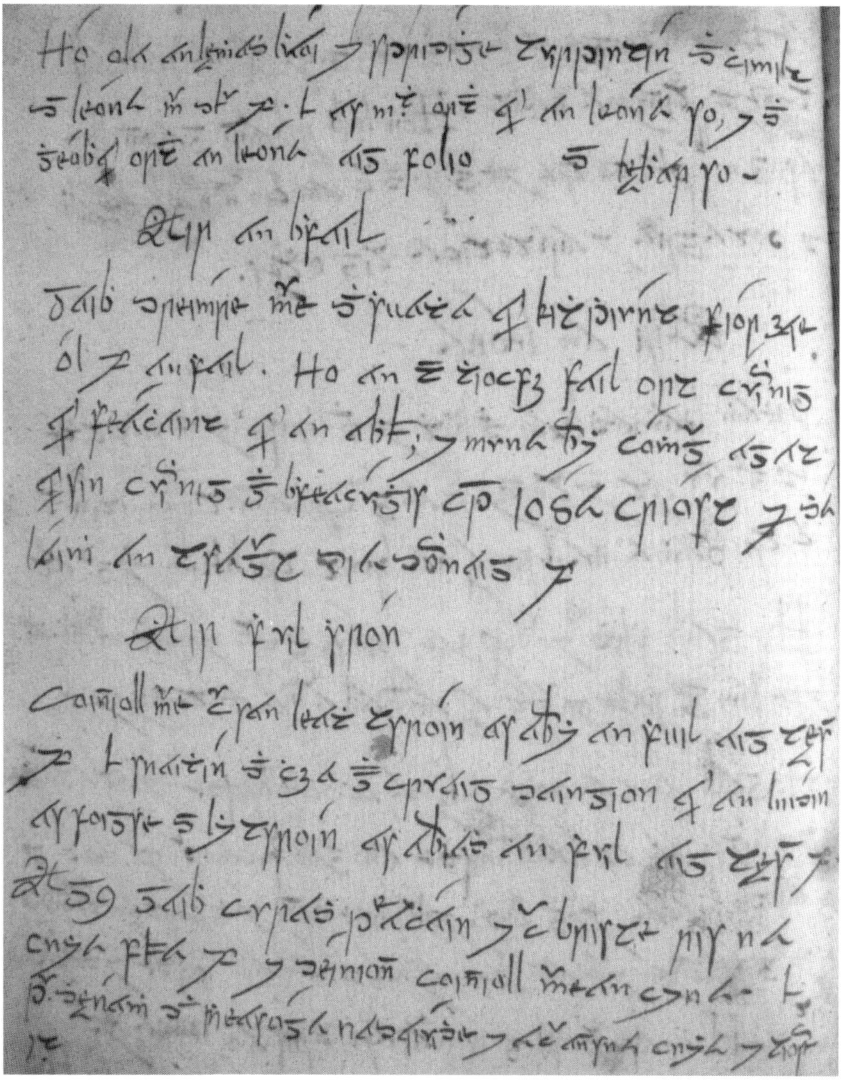

9.22 RIA MS 23 N 20, p. 21 (image courtesy of the Royal Irish Academy © RIA).

Collection now forms part of the British Library holdings. Folio 14r of our manuscript, which refers to Noah and his sons, contains Niall Ua Cuinn's colophon: *Misi Niall do scrib sin* (fig. 9.24).

Another scribal signature occurs in RIA MS 24 P 32, p. 311 – *Brian O Shiell Pattruig scripsitt*. A seventeenth-century *Materia medica*, based on either Nicholas of Salerno or Nicholas of Alexandria, this translation retains many Arabic and Greek terms, as shown on p. 325 of the manuscript (fig. 9.25).

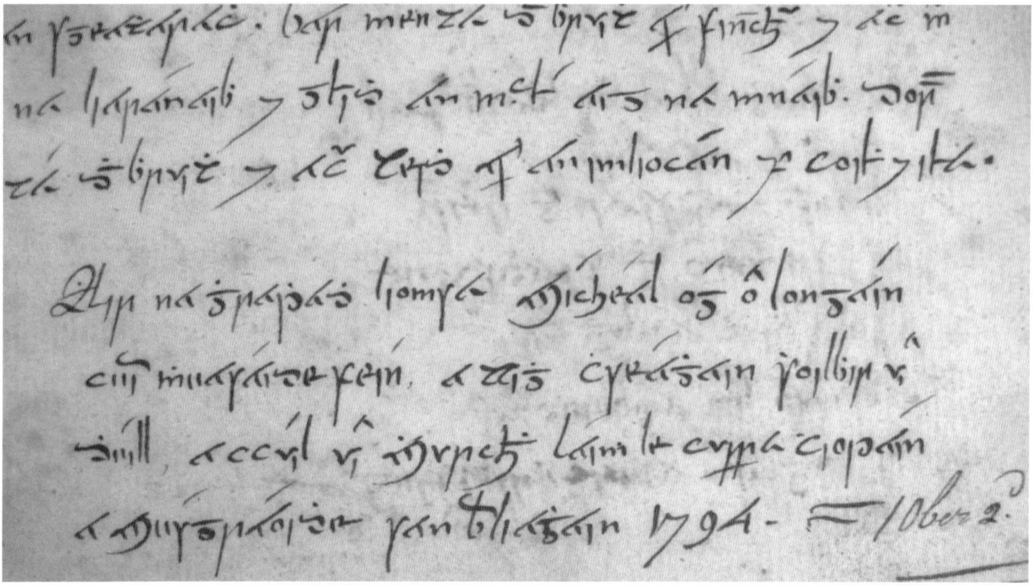

9.23 RIA MS 23 N 20, p. 134 (image courtesy of the Royal Irish Academy © RIA).

RIA MS 23 M 36, a fifteenth- to sixteenth-century compilation, deals mainly with stomach and urinary troubles. One page explains that Tadhg Ó Cuinn put Irish on the Latin text; *Uilliam O Bolgaidh do sgrib* (fig. 9.26).

RIA MS 24 P 26 is an important fifteenth-century vellum, written by Donnchadh óg Ó Híceadha, physician to the O'Briens, MacNamaras and other families of Thomond; it contains Geraldus de Solo's commentary on Rhazes, Pietro de Argelata's *De chirurgia* and Arnaldus de Villa Nova's *Regimen sanitatis*. Owned by MacAdam, it was previously owned by Edward O'Reilly (sale catalogue, lot no. 4)[15] and latterly was the property of Reeves. Reeves cites the O'Reilly sale data on the flyleaf and notes: 'I bought it in Nov. 1889 together with a large collection of Irish MSS. from Mr Robert MacAdam, & it is the gem of the lot […]'. The book had remained in O'Hickey hands for about four hundred years; evidenced by the notes therein. The work is written in a fine hand shown here (fig. 9.27). This image also illustrates a repair to the vellum carried out by John Gillis, conservator, who later worked on the Faddan More Psalter.

MS 23 P 20 is a sixteenth-century copy of John of Gaddesden's *Rosa Anglica*, which includes material not found in the printed version of the text (pl. 11).

15 *Bibliotheca Hibernica: Catalogue of the library of the late Edward O'Reilly … to be sold by auction … at … no. 33 Anglesea Street, Dublin on … November 30, 1830 …* by Charles Sharpe … (Dublin, 1830). Copy held by RIA Library.

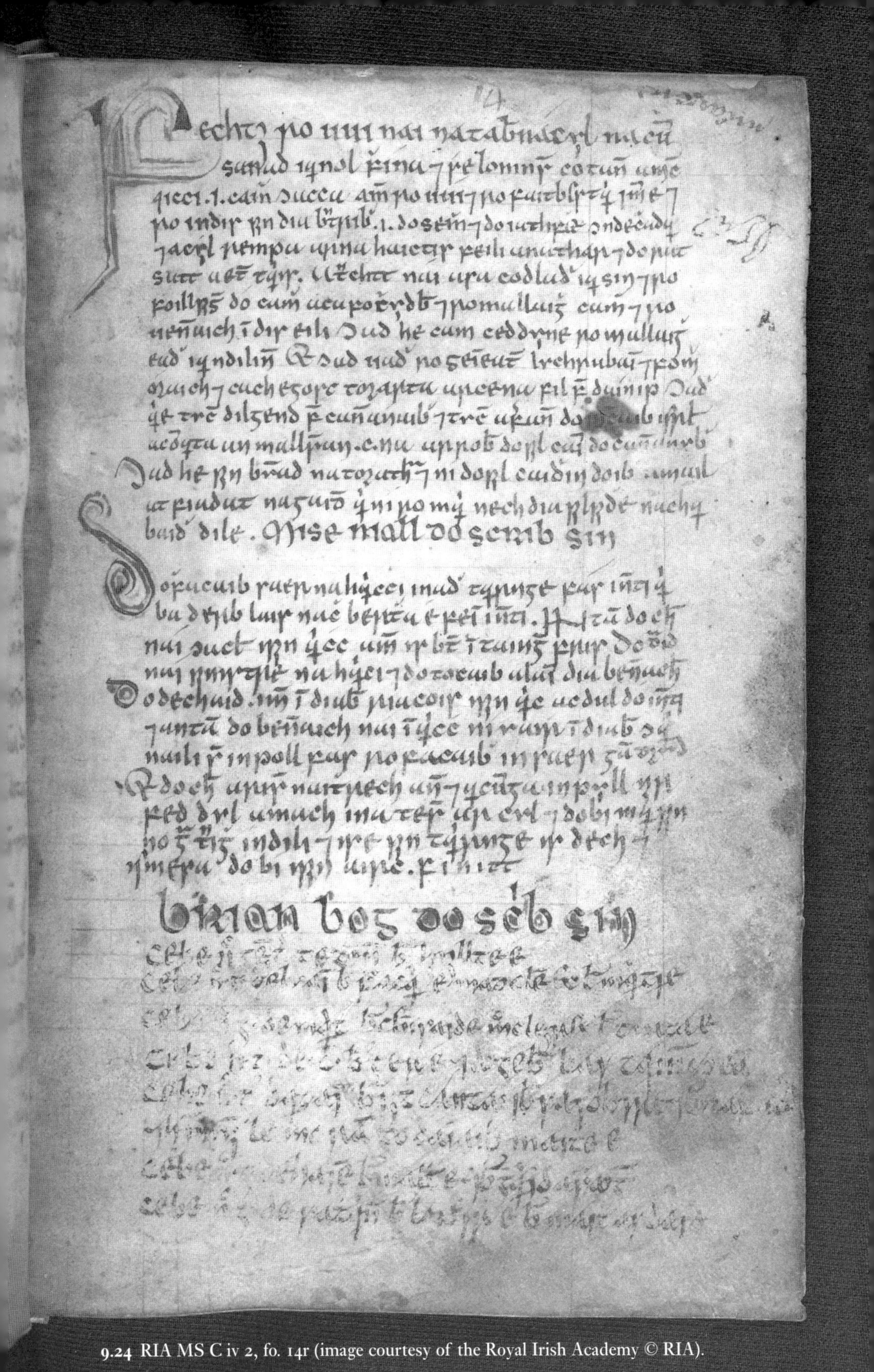

9.24 RIA MS C iv 2, fo. 14r (image courtesy of the Royal Irish Academy © RIA).

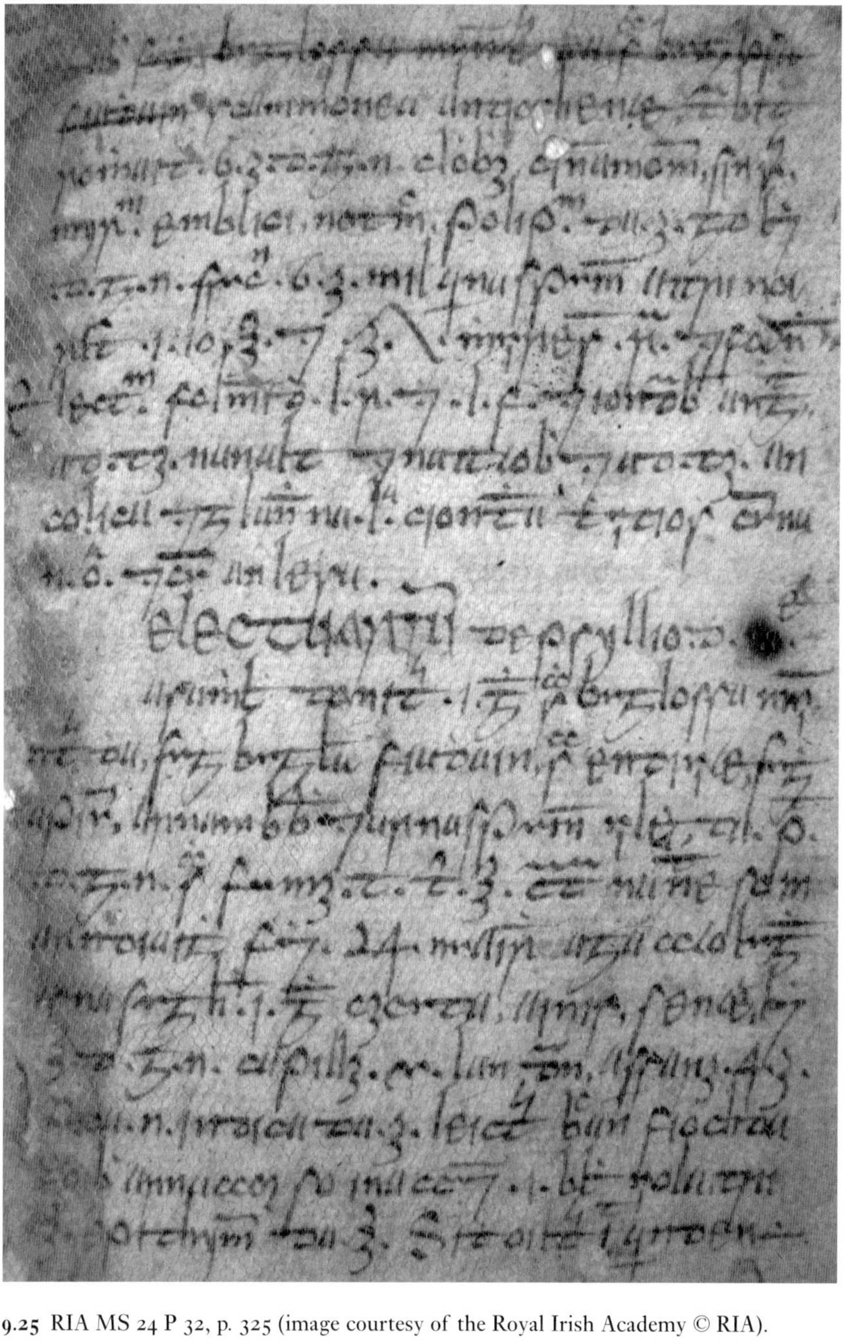

9.25 RIA MS 24 P 32, p. 325 (image courtesy of the Royal Irish Academy © RIA).

Among the jewels in the crown of the Academy's collections are MSS 23 P 10 (i)–(iii), three medical tracts bound together by Sir John T. Gilbert (1829–98)[16] in the nineteenth century but actually consisting of three separate texts, with different scribes and compilation dates. The first is a fifteenth-century compilation of medical treatises including Hippocrates' Aphorisms and Prognostics; the scribes of this tract are unidentified. A detail from p. 1 (pl. 12) and a marginal note on p. 4 (fig. 9.28) are shown here.

The most renowned of the medical books is RIA MS 23 P 10 (ii), known alternatively as the Book of O'Lees or Book of Hy Brassil. The fortunes of this book have given rise to the 'imagined islands' of my title, referring to the mythical isle of O'Brazil. Roderic O'Flaherty, writing in 1684, identified O'Brazil as an imaginary isle to the west of the Aran Islands, and noted that one Morogh O'Ley (Murchadh Ó Laidhe) claimed to have spent two days there. The story goes that O'Ley was wandering around the Carna peninsula 'going alone from one village to another, in a melancholy humour, upon some discontent of his wife' when he was kidnapped and taken off to the island. A few days later he was deposited back on the mainland where he lay ill at the house of a friend for some time, before returning to Galway. About seven or eight years later, O'Ley began to practise 'both chirurgery and physick, and so continues ever since to practise, tho' he never studied nor practised either all his life time before, as all we that knew him since he was a boy can averr'.[17] Hardiman, who edited O'Flaherty's text, noted that O'Ley 'was held to have received a book from one of the island's inhabitants, with an injunction not to look into it for seven years'. Upon so doing at the end of this period, O'Ley was 'imbued with the gift of healing'.[18] Hardiman's interpretation of this story was that it permitted O'Ley, whose family lands had been confiscated in the seventeenth century, and who had since fallen on hard times, to re-invent himself as a physician in order to earn a living. The O'Leys (O'Lee) were hereditary physicians in Iar Connacht.

The contents of the book are no less fascinating than its fortunes in the hands of the O'Lees. The book (*c.* fifteenth century) had probably been in the possession of the O'Lees for a considerable time, as attested by the signature – 'P. Lee' on folio 1:

> It is [...] the product of an academic milieu. It comprises a learned, Latinate, and highly organised pathological treatise, which is arranged in a series of forty-four tables, preceded by a table of contents. The opening three tables treat of diseases that affect the entire body, namely fever and apostemes [...].[19]

16 Gilbert was Librarian to the Academy for most of the period 1861–98. 17 Roderic O'Flaherty, *A chorographical description of West or H-Iar Connaught written* AD1684, ed. James Hardiman (Dublin, 1846), pp 68–72. 18 Nic Dhonnchadha, 'The "Book of the O'Lees"', p. 82. 19 Ibid., pp 82–3.

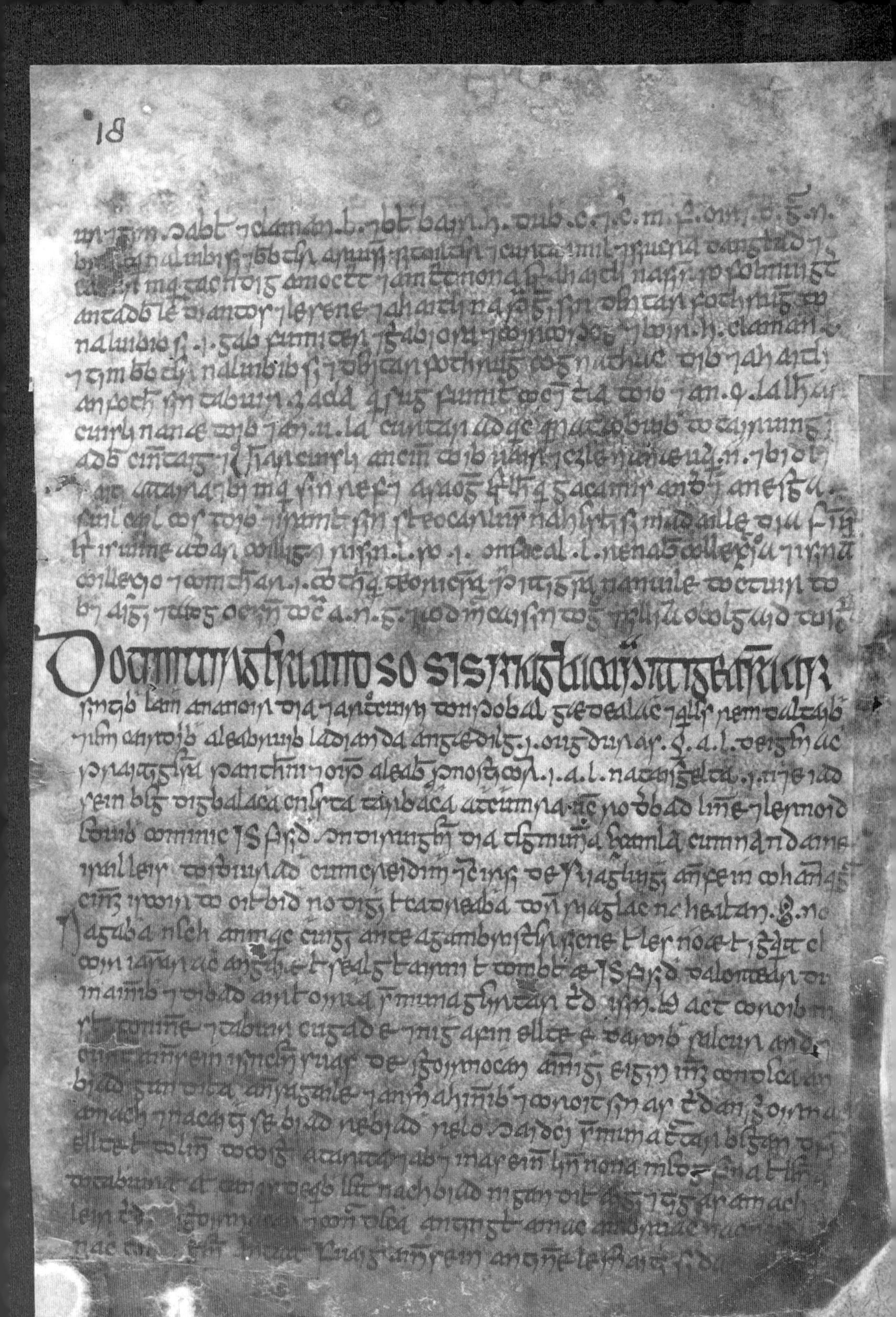

9.26 RIA MS 23 M 36, p. 18 (image courtesy of the Royal Irish Academy © RIA).

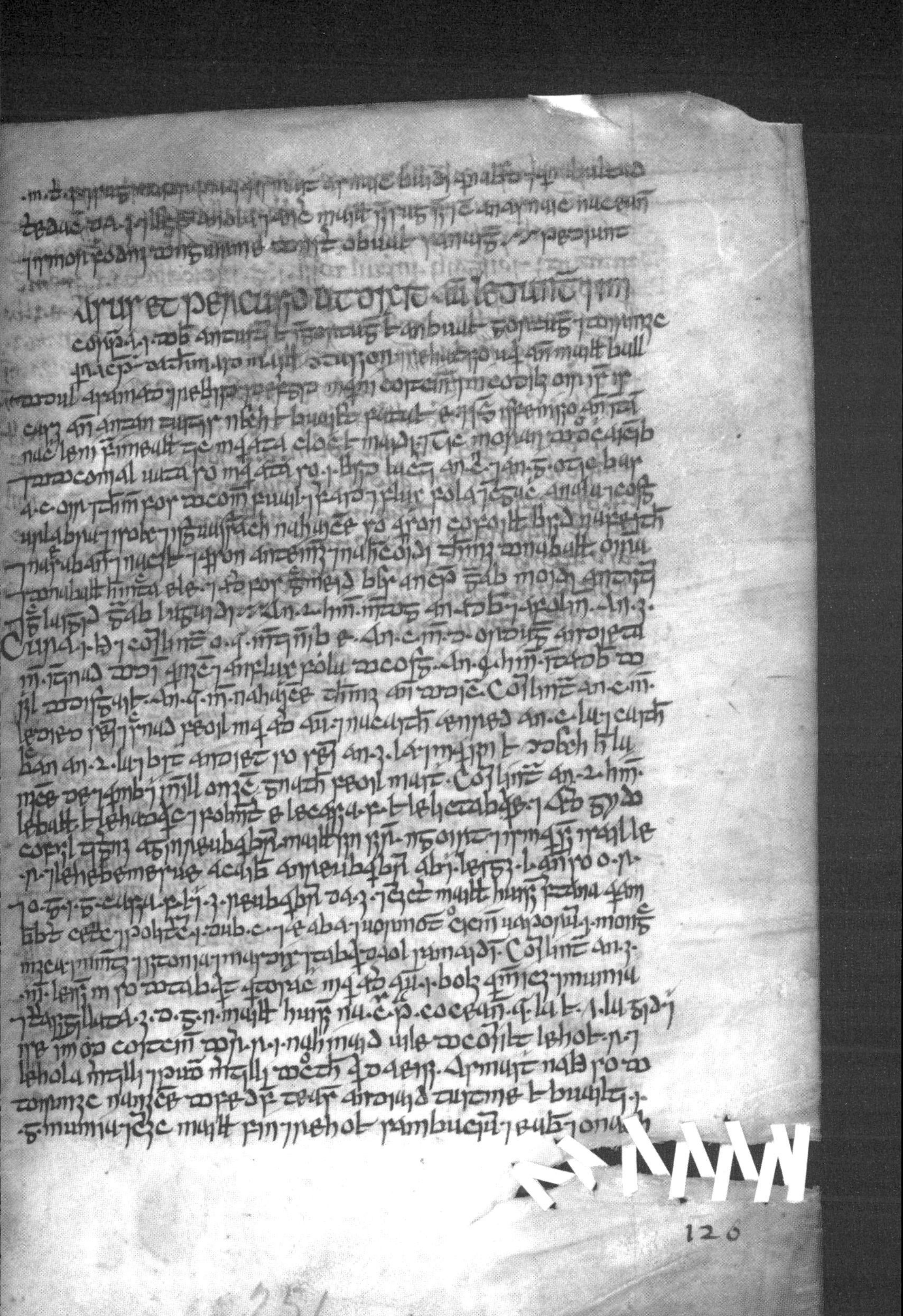

9.27 RIA MS 24 P 26, p. 251 (image courtesy of the Royal Irish Academy © RIA).

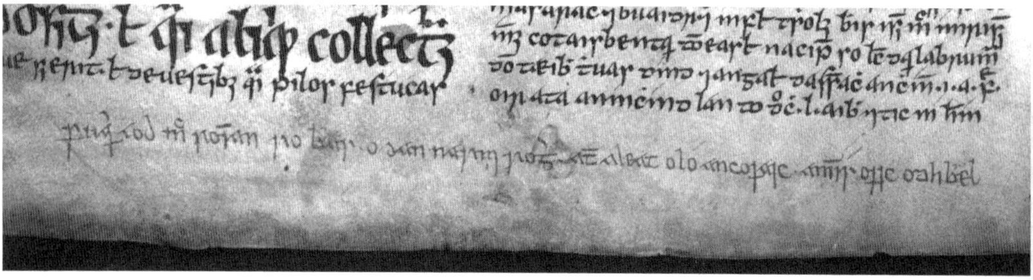

9.28 RIA MS 23 P 10 (i), p. 4 (image courtesy of the Royal Irish Academy © RIA).

Diseases of the exterior of the body, the internal members, the brain, the spinal cord and the organs of sense are then treated, followed by the organs of respiration, those of nutrition and those of generation and so on. Unfortunately, the scribal colophon has been lost due to the fact that leaves are missing from the end of the book.

The painstaking detective work of Nic Dhonnchadha has led to the identification of the original work on which the Book of the O'Lees was based, as the *Tacuini aegritudinum*, 'a Latin translation from the Arabic of the Islamic physician Ibn Jazlah (d. 1100) of Baghdad, which was completed in Sicily, in 1281, by the Jewish physician and translator, Faraj ibn Salim (*fl.* 1273–82)'. The *Tacuini* first appeared in print in 1532 (Strasburg), so it is clear that the Irish scribe and translator was working from a Latin manuscript of the text.

The layout of the text is unique among our Irish manuscripts, consisting of tables (laid out in a standard format) that present diagnostic and therapeutic information arranged under eleven headings – names, complexion, age, season, region, prognosis, cause, symptom, evacuation, royal remedy, easily found remedy. A twelfth heading, 'general remedy', is to be found on the right-hand page, opposite each table. This heading is presented in great detail. Nic Dhonnchadha considers this work to be a fine example of high learning and that it was most likely the product of an established medical school. It could have been used by a teacher, a student or a practising physician (fig. 9.29 shows p. 2; pls 13 and 14 show pp 10 and 11; pl 15 shows p. 48 (detail)).

RIA MS 23 N 8 is a sixteenth-century calendar compiled by two scribes and is almost indecipherable. It is a mix of a calendar based on the lunar cycle, saints' days, the house in which the sun is to be found and so on. The interest for our purposes is the connection made with auspicious days for carrying out certain medical procedures such as blood-letting. Mention is also made of seasonable diet. Plate 16 shows folio 18v.

RIA MS 23 P 10 (iii) is a fifteenth-century treatise, formerly owned by Searga Ó Conuill. Probably written in Co. Clare, it opens with the *Rosa Anglica*, and has

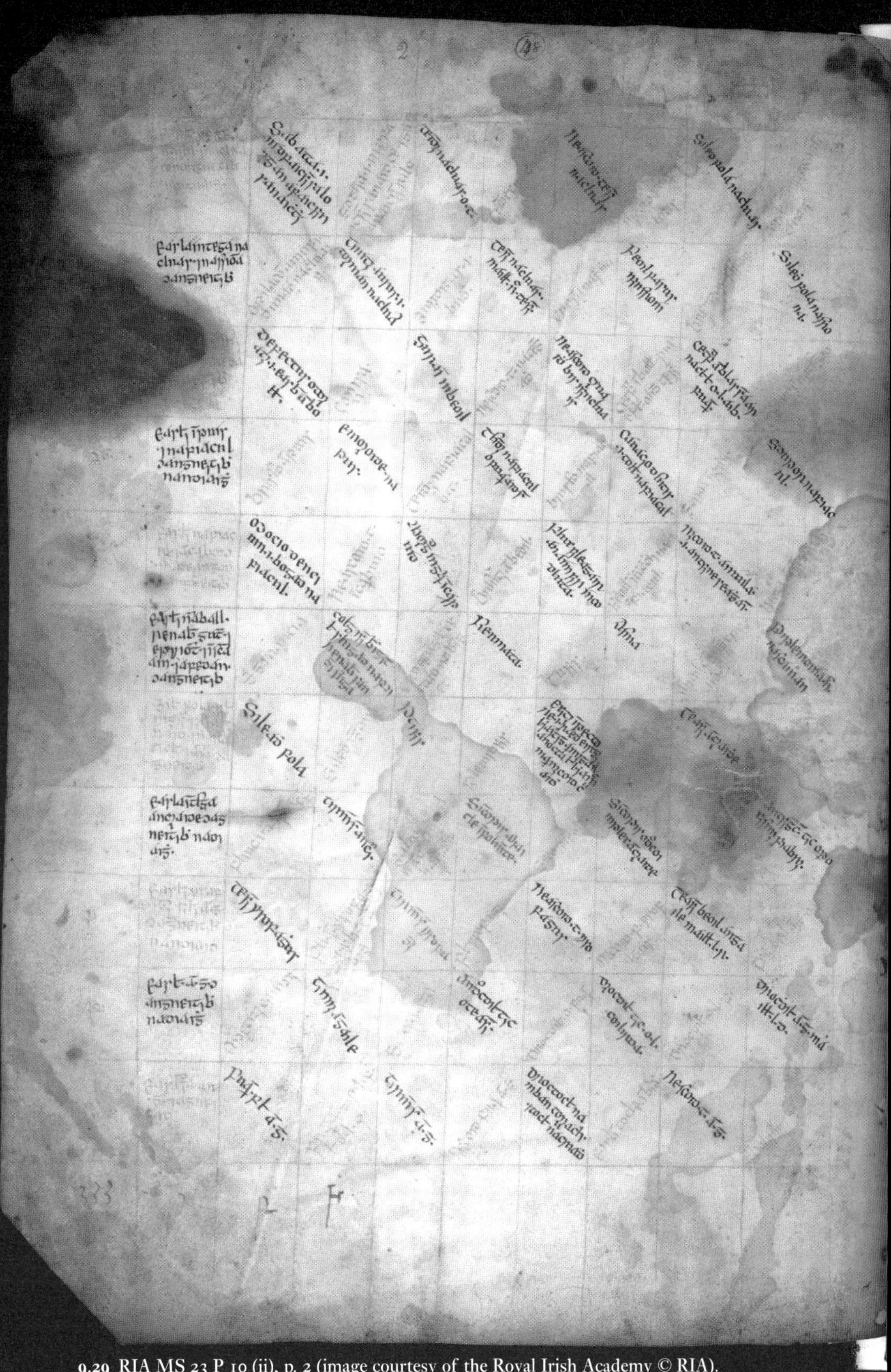

9.29 RIA MS 23 P 10 (ii), p. 2 (image courtesy of the Royal Irish Academy © RIA).

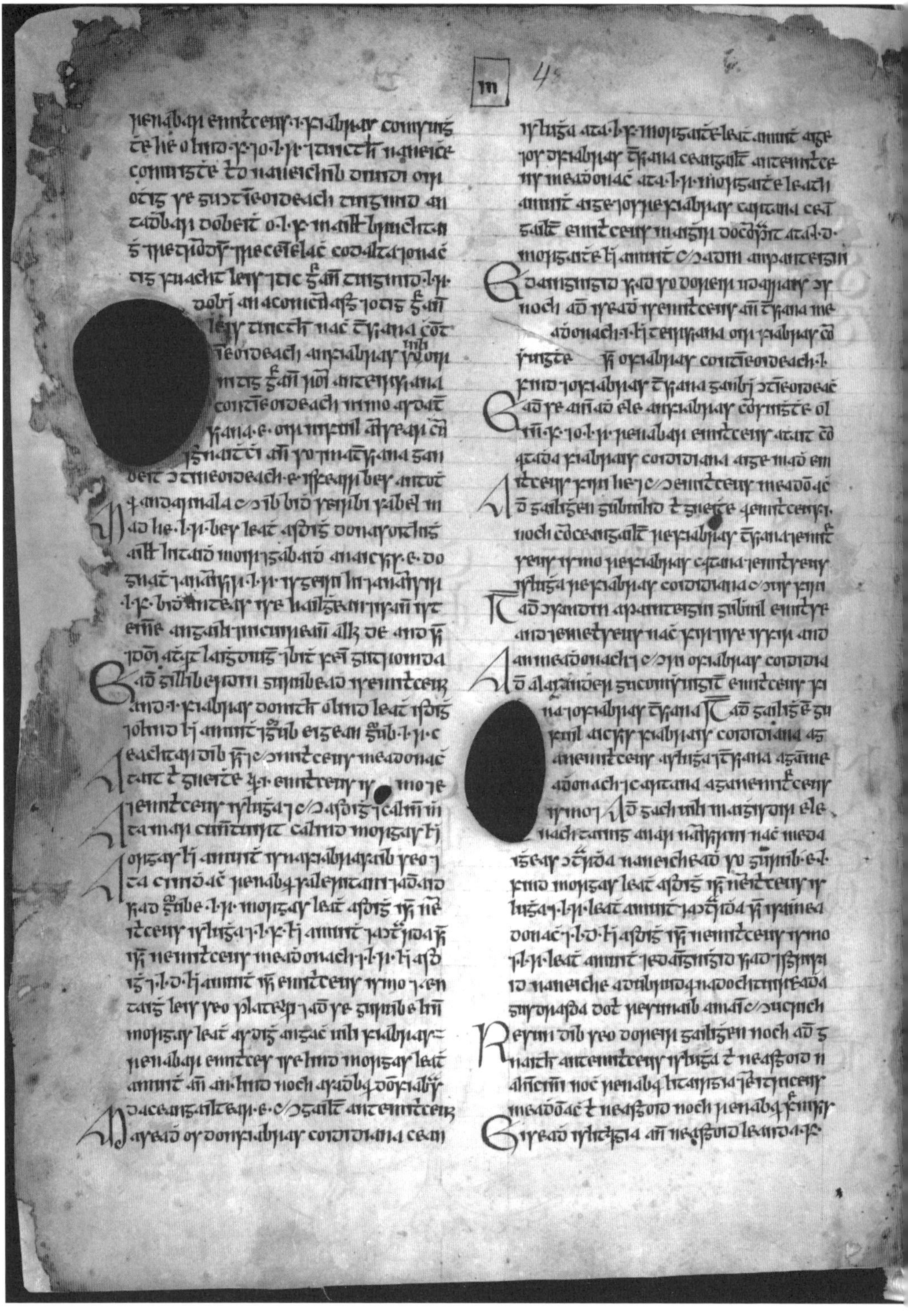

9.30 RIA MS 23 P 10 (iii), p. 48 (image courtesy of the Royal Irish Academy © RIA).

texts by another English author, Gilbert the Englishman, and the French physician, Bernard of Gordon (pl. 17).

RIA MS B ii 1, another Stowe manuscript, an astronomical tract, was edited and translated by Maura Power.[20] The commonly held belief that the stars influenced health meant that knowledge of astronomy was essential for those practising medicine. B ii 1 is a composite manuscript, the scribe of which has been identified as Aodh Buidhe Ó Leighin, member of a Munster medical family. Dating from the early fifteenth century, the manuscript unusually retains a volvelle (fo. 1v), which permits one to locate the position of the moon in the zodiac. The text is partially based on the work of Jewish astronomer, Messahalah (d. *c*.815). Over the course of forty chapters, planetary motion, the tides, the Nile floods and habitable climes are just some of the topics discussed. The manuscript is of interest also for the many diagrams illustrating the text (pl. 18).

To complete this tour of the Academy's medical and scientific manuscripts, it might be of interest to introduce a work that is not of Irish origin. It is RIA MS 23 D 43, an Icelandic medical work, described as 'a medieval miscellany of encyclopaedic character', also accessible on Irish Script on Screen, having been digitized to enable the National Library of Iceland to complete its virtual corpus of medieval Icelandic material. The text has been edited by Henning Larsen.[21] The image shown is from the beginning of a leechbook (fo. ixr) (pl. 19).

In the course of this essay, I have tried to show the range of texts used by Irish doctors, their knowledge of medicine emanating from across Europe and beyond, the ways in which the hereditary physicians made their own of the texts and the ways in which these much-used books can bring us closer to our medieval forebears.

20 Maura Power (ed.), *An Irish astronomical tract based in part on a mediaeval Latin version of a work by Messahalah* (London, [1912] 1914). Electronic version accessible at www.ucc.ie/celt. 21 Henning Larsen (ed.), *An old Icelandic medical miscellany: MS Royal Irish Academy 23 D 43 with supplement from MS Trinity College (Dublin) L–2–27* (Oslo, 1931). The volume includes an English translation.

Clocks and guns: observations on medieval science and technology

HOWARD B. CLARKE

This is a non-specialist essay composed from the point of view of a mainstream historian. The focus is on general conditions and constraints that obtained in the western half of Europe (Latin Christendom) in the period 1200 to 1500 in relation to science and technology. The range of topics has deliberately been made extremely wide, in order to present a broad perspective of the subject as a whole. For that reason, footnotes have been eschewed, since to have provided them in an adequate manner would have completely overloaded the essay, if not the whole volume! This is not an attempt to write a history of medieval science and technology, only to comment briefly on their historical significance. By the same token, just one illustration has been included, as a symbol of that practical competence which, arguably, would lead to world domination in the ensuing centuries.

I should like to begin with three observations about the times in which we ourselves live. First, of the seven billion or so human beings on planet Earth today, the majority do not possess a computer and do not know how to use one. In other words, technological expertise has always been unequally distributed between cultures and between social groups within them. Even when distributed more widely, there is the problem of tractors and other mechanical devices in the developing world that do not work for lack of spare parts and servicing expertise. Secondly, not all technological inventions are beneficial to humanity in general: examples are nuclear weapons, food preservatives and Concord. Respectively they are capable of destroying all of humanity and most animal life, of causing chronic obesity, and of polluting the upper atmosphere at completely uneconomic cost. Thirdly, for a variety of reasons, not everyone is psychologically disposed to adopt new techniques.

Next I should like to make three observations about the medieval world, in its extended sense of the period AD400–1500. First, most western Europeans, even at the height of the urban boom in the thirteenth century, lived in the countryside. They were farmers, hunters and fishermen among the males, and fieldworkers, domestics and kitchen gardeners among the females. They were also craftworkers displaying an enormous variety of inherited skills, both male and female. They were members of André Varagnac's *Civilisation traditionelle et genres de vie* (1948). Secondly, as Marc Bloch observed in an essay entitled (in

English) 'Technical change as a problem of collective psychology', they were conservative. One major reason is that large numbers of children were 'educated' in the rural economy, not by their parents but by their grandparents. Grandparents by their nature tend to nurture the ways of the past; they look backwards rather than forwards. It has been argued that human beings live longer than most animal species because it takes so long to educate them in any sense, while parents must work in order to provide the essential means of existence. Thirdly, the medieval period forms part of Lewis Mumford's eotechnic phase of technological development. Extending down to the mid-eighteenth century, this was based essentially on the use of water and wood, with iron as a much rarer component, for example, of ploughs and carts. It reached an initial climax with Leonardo da Vinci and its main climax in the seventeenth century. It was succeeded by a palaeotechnic phase based on coal and iron, reaching a climax *c.*1870. At the time of writing (1967), people were living in a neotechnic phase based on electricity and alloys.

MEDIEVAL INVENTIONS AND INVENTIVENESS

Medieval Europeans were undoubtedly inventive, despite conventional views about what the word 'medieval' signifies. Yet the evidence is difficult to interpret, even for the better documented late Middle Ages, and often defective or non-existent. Partly this is because those medieval people who were most concerned with manual work, whether peasants or artisans, were illiterate or only partly literate. By *c.*1200, the most literate groups did little or no manual work, apart from quill-pushing, despite the fact that early medieval monastic guidelines had idealized manual labour. Medieval art tends to lack the realism required for technical devices to be understood and artists often used obsolescent conventions. Archaeological analysis is still at relatively early stages globally. A fundamental problem is dating, for a casual reference may relate to the adoption of a technique or device in a particular locality, not to its original invention. Casual references tell us nothing about the wider distribution of such techniques or devices. An interesting example is the invention of spectacles in northern Italy, probably in Venice linked to its glass and crystal industry, in the late thirteenth century. Petrarch praised them for helping him to read in old age, but their use appears to have spread only slowly.

Four main aspects of advancement are here posited: the continuance of techniques inherited from the ancient world; modifications to existing devices and methods; the adoption of techniques borrowed from other cultures; and the invention of new techniques. Some examples of each of these processes will be given.

THE CONTINUANCE OF TECHNIQUES INHERITED FROM THE ANCIENT WORLD

Apart from sail power, the greatest source of energy in medieval Europe was the watermill. A description of the Porta Pontemolino at Padua dating to between 1314 and 1318 cites thirty-four mill wheels floating in the river beneath the bridge. In 1493 the district of Troyes had 20 corn mills, 14 paper mills, 4 fulling mills, 2 tanning mills and 1 cloth mill (41 in all). The vertical wheel is usually attributed to the Roman architect and engineer Marcus Vitruvius Pollio and vernacular words for 'mill' tended to be derived from late Latin *molinum*. A vertical mill wheel necessitated the use of cogs, made of wood, which transmitted power in a horizontal plane and could speed up millstones, creating an advantage over the horizontal type of mill. Experimentation with gearing was important, eventually for the mechanical clock and its derivatives.

Mill construction and operation raised a number of technical problems. First, there was a need for dams and mill races for an overshot drive. In Spain and other parts of Mediterranean Europe, many mills were situated on irrigation canals, requiring a well-regulated supply and sometimes sharing. There was conflict between farmers and owners of orchards on the one hand and millers on the other. In some places, irrigation of land was restricted to weekends when mills were not working. In northern parts of Europe, of course, watermills froze up during the winter months. Secondly, there was the problem of the transportation and installation of millstones, a pair of which might weigh up to two tonnes. Thus the carriage of one stone from Southampton to Cuxham in Oxfordshire in 1305 took three horses, a carter and the miller five full days travelling at a cost of 6*s.* 9*d.* At Dublin in 1455, the bridge-toll across the River Liffey was ½*d.* for a load of hay or wood but 4*d.* for a millstone.

The prime use of watermills was to grind grain, thereby saving labour, especially of women. North of the Alps, mills tended to be seigneurial, feudal lords building them as sources of revenue. Manorial tenants were compelled to grind at the lord's mill and multure was commonly one-sixteenth of the grain brought to the mill. There was constant resort to secret handmills and friction between landlords and tenants: at St Albans, these illegal handmills were confiscated by the monks and used by them as flooring for their buildings, only to be dug up by rioters in the Peasants' Revolt of 1381.

MODIFICATIONS TO EXISTING DEVICES AND METHODS

A major advance on the classical world was to adapt the waterwheel to a wide range of industrial processes. Examples are mashing beer (itself a late medieval invention), fulling cloth, tanning leather, beating metal, turning winches in mines and making paper. All of these processes were increasingly commonplace by the fourteenth and fifteenth centuries. They mostly required a hammer action, achieved by means of cams and trip- or tilt-hammers. The cam, incidentally, was not a medieval invention in itself, but had never been used productively before.

The earliest documented paper mill was at Xativa, near Valencia, in 1238. This technique appears to have spread only slowly, for the first recorded German example was built near Nürnberg in 1389. On the other hand, the fulling mill appears to have spread rapidly in thirteenth-century England. Again, this became a seigneurial monopoly there and women were often hauled before manorial courts for evasion. Mills were also centres of social life and we even hear of prostitutes seeking customers at queues outside grain mills. In 1192, mills being used as Cathar meeting-places at Modena were demolished.

THE ADOPTION OF TECHNIQUES BORROWED FROM OTHER CULTURES

The question of borrowing from other cultures, especially Arabic and Chinese ones, is fraught with difficulty and controversy. As in the case of archaeological artefacts or art-historical motifs, links between places and peoples are often hard to prove. There are disagreements between 'creationists' and 'diffusionists'. Thus, for example, it may or may not be the case that the magnetic compass had reached western Europe overland from China by the late twelfth century. Another instance is the invention of printing by moveable metal type, which the Chinese and Koreans had by the fourteenth century. Gutenberg could have been working independently at Mainz in the 1440s, for it is hard to establish an intermediate link at that time. More certain examples are the adoption of exotic crops, especially in Spain, such as oranges, lemons, melons and sugar.

Most important of all is the system of Hindu-Arabic numerals, including zero, which became known in western Europe in the twelfth century. Yet their use spread only slowly outside Italy and mathematical knowledge remained weak. Most late medieval manuscripts contain roman numerals and are foliated in the same manner. To take Dublin as an example, roman numerals were still being used in the municipal records as late as the early seventeenth century. When the English language came into widespread use there after 1452, a mixture of roman numerals and completely spelt out words was the norm. The first large-scale usage of Hindu-Arabic numerals appears to have been in Perrot's survey of the walls, towers and gates of the city in 1585, which includes many statements of distance and size.

THE INVENTION OF NEW TECHNIQUES

One of the most striking medieval inventions was the windmill, which first appeared in coastal districts of Normandy and England in the 1180s. There is some speculation about whether the basic idea was a borrowing from the Near East in the course of the first crusades. As early as the 1190s, Pope Celestine III was making them subject to tithe. They were apparently more popular in northern parts of Europe, where rivers often froze over in winter. They could also be used in castles and towns when under siege. These early mills were all post-mills that had to be turned bodily to face the wind, their sails being made of canvas as in ships. The much more complex tower-mills, whose sails alone were turned to the wind, were invented probably in the early fourteenth century, though they are not illustrated until the end of that century. Their construction and maintenance required considerable technical expertise.

Windmills in general were used exclusively for grinding grain until the Dutch evolved a hollow post-mill for drainage purposes in the 1430s. Here, ingeniously, the shaft operating the pump passed through the centre of the main post. Windmills were put to another purpose, namely technological speculation. In 1326, the author of a book presented to King Edward II of England demonstrated the use of a post-mill for flinging beehives into a town under siege, while in 1335 an Italian sketched a kind of 'tank' driven by windpower. All over Latin Christendom, watermills and windmills generating five to ten horsepower dominated European technology down to the end of the eighteenth century.

MEDIEVAL MACHINES

Many medieval Europeans – men, women and children – were living in close proximity to machines by the late Middle Ages. They may have been more familiar with the workings of their machines than we are. On the other hand, peripheral parts of Europe used hand-mills down to the late nineteenth century, partly because of hard winters when neither type of power mill could function. With regard to waterpower, northern Europe had an advantage over the Mediterranean region, especially in summertime. On balance, the greatest medieval concentration of industrial activity was probably the Low Countries. The constant availability of water may help to explain the greater use of watermills than in Roman times. Machines were made essentially of wood, including the cogs, so that they were relatively quiet when in operation.

The best known medieval machine was ultimately the spinning wheel, first demonstrated at Speyer *c.*1280. A belt-driven spindle was turned very fast by the large, hand-operated wheel. Late in the fifteenth century, a U-shaped flyer driven at a different speed by a separate belt made it possible to wind the spun thread on to a bobbin. By that time the productivity of female labour may have increased dramatically, if not measurably.

THE INVENTION OF THE MECHANICAL CLOCK

In earlier parts of the Middle Ages, various techniques were tried out for measuring the passage of time. The commonest was the sundial; generally positioned on the south wall of churches, it had the obvious disadvantage of needing sunshine. A famous alternative for those who could afford it was to use candles of a given length or with a graduated scale, as apparently did King Alfred the Great of Wessex in the late ninth century.

Another technique involved the use of waterpower, as in the palace of King Louis IX of France *c.*1250. The principal part was a wheel fitted with fifteen metal cones, arranged equidistantly around its perimeter so that water ran out of one cone into the next every four minutes, thus making one revolution per hour. The axle was powered by a weight hanging from a cord wound round it, so that the weight of water acted as a brake on the rotation of the wheel. Apart from the trebuchet, this was the first machine to be driven by the force of gravity. It had no dial, but rang the hours, except of course when it froze up during harsh winters.

Because of the defects of water-clocks and sand-clocks, it was desirable to find a purely mechanical alternative. By the mid-thirteenth century this problem was perceived in terms of inventing a mechanical (metal) device to regulate the flow of energy derived from the fall of the weight, that is, an escapement system. A treatise written by Robert the Englishman in 1271 shows an awareness of the problem. A book written in the 1270s for King Alfonso X of Castile shows a weight-driven astronomical clock regulated by mercury leaking from a revolving drum.

Unfortunately, we do not know who solved the problem, or exactly when or where. Most experts place the solution in the last quarter of the thirteenth century, that is, at the same time as the invention of the gun. In fact, two solutions were found. One was an oscillating foliot bar north of the Alps, which remained the standard mechanism until the pendulum was invented in the 1650s. The first fairly certain example was operating at Strasbourg in 1352, an elaborate device with a calendar and astrolabe showing the movement of the sun, moon and planets while also ringing the hours. It incorporated a statue of the Virgin Mary, to whom the three Magi would bow at noon while a carillon played a tune, after which a cock on the top would open its beak and flap its wings. The other solution was an oscillating balance wheel invented in Italy, the first fairly certain example being recorded at Milan in 1335. By 1488, Ludovico Sforza, the Milanese despot, was commissioning three pendant watches to be hung around a person's neck.

The mid-fourteenth century saw mechanical clocks become suddenly popular. In towns especially, churches and town halls acquired public clocks built at great expense. Many had decorative mechanical features to impress and amuse the public, such as angels with trumpets and processions of royal figures. As far as Ireland is concerned, the first indication of a mechanical clock is a reference to a clockmaker called Walter le Gret living in Dublin's Patrick Street in the 1330s. He was obligated to the authorities at St Patrick's Cathedral and may have been commissioned to build a clock for the tower. If he succeeded, it was presumably destroyed when the tower collapsed in 1362. The first clear reference to a working clock in Dublin (and Ireland) comes in 1466, when a cutler named William Meyler was being paid 13*s*. 4*d*. a year to maintain the city's clock atop the tholsel opposite Christ Church Cathedral. As elsewhere, time-keeping in Dublin had been urbanized and laicized.

By far the most impressive of all medieval clocks was that built by Giovanni di Dondi and installed in the library of the Visconti castle at Pavia (pl. 20). Giovanni is stated to have done nothing else for sixteen years (1348–64), at the time of the initial outbreak of plague! The known details come from a long treatise written on its completion, provided with numerous diagrams, hence the possibility of modern reconstruction. This is essentially an astronomical clock (astrarium), with separate dials for the sun, moon and five planets. Hence the main structure is septagonal, the entire mechanism being powered by a single weight. The moon and mercury were given eliptical orbits, necessitating eliptical gears. The lower part of the machine had a 24-hour dial, another showing fixed ecclesiastical feasts, and an even more complex arrangement showing moveable feasts related to Easter. By any standard, this was an astounding feat of mechanical engineering, representing a challenge even now to the Smithsonian Institute in Washington.

THE INVENTION OF THE GUN

Human ingenuity has always been applied, regrettably, to warfare and weaponry. Virtually all cultures worldwide have engaged in the development of more efficient means of killing other humans. As far away as the North Island of New Zealand, Maori tribesmen who migrated there in the second half of the fourteenth century equipped themselves with ferocious weaponry. Medieval Europeans spent enormous amounts of time and resources on offensive and defensive weaponry. As today, military technology was in some respects in advance of civilian technology. One of the social elites, the *bellatores*, 'warriors' of a particular early medieval sociological construct, deployed vast resources. Even their favourite sports, such as hunting and falconry or hawking, were violent and the occasion of accidents, as in the famous case of King William

Rufus in 1100. Warfare came to be codified by laws of war and ritualized by a code of honour known as chivalry.

'Traditional' medieval warfare involved mainly hand-to-hand fighting, with the addition of missiles propelled by bows (crossbows and longbows). The longbow was revolutionary in that it killed or wounded men and horses at a fast and terrifying rate, whereas the earlier but much slower crossbow, though widely used, was rarely decisive. The longbow could be fired at ten to twelve arrows a minute, with an effective range of 200m, and it was simpler and cheaper to make. It was first used with decisive effect against Scottish schiltrons at Falkirk in 1298.

Before then, however, a substance had been invented with more decisive properties – gunpowder capable of explosive force. The mixture of saltpetre, sulphur and charcoal is first referred to in a letter-collection of Roger Bacon dedicated to the bishop of Paris in 1249. Bacon was discussing its scientific properties, but foresaw military application. Probably early in the fourteenth century, gunpowder was adapted to propelling missiles by another unknown inventor. Early hand-guns shot bolts or pellets and were attached to a wooden stock set in the ground. They were used during Edward III's first Scottish campaign in 1327, only a generation after the Battle of Falkirk. By the second half of the fourteenth century, hand-guns were used bound together in rows (ribauldiquins) that could be fired simultaneously, though they were slow to operate. In the early fifteenth century, a change to firing from the shoulder occurred. The invention of a trigger operating the match enabled the user to hold the gun with both hands. Herstmonceux Castle in Sussex, built of brick in the 1440s, was equipped with gun-ports for fear of a French invasion.

Hand-guns were first used with real effect by Hussites, for John Žižka saw the necessity of inventing new techniques. Coupled with waggon squares, offensive and tactical possibilites were exploited from 1421 onwards. Gunfire from waggons created noise and fear as well as causing casualties. Nevertheless, the waggon square had no future in Europe as a military technique, following the defeat of the Taborites by Bohemian forces at Lipany in 1434. Incidentally, the first recorded use of firearms in Ireland occurred in 1487 when they were used by troops of Aodh Ruadh Ó Domhnaill. The first Irishman to die from gunshot wounds was Brian, son of Cathal Ó Ruairc, at Glencar, Co. Leitrim, that same year.

Another major offensive development was that of the field cannon, first illustrated in an Oxford manuscript of 1327, which shows a 'quarrel' in the barrel. The earliest cannons comprised iron bars welded together around a hardwood core and surrounded by hoops, the core then being burned or bored out. An example in the Tower of London dated 1464 is made of brass, is 5.1m long with a bore of 63cm and weighs 18 tonnes. Early cannons were very dangerous to their users, causing the death of King James II of Scotland in 1460 at the siege of Roxburgh. Wheels began to be applied in the fifteenth century, which may have helped to absorb the recoil. Such weapons, firing stones for cheapness, played a part in French successes from 1369. On the other hand, at Aljubarrata in 1385 Castilians equipped with sixteen cannons were defeated by a smaller Portuguese army with no artillery. The first really dramatic success was achieved by Turks at Constantinople in 1453, prior to their final night-time attack. Cannons also played a part in the rapid ending of the Hundred Years War, the English commander at Castillon being killed by a cannonball.

MATHEMATICS AND SCIENTIFIC EXPERIMENTATION

The classical revival of the twelfth century concerned both Greek and Roman authors. As far as science is concerned, the key development was the translation of Aristotle's works into Latin. Many of these translations were made in Spain, especially at Toledo, via an Arabic intermediary, thereby introducing many inaccuracies. By the thirteenth century, western Europe had for the first time a body of 'scientific' literature, essentially on mathematics, astronomy and medicine.

In the early universities, the *quadrivium* comprised arithmetic, geometry, astronomy and music. In the late twelfth century, Paris replaced Chartres as the leading centre of higher education in northern Europe. The influence of Aristotle's thought eventually led to the condemnation of 219 propositions by Bishop Tempier of Paris and Archbishop Peckam of Canterbury in 1277. What were most strongly opposed were elements of Averroist determinism derived from Abū'l-Walīd ibn Rushd (1126–98), the Spanish Muslim philosopher. In 1278, circulation of works of Roger Bacon (*c.*1214–92) was restricted by the general of the Franciscan order. Thus, one of medieval Europe's most inventive and speculative minds was obstructed by ecclesiastical censure. This condemnation had important effects on the scientific world, for thinkers no longer believed that natural causes and laws could be ascertained with confidence. Ideas were thenceforth expressed tentatively in the form of hypotheses. The aim of science was not to solve real problems, but to satisfy intellectual curiosity about God and the universe.

Two of the most original thinkers in the thirteenth century gave mathematics a central place in their concept of scientific thought. One was Robert Grosseteste (*c.*1175–1253), a man of humble origins who became chancellor of Oxford University in 1224 and bishop of Lincoln in 1235. An essential asset was a competent knowledge of Greek, a rare accomplishment in northern Europe; accordingly he ordered Greek grammars for the university and teachers from Byzantium. Grosseteste regarded light as the prime element, which then generated the universe. The other thinker was Roger Bacon, who was also educated at Oxford and later at Paris. Like Grosseteste, he grasped the essence of the inductive method, by which a scientist eliminates hypotheses contradicted by the facts to discover the 'true' cause. He pointed out a defect in the Julian calendar that was first corrected only in the late sixteenth century.

A fundamental limitation was the continued recourse to roman numerals, which made even simple calculations difficult, hence the reliance on the abacus. In universities, arithmetic was very elementary and geometry based on the first three books of Euclid. The most advanced notion was that base angles of an isosceles triangle are equal. Hindu-Arabic numerals reached Europe gradually from the twelfth century onwards, but were not widely known until the thirteenth century. A key link was the *Liber abaci* of Leonardo Fibonacci, the son of a Pisan merchant, published in 1202. Yet their use spread only slowly outside Italian mercantile and intellectual circles. Even Leonardo da Vinci (1452–1519), who referred to himself as an engineer, was hampered by inadequate mathematics. His experiments on mechanical stress relied on wooden beams, though he also considered the loading strength of iron wire.

PHYSICS

The purpose of physics in the Middle Ages was to determine causes and the fundamental principles governing them. In general, medieval Europeans remained within the framework of Aristotle's theory of nature, though some remarkably accurate observations were made. One was by Dietrich of Freiberg, who in 1307 wrote a treatise called *De iride et radialibus impressionibus*. In this, he explained colours of a rainbow as a consequence of the refraction of sunlight into each water droplet, prior to reflection at the inner surface and refraction out again. His experimental apparatus consisted simply of a spherical glass vessel and a light-source. Another was by Thomas Bradwardine (d. 1349, of plague), archbishop of Canterbury. In an argument of fundamental importance later to Newtonian physics, he stated that objects of unequal weight fall at the same speed in a vacuum. Then there was Jean Buridan (d. 1359), who deduced correctly that the speed of an object in motion is a function of its mass, that is, that objects with greater mass travel farther under the same impetus. This may be compared with the discovery in the fifteenth century that small iron cannonballs were more effective than large stone ones.

Thus Galileo had several predecessors in the field of mechanics in the fourteenth and fifteenth centuries. But what were left as intellectual exercises were arranged by Galileo into a logical and ordered system. Medieval speculators and experimentalists, many of them churchmen, were content to remain within an Aristotelian-cum-Christian framework. They had inherited the Aristotelian concept of a fixed Earth, round which the sun, moon, planets and stars revolved, while the main practical application of astronomy was to determine the date of Easter. The heliocentric view of the solar system was not proposed until Nicolaus Copernicus' work was published in 1543, the year of his death. Like Martin Luther, he was a product of the late Middle Ages, studying mathematics and science at Kraków University, followed by astronomy at Bologna. What he demonstrated was less a new idea than a new attitude, searching for fundamental truths irrespective of their implications for religion, philosophy or even physics.

MEDICAL THEORY

The writer who dominated medieval medicine was Galen, a Greek from Asia Minor (AD131–201). Until Leonardo da Vinci, his works were generally accepted without question, including anatomical errors. Galen and other Greek writers were known in the West in the form of Latin translations of Arabic–Jewish translations from the Greek. Only from the late fifteenth century were texts retranslated directly into Latin more accurately. The earliest complete medical text from England, for example, is the *Compendium medicinae* of Gilbert Anglicus, written *c.*1240. He had studied at Salerno and the book contains a surprising amount on surgery as well as on a wide range of complaints such as bad breath, seasickness, sexual impotence and warts. A later standard treatise on surgery was the *Chirugia magna* by Guy de Chauliac, resident physician to the Avignon popes, written in 1363.

A major weakness in medieval theory was the lack of anatomical research, which was discouraged by the Church because of belief in the resurrection. In 1300, Pope Boniface VIII issued a bull prohibiting the boiling and dismembering of the corpses of dead crusaders, though this measure was aimed at relic worshippers. In a later, humanistic environment, Sixtus IV (1471–84) did allow dissections, provided that ecclesiastical permission was sought. At Montpellier, there was an anatomy lesson roughly every two years, using the bodies of criminals. Thus great ignorance remained, such as the notion of the stomach as a cauldron in which food was cooked by heat from the liver. In general, the Church advocated supernatural cures and so there was more emphasis on care than on cure. For example, leprosy was incurable and a special church service accompanied the committal of a leper, in which a priest gave the victim a bell or clapper and a dish for alms.

EUROPEAN SCIENCE AND TECHNOLOGY IN 1500

By the end of the Middle Ages, there had been no dramatic improvements in agrarian technology, with the result that the ox remained the standard plough-beast in most parts of Europe. Only because of depopulation did the diet of some people improve, especially the eating of more meat. According to certain scholars, this may have been of crucial importance in giving western Europeans more physical energy than was available in other world cultures, such as rice-based ones. It has been pointed out that the social status of butchers in some places rose, even to the extent of serving on town councils. The main developments had been in the fields of metallurgy, mining, shipping and warfare. By the later thirteenth century, western Europeans had invented the idea and practice of invention: examples are the search for an escapement mechanism for clocks and for a perpetual motion machine.

The mathematical, medical and general scientific knowledge of western Europeans was still relatively backward. Where they scored was in their capacity to manufacture useful things, represented most practically by their clocks, their guns and their ocean-going ships. Along with the ships themselves were navigational skills honed by sailors over many centuries, ships being differentiated as early as the Viking Age. In the fourteenth and fifteenth centuries, the capacity of western Europeans to voyage farther and farther was enhanced by the making of portolan charts. Such charts could be used for repeated voyages to the same places and then beyond them; sailors could literally chart a course at sea.

Accordingly, *c.*1500 western Europeans were poised for world domination more thanks to the abilities of their craftworkers than to those of their intellectuals. Western Europeans, and in mining contexts central Europeans, were on the road to the replacement of human labour by machines. At the 2012 science convention held in Dublin, an American panellist, Brian Christian, described how, since the time of Greek philosophers, humans have drawn comparisons of their capabilities against those of animals. Now, he observed, they are much more likely to measure themselves against machines. By the end of the Middle Ages, western Europeans were indeed on that road; we know with hindsight, however, how much farther there was to go.

EPILOGUE

As a postcript I should like to make two final points about inventiveness in the Middle Ages. First, this collection of essays contains a good deal about scientific enquiry and rationality and about applied mathematical thought in early medieval Ireland. There is no doubt that sophisticated mathematical calculations were being made throughout the island of Ireland. Large numbers of churchmen were involved and over a considerable period of time. But there was a definite reason that this was so – the need to calculate the date of Easter; mathematical speculation was responding to a perceived requirement. Once the Easter controversy had been settled, however, such speculation ceased, although Irish scholars' interest in the most innovative mathematics of the continent of

Europe (ultimately through the work of Gerbert of Aurillac) may be seen in textbooks in use in the school at Glendalough in the early twelfth century.

Secondly, this phenomenon is of a piece with other scattered details about those early centuries. The Utrecht Psalter of *c*.830 shows a crank-driven grindstone on the side of the godly, in contrast to a conventional whetstone on the side of the evildoers. In the first decade of the eleventh century, a monk named Eilmer at Malmesbury Abbey in Wiltshire experimented with the possibility of human flight. Having ascended the tower, he strapped rigid wings on to his arms and legs and took off. He travelled about 200m before crash-landing, breaking both legs and concluding that he lacked an adequate tail-plane unit. When an Italian tried a similar experiment in 1507 at Stirling Castle in central Scotland, he also broke a leg; no significant progress was possible without a portable power source. My question is: did greater church control and stultified university curricula inhibit this early medieval spirit of free enquiry?

ACKNOWLEDGMENT

I wish to thank Jennifer Moore of the Royal Irish Academy's Irish Historic Towns Atlas project for invaluable technical assistance with regard to the image of the astrarium.

FURTHER READING

Bartlett, Robert, *The natural and the supernatural in the Middle Ages* (Cambridge, 2008).

Biller, Peter, and Joseph Ziegler (eds), *Religion and medicine in the Middle Ages* (Woodbridge, 2001).

Bloch, Marc, 'Technical change as a problem of collective psychology' in *Land and work in mediaeval Europe*, trans. J.E. Anderson (London, 1967), pp 124–35.

Cipolla, Carlo Maria, *Before the industrial revolution: European society and economy, 1000–1700* (3rd ed., London & New York, 1993).

Cipolla, Carlo Maria, *Clocks and culture, 1300–1700* (repr. New York, 2003).

Clagett, Marshall, *The science of mechanics in the Middle Ages* (Madison, WI, 1959).

Clagett, Marshall, *Studies in medieval physics and mathematics* (London, 1979).

Crombie, Alistair Cameron, *Augustine to Galileo* (2 vols, 2nd ed., London, 1969).

Dales, Richard C., *The scientific achievement of the Middle Ages* (Philadelphia, 1973).

Daumas, Maurice (ed.), *A history of technology and invention: progress through the ages*, 1: *The origins of technological civilization to 1450*, trans. E.B. Hennessy (London, 1980).

Derry, Thomas Kingston, and Trevor Illtyd Williams, *A short history of technology: from the earliest times to AD1900* (Oxford, 1960).

Gimpel, Jean, *The medieval machine: the industrial revolution of the Middle Ages* (2nd ed., London, 1992).

Glick, Thomas F., Steven John Liversey, and Faith Wallis (eds), *Medieval science, technology and medicine: an encyclopedia* (London, 2005).
Grant, Edward (ed.), *A source book in medieval science* (2 vols, Cambridge, MA, 1974).
Grant, Edward (ed.), *Studies in medieval science and natural philosophy* (London, 1981).
Grant, Edward (ed.), *The foundations of modern science in the Middle Ages: their religious, institutional and intellectual contexts* (Cambridge, 1996).
Grant, Edward, and John Emery Murdoch (eds), *Mathematics and its applications to science and natural philosophy in the Middle Ages: essays in honour of Marshall Clagett* (Cambridge, 1987).
Hackett, Jeremiah (ed.), *Roger Bacon and the sciences: commemorative essays* (Leiden & New York, 1997).
Hall, Bert S., and Delno C. West (eds), *On pre-modern technology and science: a volume of studies in honor of Lynn White, jr* (Malibu, CA, 1976).
Hannam, James, *God's philosophers: how the medieval world laid the foundations of modern science* (London, 2009).
Haskins, Charles Homer, *Studies in the history of mediaeval science* (New York, 1927).
Hilton, Rodney Howard, and Peter Hayes Sawyer, 'Technical determinism: the stirrup and the plough', *Past & Present*, 24 (1963), 90–100.
Kibre, Pearl, *Studies in medieval science: alchemy, astrology, mathematics and medicine* (London, 1984).
Lilley, Samuel, *Men, machines and history: a short history of tools and machines in relation to social progress* (2nd ed., London, 1965).
Lindberg, David C., *The beginnings of western science: the European scientific tradition in philosophical, religious and institutional context, prehistory to AD1450* (2nd ed., Chicago & London, 2007).
Long, Pamela Olivia (ed.), *Science and technology in medieval society* (New York, 1985).
Mann, Nicholas, and Birger Munk Olsen (eds), *Medieval and Renaissance scholarship: proceedings of the second European Science Foundation workshop on the classical tradition in the Middle Ages and the Renaissance*, London, Warburg Institute, 27–8 November 1992 (Leiden, 1997).
Moody, Ernest Addison, *Studies in medieval philosophy, science and logic: collected papers, 1933–1969* (Berkeley, CA, 1975).
Mumford, Lewis, *The myth of the machine: technics and human development* (New York, 1967).
Murdoch, John Emery, and Edith Dudley Sylla (eds), *The cultural context of medieval learning: proceedings of the first International Colloquium on Philosophy, Science and Theology in the Middle Ages – September 1973* (Dordrecht, 1975).
Pines, Shlomo, *Studies in Arabic versions of Greek texts and in mediaeval science* (Jerusalem & Leiden, 1986).
Singer, Charles Joseph et al. (eds), *A history of technology*, 2: *The Mediterranean civilizations and the Middle Ages*, c.*700BC to AD1500* (Oxford, 1956).
Thorndike, Lynn, 'Invention of the mechanical clock about 1271AD', *Speculum*, 16 (1941), 242–3.
Thorndike, Lynn, *Science and thought in the fifteenth century: studies in the history of medicine and surgery, natural and mathematical science, philosophy and politics* (New York, 1963).

White, Lynn Townsend, jr, *Medieval technology and social change* (Oxford, 1962).
White, Lynn Townsend, jr, 'The expansion of technology, 500–1500' in C.M. Cipolla (ed.), *The Fontana economic history of Europe*, 1: *The Middle Ages* (London, 1972), pp 143–74.
White, Lynn Townsend, jr, 'The medieval roots of modern technology and science' in idem, *Medieval religion and technology: collected essays* (Berkeley, CA, 1986), pp 75–91.
Unger, Richard W., *The art of medieval technology: images of Noah the shipbuilder* (New Brunswick, NJ, 1991).
Wiet, Gaston et al., *History of mankind: cultural and scientific development*, 3: *The great medieval civilizations* (London, 1975).
Wilson, Curtis, *William Heytesbury: medieval logic and the rise of mathematical physics* (Madison, WI, 1956).
Wright, John Kirtland, *The geographical lore of the time of the crusades: a study in the history of medieval science and tradition in western Europe* (rev. ed., New York, 1965).
Wu, Nancy Yee Sin, *Ad quadratum: the practical application of geometry in medieval architecture* (Aldershot, 2002).

Music *and* the stars – the model for the present!

ANTHONY HARVEY

Readers of this volume, like those who participated in the two days of the conference that it reports, will probably be increasingly aware of a tension between two essentially opposed views that underlie the substance of the contributions. One is that the biblical, Christian background to the western Middle Ages was a help, if not a *sine qua non*, to scientific progress in the period. The other view is that this background actually constituted a constraint or impediment to such progress. I wish to suggest that the tension arises because, between the essays in the book, the concept of science is understood in no fewer than four different, not necessarily compatible modern senses.

One contributor evaluates scientific progress in terms of *technological advances* in medieval Ireland. Others judge a medieval worldview as 'scientific' or not, depending on the extent to which it arrived at *conclusions that we now believe to be true*. For example, the realization that the tides are related to the phase of the moon would on this reading be considered scientific, whereas the notion that at high tide the sea filled up from the 'great deep' below it would not be. There is nothing to do with technology as a criterion here. A third approach calls medieval thought scientific if it has been *logical*; that is, if it has proceeded rationally from what were considered at the time to be given data. And the fourth criterion to be applied measures the degree of science in medieval thinking by the amount of *original, empirical observation* it did or did not contain.

The contradictions set up by these varying approaches include the following. On the third criterion, that of logicality, medieval computistics was very scientific (since the seventh-century Irish not only were the first to analyse how the 19-year cycle worked, but had even been able to begin by deconstructing the distortions in the transmission of the tradition as they had received it); but, in view of its preference for tradition over something as obvious as the visible phases of the moon, computistics would score very low by the fourth criterion, as it set little store by original observation. Again, on outlook number two, the 'great deep' explanation for the tides is not at all scientific, since it began from a statement in the book of Genesis; but on outlook number three it is entirely so, since it was logically based upon that statement, which was taken as a legitimate piece of data to start from.

Before prescribing the medieval approach to science as one that should be commended to policy-makers today, we therefore need to think more, and clearly, about what we mean by it. Something valuable has indeed been lost between those days and these; but what is it? At a more over-arching level than the implicit disagreements about what medieval science consisted of, however, a clear and urgent unanimity is discernible throughout this volume on that question. It is the realization that, in the Middle Ages, it was still recognized that knowledge is the means by which humanity gains access to what is a unified field of truth. To draw a fundamental distinction between science and humanities is therefore out of line with this recognition. In that case, it is not merely important to bridge the chasm between the two; rather, it should be denied that any such chasm should really be seen as existing in the first place.

Unfortunately, modern English usage, in particular, is unhelpful in this regard. The Royal Irish Academy was established by royal charter in 1785 on sound Enlightenment principles to investigate … what? It is said that, shamefully few years ago, an engineering member of the Academy asked why the organization was involved in running research projects in the humanities when its sister institutions in other countries tended to bear the title Academy of Sciences. It had to be pointed out to him that the English language is more or less alone in interpreting the word *science* as referring solely to the empirical investigation of nature; its counterparts *la science* in French, or *Wissenschaft* in German, or the Slavic *nauka*, as in the names of those organizations, all cover the humanities as well. It is a delight to me, as a professional dictionary editor, to know that my counterpart in Paris works in a *laboratoire*, and that the assistant editors on the corresponding project there have the title of *ingénieur*. Fortunately, when our own Academy began in the eighteenth century, the scholarly air of Ireland was still free from the artificial distinction between (what are perceived as) the two types of learning; its founding principles were and still are those of the Renaissance – and, we can now see, of much longer ago than that – in recognizing that the one cannot flourish without the other, and in seeing the need to foster both. However they may disagree even in substance, the essays in this volume are at one in pointing to the need for modern scholarship and its promoters to regain and reassert this essential insight.

Glossary

agrimensores	landsurveyors, noted for their use of roman fractions.
argumentum, pl. *a*	a formula for computistical calculation, evidence, support, proof.
bissextile	*bissextilis* < *bis* 'two' + *sextus*, 'sixth'. In leap years, an intercalary day was added – originally on the sixth calends of March, that is 24 February. Therefore it was the second 'sixth' day.
brachiopod	marine animal with hard shell above and below, hinged at rear end.
calends	< Lat. *kalendae*: first day of each month in Roman calendar.
Chalcidius	fourth-century philosopher who translated the first part of Plato's *Timaeus* into Latin and made a commentary on it.
computus	'computation' – science concerning the calculation of the date of Easter.
concurrentes	weekday of 24 March.
De nuptiis Philologiae et Mercurii	'On the marriage of Philology and Mercury': an allegory in prose and verse by fifth-century Martianus Capella. It defined the standard of academic learning during the early Middle Ages.
embolism	< Latin *embolismus* < Greek *embolismos* < *emballein* 'insert': used particularly for the intercalation of a month.
epacts	< Latin *epactae* < Greek *epaktai hèmerai*, 'added days': the age of the moon; i.e., the day of the lunar month on any given year in the solar calendar, important for computistics are 22 March (Dionysius Exiguus' 19-year cycle) and 1 January (84-year cycle and Tables of Victorius of Aquitaine).
exegesis	(seventeenth century) < Greek *exegesis* < *exegeisthai* 'to interpret', specifically bible studies.

feria, pl. *ae*	day of the week of any given date.
foliot bar	horizontal bar with weights on each end as part of a primitive balance wheel < 'decoy bird', ducks' heads and waddling motion.
hendecas	< Greek *hendeka*, 'eleven', the final eleven years of the 19-year lunar cycle.
Hindu-Arabic numerals	decimal numeral system developed by Indian mathematicians (first–fourth centuries) adopted by Arabic mathematicians in ninth century and spread to Western world by high Middle Ages.
indiction	< Latin *indicere*. Any year in 15-year cycle (fiscal period). Used in dating since its inception by Constantine in AD313.
intercalation	the insertion of an extra day, week or month to a year to adjust a calendar to the sun or moon.
lambda diagrams	graphical notation for closed lambda terms, in which abstractions (lambdas) are represented by horizontal lines, variables by vertical lines emanating down from their binding lambda (< eleventh letter of Greek alphabet).
latercus	the 84-year Insular Paschal cycle. The name derives from *laterculus*, diminutive form of the Latin *later, -eris*, brick, tile, used to describe a mathematical table divided into cells. Insular computists incorrectly understood *laterculus* to be the diminutive form of *latercus*. Also *latercus* as < *latens cultus* 'hidden knowledge' / 'secret education', an etymology attributed to Isidore in the sixth-century Irish computistical tract *De ratione conputandi*.
ledones tides	< Greek 'dead waters', weak tides / low tides.
luni-solar calendar	a lunar calendar adjusted to coincide with the solar seasons by the intercalation of lunar months.
Macrobius	Macrobius Ambrosius Theodosius, fifth-century author of *Commentarii in somnium Scipionis* ('Commentary on the Dream of

	Scipio'). The dream was narrated by Cicero at the end of his *Republic*. The discussion is from a Neo-Platonic point of view and was an important source of Platonism in the medieval Latin West.
malinae tides	< Greek 'living waters', strong tides / high tides.
Martianus Capella	author of *De nuptiis* (above). Active during the fifth century and seems to have practiced as a jurist at Carthage. His work (like that of Macrobius) introduced Neo-Platonism to the medieval West.
metrum	'metre'. The basic rhythmic structure of a verse.
ogdoas	< Greek *ogdoas* 'eightfold', the first eight years of the 19-year lunar cycle.
ontogeny	< Greek *ontos* 'to be' + *geny* 'to be born', the history of structural change in a unity.
portolan charts	< Italian *portolano* < *porto* 'harbour'. Origin in thirteenth century. Books of sailing directions and charts with descriptions of harbours and ports.
prognostica	< Greek *prognostikos* 'advanced indication of future events'.
quarrel	< French *carré* 'square', short squared-headed arrow for use in crossbow or bolt.
quartodecimanism	< Vulgar Latin *quarta decima* 'fourteen'. The *Quartodecimani* were so called because they ended the pre-Paschal fast and celebrated the feast on the 14th day of the Jewish lunar month of Nisan.
ribauldiquins	primitive machine gun with a number of barrels.
saltus lunae	'leap of the moon'. The omission of a lunar day at the end of a 19-year lunar cycle which, following the intercalation of 7 lunar months throughout the cycle, allows the end of the lunar and solar years to coincide.
schiltrons	Old English *scildtruma*, *sceldtroman*, *scheltron* < 'shield troup', 'shield wall'. A defensive formation of up to 2,000 men wielding

	outward-pointing pikes or polearms against cavalry attacks.
sesquialter	< Latin *sesqui* 'half' + *alter* 'other': ratio of 3:2 or as in an interval of a fifth; a term found in Boethius' *De institutione arithmetica*, the basis for medieval mathematics: in this instance *sesqui-* is a number which in relation to a smaller number contains that number + a fractional part of it, e.g., *sesquialter* is the ratio of 3:2 where 3 contains 2 + one half of 2.
sesquioctave	the ratio of 9:8, where 9 contains 8 + one eighth of 8.
sesquitertial	the ratio of 4:3, where 4 contains 3 + one third of 3.
trebuchet	< Middle English < French *trebucher* 'overthrow'. A siege engine (sling) for throwing large stones or objects at enemy.
volvelle	< Latin volvere 'to roll', 'turn'. A wheel chart, usually of paper with rotating parts.

Bibliography

PRIMARY SOURCES

Alberto, P.F. (ed.), *Eugenius Toletanus opera omnia* (Turnhout, 2005).

Barney, S.A., W.J. Lewis, J.A. Beach, and O. Berqhof (trans.), *The etymologies of Isidore of Seville* (Cambridge, 2006).

Benson, L. (ed.), *The Riverside Chaucer* (3rd ed., Boston, 1987).

Bieler, L. (ed.), *Anicii Manlii Seuerini Boethii Philosophiae consolatio* (Turnhout, 1957).

Borst, A. (ed.), *De quatuor quaestionibus compoti* (unpublished).

Bubnov, N. (ed.), *Gerberti, postea Silvestri II papae, opera mathematica (972–1003)* (Berlin, 1899).

Burnett, C., 'H. Ocreatus, *Algorismi vel helcep decentior est diligentia*: the arithmetic of Adelard of Bath and his circle' in Folkerts, *Mathematische Probleme*, pp 221–31.

Calder, G., *Auraicept na n-éces: the scholars' primer* (Edinburgh, 1917; repr. Dublin, 1995).

Díaz y Díaz, M. (ed.), *Liber de ordine creaturarum: un anónimo irlandés del siglo VII*, Monografias de la Universidad de Santiago de Compostela 10 (Santiago de Compostela, 1972).

Dyson, R.W. (trans.), *Augustine: The City of God against the pagans* (Cambridge, 1998).

Ehwald, R. (ed.), *Aldhelmi opera omnia* (Berlin, 1919; repr. 1961).

Flint, V.I.J., 'Honorius Augustodunensis Imago mundi', *Archives d'histoire doctrinale et littéraire du moyen âge*, 49 (1982), 7–153.

Friedlein, G. (ed.), *Anicii Manlii Torquati Severini Boetii De institutione arithmetica libri duo, De institutione musica libri quinque. Accedit geometria quae fertur Boetii* (Leipzig, 1867).

Goméz Pallarés, J., 'Hacia una nueva edición de los argumenta paschalia de Dionisio el Exiguo', *Hispania Sacra*, 49 (1994), 13–31.

Herren, M. (ed.), *Hisperica Famina* (2 vols, Toronto, 1974, 1987).

Hervieux, L. (ed.), *Les fabulistes latins depuis le siècle d'Auguste jusqu'à la fin du moyen age* (5 vols, Paris, 1893–9).

Q. Julius Hilarianus, *Expositum de die paschae et mensis*, *PL* 13, cols 1105–14 (Paris, 1845).

Hofman, R. (ed.), *The Sankt Gall Priscian commentary* (2 vols, Münster, 1996).

Jan, W., *Historia cycli Dionysiani cum argumentis paschalibus et aliis eo spectantibus* (Wittenberg, 1718).

Jones, C.W. (ed.), *Libri quatuor in principium Genesis usque ad natiuitatem Isaac et eiectionem Ismahelis adnotationum*, *Bedae Venerabilis opera*, *Pars I. Opera exegetica* (Turnhout, 1967).

Jones, C.W. (ed.), *Bedae opera de temporibus* (Cambridge, MA, 1943).

Jones, C.W. (ed.), *Beda Venerabilis. Opera didascalica. 2. De temporum ratione* (Turnhout, 1977).

Kautzsch, E. (ed.), *Gesenius' Hebrew grammar*, [2nd Eng. ed., A.E. Cowley] (Oxford, 1910; repr. 1985).

King, J., and P. Tax (eds), *Die Werke Notkers des Deutschen*, 7 (Tübingen, 1996), 7a (Tübingen, 2003).

Krusch, B. (ed.), *Ionae Vitae sanctorum Columbani, Vedastis, Iohannis* (Hannover, 1905).

Lapidge, M., and M. Herren (trans.), *Aldhelm: the prose works* (Cambridge, 2009).

Leone, C. (ed.), *Pietro Alfonsi: Disciplina clericalis. Sapienza orientale e scuola delle novelle* (Rome, 2010).

Liuzza, R.M., *Anglo-Saxon prognostics: an edition and translation of texts from London, British Library, MS Cotton Tiberius A.iii* (Cambridge, 2011).

Mac Airt, S., and G. Mac Niocaill (eds), *The Annals of Ulster (to AD 1131)*, pt I: text and trans. (Dublin, 1983).

McDonough, C. (ed.), Alexander Neckam: *Sacerdos ad altare*, CCCM, 227 (Leiden, 2010).

MacGinty, F.P. [= Gerard], 'The treatise *De mirabilibus sacrae scripturae*' (PhD, UCD, 1971).

Moreton, J. (ed.), *Computus ecclesiasticus* (unpublished typescript, no date).

Morin, G., B. Capelle and J. Fraipont (eds), *Hieronymus: tractatus sive homiliae in psalmos. In Marci evangelium. Alia varia argumenta* (Turnhout, 1958).

Nauta, L. (ed.), *Guillelmi de Conchis glosae super Boetium* (Turnhout, 1997).

Oosthout, H., and J. Schilling (eds), *Anicii Manlii Severini Boethii De arithmetica* (Turnhout, 1999).

Peden, A., *Abbo of Fleury and Ramsey: commentary on the Calculus of Victorius of Aquitaine* (Oxford, 2003).

Plummer, C. (ed.), *Baedae opera historica* (Oxford, 1896; repr. 1969).

Power, M., *An Irish astronomical tract based in part on a mediaeval Latin version of a work by Messahalah* (London, [1912] 1914).

Prosper of Aquitaine, 'Epitoma Chronicon' in Th. Mommsen (ed.), *Chronica minora saec. IV, V, VI, VII* (Berlin, 1892), pp 343–499.

Schenkl, K. (ed.), *Sancti Ambrosii opera*, 3 vols, CCEL (Leipzig, 1897–1902).

Sharpe, R. (trans.), *Life of Saint Columba* (London, 1995).

Silk, E.T., *Saeculi noni auctoris in Boetii Consolationem philosophiae commentarius* (Rome, 1935).

Smyth, M., 'The seventh-century Hiberno-Latin treatise *Liber de ordine creaturarum*: a translation', *Journal of Medieval Latin*, 21 (2011), 137–222.

Steele, R. (ed.), *Opera hactenus inedita Rogeri Baconi*, 6 (Oxford, 1926).

Strecker, K. (ed.), *Poetae Latini aevi Carolini* (Berlin, [1896], [1923] 1964).

Taylor SJ, J.H., *St Augustine: the literal meaning of Genesis* (New York, 1982).

Tertullian, *De pallio*, *PL*, 2, cols 1032–4.

Walker, G. (ed.), *Sancti Columbani opera* (Dublin, 1957, repr. 1971).

Wallis, F., *Bede: the reckoning of time (translated, with introduction, notes and commentary)* (Liverpool, 1999).

Walsh, M., and D. Ó Cróinín (eds), *Cummian's letter 'De controversia paschali' and the 'De ratione conputandi'* (Toronto, 1988).

Walther, H. (ed.), *Initia carminum ac versuum medii aeui posterioris latinorum* (Göttingen, 1959).

Wasserschleben, H. (ed.), *Die irische Kanonensammlung* (Giessen, 1874; Leipzig, 1885; repr. Aalen, 1966).

Wilcox, J. (ed.), *Costa Ben Luca, De differentia spiritus et animae* (in preparation).
Wulff, W. (ed.), *Rosa Anglica; seu, Rosa medicinae Johannis Anglici: an early modern Irish translation of a section of the mediaeval medical text-book of John of Gaddesden* (London, [1923] 1929).
Zangemeister, C. (ed.), *Pauli Orosii Historiarum adversum paganos libri VII* (Vienna, 1882).
Zycha, J., *De Genesi ad litteram: 'On the literal meaning of Genesis'* (Vienna, 1894).

SECONDARY SOURCES

Ahlqvist, A., 'Review: *The Insular Latin grammarians*, by Vivien Law, Woodbridge 1982', *CMCS*, 6 (winter 1983), 100–1.
Beeson, C.H., *Lupus of Ferrières as scribe and text critic: a study of his autograph copy of Cicero's 'De oratore' with a facsimile of the manuscript* (Cambridge, MA, 1930).
Bergmann, W., 'Dicuils Osterfestalgorithmus im Liber de astronomia' in Warntjes and Ó Cróinín (eds), *The Easter controversy of late antiquity*, pp 242–87.
Bernhard, M. (ed.), *Die Thomas von Aquin zugeschriebenen Musiktraktate* (Munich, 2006).
Bernt, G., F. Rädle and G. Silagi (eds), *Tradition und Wertung: Festschrift für Franz Brunhölzl zum 65. Geburtstag* (Sigmaringen, 1989).
Bieler, L., 'The hymn of St Secundinus', *PRIA*, 55C (1952/3), 117–27.
Bieler, L., and Bernhard Bischoff, 'Fragmente zweier frühmittelalterlicher Schulbücher aus Glendalough', *Celtica*, 3 (1956), 211–20.
Birkhan, H. (ed.), *Kelten Einfälle an der Donau. Akten des vierten Symposiums deutschsprachiger Keltologinnen und Keltolegen Linz/Donau, 17–21 Juli 2003* (Vienna, 2007).
Bischoff, B., 'Die europäische Verbreitung der Werke Isidors von Sevilla' in idem, *Mittelalterliche Studien*, pp 171–94.
Bischoff, B., 'Wendepunkte in der Geschichte der lateinischen Exegese im Frühmittelalter', *Sacris Erudiri*, 6 (1954), 189–279.
Bischoff, B., *Mittelalterliche Studien: Ausgewälte Aufsätze zur Schriftkunde und Literaturgeschichte*, 3 vols (Stuttgart, 1966, 1967, 1981).
Blackburn, B., and L. Holford-Strevens, *The Oxford companion to the year* (Oxford, 1999).
Böhne, W., 'Beginn und Dauer der römischen Fastenzeit im sechsten Jahrhundert', *Zeitschrift für Kirchengeschichte*, 77 (1966), 224–37.
Borst, A., 'Ein Forschungsbericht Hermanns des Lahmen', *Deutsches Archiv für Erforschung des Mittelalters*, 40 (1984), 379–477.
Borst, A., *Das Buch der Naturgeschichte: Plinius und seine Leser im Zeitalter des Pergaments*, Abhandlungen der Heidelberger Akademie der Wissenschaften, philosophisch-historische Klasse, Jahrgang 1994, 2. Abhandlung (Hannover, 1995).
Borst, A., *Die karolingische Kalenderreform*, *MGH*, Schriften 46 (Hannover, 1998).
Borst, A., *Ein Forschungsbericht Hermanns des Lahmen* (Böhlau, 1984).
Borst, A., *Schriften zur Komputistik im Frankenreich von 721 bis 818*, 3 vols, *MGH*, *Quellen zur Geistesgeschichte des Mittelalters* 21 (Hannover, 2006).

Bouché-Leclercq, A., *L'Astrologie grècque* (Paris, 1899).

Bracken, D., 'Rationalism and the Bible in seventh-century Ireland' in Luiselli Fadda and Ó Carragáin, *Le isole britanniche*, pp 129–70.

Bracken, D., 'Virgil the Grammarian and Bede: a preliminary study', *Anglo-Saxon England*, 35 (2006), 7–21.

Breatnach, P.A., 'Review: *The Insular Latin grammarians*, Vivien Law', *Celtica*, 16 (1984), 182–6.

Brown, G.H., and L.E. Voigts (eds), *The study of medieval manuscripts of England: Festschrift in honor of Richard W. Pfaff* (Tempe, AZ, 2010).

Burnett, C., 'Learning Indian arithmetic in the early thirteenth century' in idem, *Numerals and arithmetic* (article IX).

Burnett, C., 'Learning Indian arithmetic in the early thirteenth century', *Boletín de la Asociación Matemática Venezolana*, 9 (2002), 15–26.

Burnett, C., 'The Eadwine Psalter and the western tradition of the onomancy in Pseudo-Aristotle's *Secret of Secrets*', *Archives d'histoire doctrinale et littéraire du moyen âge*, 55 (1988), 143–67.

Burnett, C., 'The Eadwine Psalter and the western tradition of the onomancy in Pseudo-Aristotle's *Secret of Secrets*' in idem, *Magic and divination*, pp 154–5.

Burnett, C., *Arabic into Latin in the Middle Ages: the translators and their intellectual and social context* (Farnham, 2009).

Burnett, C., *Magic and divination in the Middle Ages* (Aldershot, 1996).

Burnett, C., *Numerals and arithmetic in the Middle Ages* (Farnham, 2010).

Butzer, P.L., and D. Lohrmann (eds), *Science in Western and Eastern civilization in Carolingian times* (Basel, 1993).

Byrne, F.J., *A thousand years of Irish script: an exhibition of Irish manuscripts in Oxford libraries arranged by Francis John Byrne* (Oxford, 1979), pp 14–15.

Byrne, F.J., 'Seventh-century documents', *Irish Ecclesiastical Record*, 108 (1967), 164–82.

Cahill, T., *How the Irish saved civilization* (New York, 1995).

Carey, J., 'The obscurantists and the sea-monster: reflexions on the Hisperica Famina', *Peritia*, 17–18 (2003–4), 40–60.

Carey, J., *A single ray of the sun: religious speculation in early Ireland* (Aberystwyth, 1999).

Carey, J., *King of mysteries: early Irish religious writings* (Dublin, 2000).

Carney, J., *The problem of Saint Patrick* (Dublin, 1961).

Chardonnens, L.S., *Anglo-Saxon prognostics, 900–1100: study and texts* (Leiden, 2007).

Charles-Edwards, T., *Early Christian Ireland* (Cambridge, 2000).

Chenu, M.-D., 'Platon à Cîteaux', *Archives d'histoire doctrinale et littéraire du moyen âge*, 21 (1954), 99–106.

Clark, S., and E.M. Leach, *Citation and authority in medieval and Renaissance musical culture* (Woodbridge, 2005).

Clarke, H., and M. Brennan (eds), *Columbanus and Merovingian monasticism* (Oxford, 1984).

Connolly, S., and J.-M. Picard, 'Cogitosus's "Life of Brigit" content and value', *JRSAI*, 117 (1987), 5–27.

Cordoliani, A., 'Le comput de Dicuil', *Cahiers de civilisation médiévale*, 4 (1960), 325–37.

Cordoliani, A., 'Le computiste Hermann de Reichenau', *Miscellanea Storica Ligure*, 3 (1963), 165–90.

Corning, C., *The Celtic and the Roman traditions: conflict and consensus in the early medieval church* (New York, 2006).

Corthals, J., 'Stimme, Atem und Dichtung: aus einem altirischen Lehrbuch für die Dichterschüler (*Uraicept na Mac Sésa*)' in Birkhan, *Kelten Einfälle*, pp 127–47.

Courcelle, P., *La Consolation de philosophie dans la tradition littéraire: antécédents et postérité de Boèce* (Paris, 1967).

Cunningham, B., and S. Fitzpatrick (eds), *Treasures of the Royal Irish Academy Library* (Dublin, 2009).

de Paor, A., 'The Life and work of Robert MacAdam of Belfast, 1808–95' in Ní Chatháin et al., *Pathfinders to the past*, pp 72–85.

Dekkers, E., and J. Fraipont (eds), *Enarrationes in Psalmos CI–CL* (Turnhout, 1990 [1956]).

Doherty, C., L. Doran and M. Kelly (eds), *Glendalough: City of God* (Dublin, 2011).

Draak, M., 'A Leyden Boethius-fragment with Old-Irish glosses', *Mededelingen der Koninklijke Nederlandsche Akademie van Wetenschappen*, n.s. 11:3 (1948), 115–27.

Duemmler, E. (ed.), *Epistolae Karolini aevi II* (Berlin, 1895).

Duhem, P., *Le système du monde: histoire des doctrines cosmologiques, de Platon à Copernic* (Paris, 1913–59).

Eastwood, B.S., *Ordering the heavens: Roman astronomy and cosmology in the Carolingian Renaissance* (Leiden, 2007).

Eastwood, B.S., and Gerd Grasshoff, *Planetary diagrams for Roman astronomy in medieval Europe, c.800–1500* (Philadelphia, 2004).

Esposito, M., 'An unpublished astronomical treatise by the Irish monk Dicuil', *PRIA*, 26C (1907), 387–446.

Esposito, M., 'An unpublished astronomical treatise by the Irish monk Dicuil' in Lapidge, *Irish books and learning*, ch. 7.

Feehan, J., 'The heritage of the rocks' in Foster and Chesney, *Nature in Ireland*, pp 3–22.

Finan, T., and V. Twomey (eds), *Scriptural interpretation in the Fathers: letter and spirit* (Dublin, 2008).

Flower, R., 'Quidam Scotigena .i. discipulus Boëthii', *Ériu*, 8 (1916), 150–4.

Folkerts, M. (ed.), *Essays on early medieval mathematics: the Latin tradition* (Aldershot, 2003).

Folkerts, M. (ed.), *Mathematische Probleme im Mittelalter: Der lateinische und arabische Sprachbereich* (Wiesbaden, 1996).

Folkerts, M., *'Boethius' Geometri II: Ein mathematisches Lehrbuch des Mittelalters* (Wiesbaden, 1970).

Folkerts, M., 'Die Altercatio in der Geometrie I des Pseudo-Boethius. Ein Beitrag zur Geometrie im mittelalterlichen Quadrivium' in Keil et al., *Fachprosa-Studien*, pp 84–114.

Folkerts, M., 'The importance of the pseudo-Boethian geometria during the Middle Ages' in Masi, *Boethius and the liberal arts*, pp 187–209.

Foster, J.W., and H.C.G. Chesney (eds), *Nature in Ireland: a scientific and cultural history* (Dublin, 1997).

Fotheringham, J.K. (ed.), *The Bodleian manuscript of Jerome's version of the Chronicles of Eusebius* (Oxford, 1905).

Germann, N., *De temporum ratione: Quadrivium und Gotteserkenntnis am Beispiel Abbos von Fleury und Hermanns von Reichenau* (Leiden, 2006).

Gneuss, H., 'A preliminary list of manuscripts written or owned in England up to 1100', *Anglo-Saxon England*, 9 (1981), 1–60.

Gneuss, H., *Handlist of Anglo-Saxon manuscripts: a list of manuscripts and manuscript fragments written or owned in England up to 1100* (Tempe, AZ, 2001).

Goméz Pallarès, J., *Studia Chronologica: estudios sobre manuscritos latinos de cómputo* (Madrid, 1999).

Gorman, M., 'The myth of Hiberno-Latin exegesis', *Revue Bénédictine*, 110 (2000), 44–85.

Grosjean, P., 'Sur quelques exégètes irlandais du VIIe siècle', *Sacris Erudiri*, 7 (1955), 67–98.

Hardiman, J. (ed.), *Roderic O'Flaherty, A chorographical description of West or H-Iar Connaught written AD1684* (Dublin, 1846).

Häring, N.M., 'Four commentaries on the *De consolatione philosophiae* in MS Heiligenkreuz 130', *Mediaeval Studies*, 31 (1969), 287–316.

Haskins, C.H., *Studies in the history of mediaeval science* (New York, 1924).

Haskins, C.H., *The renaissance of the 12th century* (Cambridge, MA, 1927; repr. 1993).

Hein, H., *Die Mathematik im Mittelalter: von Abakus bis Zahlenspiel* (Darmstadt, 2010).

Henry, F., and G.L. Marsh-Micheli, 'A century of Irish illumination (1070–1170)', *PRIA*, 62C (1962), 101–64.

Herren, M., 'Review: *Insular grammarians* (Vivien Law, *The Insular Latin grammarians* (Studies in Celtic History 3))', *Peritia*, 2 (1983), 312–16.

Hillgarth, J.N., 'Ireland and Spain in the seventh century', *Peritia*, 3 (1984), 1–16.

Holford-Strevens, L., 'Paschal lunar calendars up to Bede', *Peritia*, 20 (2008), 165–208.

Holtz, L., 'Irish grammarians and the Continent in the seventh century' in Clarke and Brennan *Columbanus*, pp 135–52.

Howlett, D., 'Five experiments in textual reconstruction and analysis', *Peritia*, 9 (1995), 1–50.

Howlett, D., 'Gematria, number and name in Anglo-Norman', *French Studies Bulletin*, 27 (101) (2006), 90–2.

Howlett, D., 'Hibero-Latin, Hiberno-Latin and the Irish foundation legend', *Peritia*, 19 (2005), 44–60.

Howlett, D., 'Insular Latin writers' rhythms', *Peritia*, 11 (1997), 53–116.

Howlett, D., 'Seven studies in seventh-century texts', *Peritia*, 10 (1996), 1–70.

Howlett, D., 'The Brigitine hymn *Xpistus in nostra insula*', *Peritia*, 12 (1998), 79–86.

Howlett, D., 'The earliest Irish writers at home and abroad', *Peritia*, 8 (1994), 1–17.

Howlett, D., 'The polyphonic colophon to Cormac's Psalter', *Peritia*, 9 (1995), 81–91.

Howlett, D., '*Tres linguae sacrae* and threefold play in Insular Latin', *Peritia*, 16 (2002), 94–115.

Howlett, D., *British books in biblical style* (Dublin, 1997).

Howlett, D., *Caledonian craftsmanship: the Scottish Latin tradition* (Dublin, 2000).

Howlett, D., *Muirchú Moccu Macthéni's 'Vita Sancti Patricii': Life of Saint Patrick* (Dublin, 2006).

Howlett, D., *The Celtic Latin tradition of biblical style* (Dublin, 1995).

Hughes, A.J., *Robert Shipboy MacAdam: his life and Gaelic proverb collection* (Belfast, 1998).

Hughes, K., *Early Christian Ireland: introduction to the sources* (London, 1972).

Jones, C.W., 'The "lost" Sirmond manuscript of Bede's computus', *English Historical Review*, 52 (1937), 204–19.

Juste, D., *Les* Alchandreana *primitifs: étude sur les plus anciens traités astrologiques latins d'origine arabe (Xe siècle)* (Leiden, 2007).

Keil, G., P. Assion, W.F. Daems and H.-U. Roehl (eds), *Fachprosa-Studien: Beiträge zur mittelalterlichen Wissenschafts- und Geistesgeschichte* (Berlin, 1982).

Kelly, D., 'A sense of proportion: the metrical and design characteristics of some Columban high crosses', *JRSAI*, 126 (1996), 108–46.

Kelly, D., 'The heart of the matter: models for Irish High Crosses', *JRSAI*, 121 (1991), 105–45.

Kelly, M., 'Twelfth-century ways of learning: from Worcester or Cologne to Glendalough', *JRSAI*, forthcoming.

Kenney, J.F., *The sources for the early history of Ireland: ecclesiastical. An introduction and guide* (Dublin, [1929] 1997).

Kiesel, G., and J. Schroeder (eds), *Willibrord. Apostel der Niederlande, Gründer der Abtei Echternach. Gedenkgabe zum 1250. Todestag des angelsächsischen Missionars* (Luxembourg, 1989).

Knappe, G., 'On rhetoric and grammar in the Hisperica Famina', *Journal of Medieval Latin*, 4 (1994), 130–62.

Krusch, B., 'Die Einführung des griechischen Paschalritus im Abendlande', *Neues Archiv der Gesellschaft für ältere deutsche Geschichtskunde*, 9 (1884), 99–169.

Krusch, B., *Studien zur christlich-mittelalterlichen Chronologie: Der 84jährige Ostercyclus und seine Quellen* (Leipzig, 1880).

Krusch, B., *Studien zur christlich-mittelalterlichen Chronologie: Die Entstehung unserer heutigen Zeitrechnung* (Berlin, 1938), pp 59–86.

Kruse, N., 'Eine neue Schrift Notkers des Deutschen: Der althochdeutsche Computus', *Sprachwissenschaft*, 28 (2003), 123–55.

Lalor, B., *The Irish round tower: origins and architecture explored* (Dublin, 1999).

Landes, R., 'Computistical and chronological activity at the approach of the millennial years 6000/800 and 1000' in *Fourth International Conference on the Science of Computus, Galway, 13–15 July, 2012*, forthcoming.

Landes, R., 'Lest the millennium be fulfilled: apocalyptic expectations and the pattern of Western chronography, 100–800 CE' in Verbeke et al., *The use and abuse of eschatology*, pp 137–211.

Lapidge, M. (ed.), *Columbanus: studies on the Latin writings* (Woodbridge, 1997).

Lapidge, M. (ed.), *Irish books and learning in medieval Europe* (Aldershot, 1990).

Lapidge, M. (ed.), *The Anglo-Saxon library* (Oxford, 2006).

Lapidge, M., and R. Sharpe, *A bibliography of Celtic–Latin literature, 400–1200* (Dublin, 1985).

Larsen, H. (ed.), *An old Icelandic medical miscellany: MS Royal Irish Academy 23 D 43 with supplement from MS Trinity College (Dublin) L–2–27* (Oslo, 1931).
Law, V., *The Insular Latin grammarians* (Woodbridge, 1982).
Lehmann, P., 'Cassiodorstudien' in *Erforschung des Mittelalters: Ausgewählte Abhandlungen und Aufsätze*, 2 (Stuttgart, 1959), pp 38–108.
Lehmann, P., 'Cassiodorstudien', *Philologus*, 71 (1912), 278–99; 72 (1913), 503–17; 73 (1914–16), 252–73; 74 (1917), 351–83.
Liuzza, R.M., 'Anglo-Saxon prognostics in context: a survey and handlist of manuscripts', *Anglo-Saxon England*, 30 (2001), 181–230.
Lowe, E.A., *Codices Latini Antiquiores IV* (Oxford, 1947).
Luiselli Fadda, A., and É. Ó Carragáin (eds), *Le isole britanniche e Roma in età romano-barbarica* (Rome, 1998).
MacNeill, J., and H.M. Gamer, *Medieval handbooks of penance* (New York, 1938; repr. 1990).
Marenbon, J. (ed.), *The Cambridge companion to Boethius* (Cambridge, 2009).
Martin, C.J., 'The logical textbooks and their influence' in Marenbon, *The Cambridge companion to Boethius*, pp 56–84.
Masi, M. (ed.), *Boethian number theory: a translation of the De institutione arithmetica* (Amsterdam, 1983).
Masi, M. (ed.), *Boethius and the liberal arts: a collection of essays* (Bern, 1981).
Mc Carthy, D., 'Bede's primary source for the Vulgate chronology in his chronicles in *De temporibus* and *De temporum ratione*' in Warntjes and Ó Cróinín, *Computus and its cultural context*, pp 159–89.
Mc Carthy, D., 'Easter principles and a fifth-century lunar cycle used in the British Isles', *Journal for the History of Astronomy*, 24 (1993), 204–24.
Mc Carthy, D., 'On the arrival of the *latercus* in Ireland' in Warntjes and Ó Cróinín, *The Easter controversy of late antiquity*, pp 48–75.
Mc Carthy, D., 'The chronology of the Irish annals', *PRIA*, 98C (1998), 203–55.
Mc Carthy, D., 'The origin of the *latercus* paschal cycle of the Insular Celtic churches', *CMCS*, 28 (1994), 25–49.
Mc Carthy, D., 'The study and use of numbers in early Irish monasteries' in Doherty et al., *Glendalough*, pp 223–37.
Mc Carthy, D., and D. Ó Cróinín, 'The "lost" Irish 84-year Easter table rediscovered', *Peritia*, 6–7 (1987–8), 227–42.
Mc Carthy, D., *The Irish Annals: their genesis, evolution and history* (Dublin, 2008; repr. 2010).
McCluskey, S.C., *Astronomies and cultures in early medieval Europe* (Cambridge, 1998).
McCormick, F., Thom Kerr, Meriel McClatchie and Aidan O'Sullivan, *The archaeology of livestock and cereal production in early medieval Ireland, AD400–1100*, EMAP Report 5.1 (Dec. 2011).
McErlean, T.C., Caroline Earwood, Dermot Moore and Eileen Murphy, 'The sequence of Early Christian period horizontal tide mills at Nendrum monastery: an interim statement', *Historical Archaeology*, 41:3 (2007), 63–75.
McNamara, M., 'Patristic background to medieval Irish ecclesiastical sources' in Finan and Twomey, *Scriptural interpretation*, pp 253–81.
McNamara, M., *The Apocrypha in the Irish Church* (Dublin, 1975).

Meyer, C., 'Le diagramme lambdoide du MS Oxford Bodleian Library Auct. F.III.15 (3511)', *Scriptorium*, 49 (1995), 228–37.
Meyer, G., 'Die sieben freien Künste im Mittelalter, Teil 2' in *Jahresbericht über die Lehr- und Erziehungsanstalt des Benediktiner-Stiftes Maria-Einsiedeln der Studienjahre 1886/87* (Einsiedeln, 1887), pp 3–36.
Moore, R.C., A. O'Sullivan and V. Ginn, 'Killoteran 9: early medieval vertical water-mill' in James Eogan and Elizabeth Shee Twohig (eds), *Cois tSiúire: nine thousand years of human activity in the Lower Suir Valley* (Dublin, 2011).
Morel, G., 'Handschriften der Klosterbibliothek zu Einsiedeln', *Archiv der Gesellschaft für ältere deutsche Geschichtskunde*, 4 (1843), 736–49.
Moriarty, C., 'The early Irish naturalists' in Foster and Chesney, *Nature in Ireland*, pp 71–90.
Moss, R. (ed.), *The art and architecture of Ireland: medieval period,* c.*400–1600* (forthcoming).
Mosshammer, A., *The Easter computus and the origins of the Christian era* (Oxford, 2008).
Naismaith, R., 'Antiquity, authority and religion in the *Epitomae* and *Epistolae* of Virgilius Maro Grammaticus', *Peritia*, 20 (2008), 59–85.
Nauta, L., 'The Consolation: the Latin commentary tradition, 800–1700' in Marenbon, *Cambridge companion to Boethius*, pp 255–78.
Neugebauer, O., 'On the Computus Paschalis of "Cassiodorus"', *Centaurus*, 25:3 (1982), 292–302.
Ní Chatháin, P., S. Fitzpatrick and H. Clarke (eds), *Pathfinders to the past: the antiquarian road to Irish historical writing, 1640–1960* (Dublin, 2012).
Ní Dhonnchadha, M., 'The beginnings of Irish vernacular literary tradition' in *L'Irlanda e Gli Irlandesi Nell'alto Medioevo, Spoleto, 16–21 aprile 2009* (Spoleto, 2010), pp 532–92.
Nic Dhonnchadha, A., 'The medical school of Aghmacart, Queen's County', *Ossory, Laois and Leinster*, 2 (2006), 11–43.
Nothaft, C.P.E., 'The date of the Passion in early medieval computistics: the strange case of the Victorian-Dionysiac "hybrid" table in the Sirmond manuscript (Oxford, Bodl. Libr., Bodley 309)' in *Fourth International Conference on the Science of Computus, Galway, 13–15 July 2012*, forthcoming.
Nothaft, C.P.E., *Dating the passion: the life of Jesus and the emergence of scientific chronology (200–1600)* (Leiden, 2012).
Nuvolone, F.G. (ed.), *Gerberto d'Aurillac, da abate du Bobbio a papa dell'anno 1000* (Bobbio, 2001).
Ó Carragáin, T., 'The architectural setting of the Mass in early medieval Ireland', *Medieval Archaeology*, 53 (2009), 119–54.
Ó Carragáin, T., *Churches in early medieval Ireland: architecture, ritual and memory* (New Haven, CT, 2010).
Ó Cróinín, D. (ed.), 'Hiberno-latin literature to 1169' in idem, *A new history of Ireland*, 1, pp 371–404.
Ó Cróinín, D. (ed.), *A new history of Ireland*, 1: *Prehistoric and early Ireland* (Oxford, 2005).
Ó Cróinín, D., 'A new seventh-century Irish commentary on Genesis', *Sacris Erudiri*, 40 (2001), 231–65.

Ó Cróinín, D., 'A seventh-century Irish computus from the circle of Cummianus', *PRIA*, 82C (1982), 405–30.
Ó Cróinín, D., 'Bischoff's "Wendepunkte" fifty years on', *Revue Bénédictine*, 110 (2000), 204– 37.
Ó Cróinín, D., 'Early Echternach manuscript fragments with Old Irish glosses' in Kiesel and Schroeder, *Willibrord*, pp 135–43.
Ó Cróinín, D., 'Early Irish annals from Easter tables: a case restated', *Peritia*, 2 (1983), 74–86.
Ó Cróinín, D., 'Mo-sinu maccu Min and the computus at Bangor', *Peritia*, 1 (1982), 281–95.
Ó Cróinín, D., 'New light on Palladius?', *Peritia*, 4 (1986), 276–83.
Ó Cróinín, D., 'Rath Melsigi, Willibrord and the earliest Echternach manuscripts', *Peritia*, 3 (1984), 17–49.
Ó Cróinín, D., 'Review: *The Insular Latin grammarians: studies in Celtic history* 3', *Studia Hibernica*, 22/3 (1982/3), 149–56.
Ó Cróinín, D., 'The computistical works of Columbanus' in Lapidge, *Columbanus*, pp 264–70.
Ó Cróinín, D., 'The date, provenance and earliest use of the works of Virgilius Maro Grammaticus' in Bernt et al., *Tradition und Wertung*, pp 13–22.
Ó Cróinín, D., 'The Irish as mediators of antique culture on the Continent' in Butzer and Lohrmann, *Science*, pp 41–52.
Ó Cróinín, D., 'The Irish provenance of Bede's computus', *Peritia*, 2 (1983), 229–47.
Ó Cróinín, D., 'Who was Palladius, "first bishop of the Irish"?', *Peritia*, 12 (2000), 205–37.
Ó Cróinín, D., *Early Irish history and chronology* (Dublin, 2003).
Ó Cróinín, D., *Early medieval Ireland* (London, 1995).
Ó Néill, P.P., 'A Middle-Irish note on Boethius's *De institutione arithmetica*', *Éigse*, 35 (2005), 1–8.
Ó Néill, P.P., 'An Irishman at Chartres in the twelfth century: the evidence of Oxford, Bodleian Library, MS Auct. F. III. 15', *Ériu*, 48 (1997), 1–35.
Ó Néill, P.P., 'Irish glosses in a twelfth-century copy of Boethius's *Consolatio philosophiae*', *Ériu*, 55 (2005), 1–17.
Ó Néill, P.P., 'Irish observance of the three Lents and the date of the St Gall Priscian (MS 904)', *Ériu*, 51 (2000), 159–80.
O'Rahilly, T.E., and Kathleen Mulchrone et al., *Catalogue of Irish manuscripts in the Royal Irish Academy*, fasc. 1–28 (Dublin, 1926–70).
O'Sullivan, A., 'Place, memory and identity among estuarine fishing communities: interpreting the archaeology of early medieval fish weirs', *World Archaeology*, 35:3 (2003), 449–68.
O'Sullivan, A., and Donal Boland, 'Underwater archaeological excavations of an early medieval bridge in the River Shannon': www.science.ulster.ac.uk/crg/nua/research/shannon/shannon.htm: accessed 5 June 2013.
O'Sullivan, T., 'Texts and transmissions of the *Scúap Chrábaid*: an Old-Irish litany in its manuscript context', *Studia Celtica Fennica*, 7 (2010), 26–47.
O'Sullivan, W., 'Manuscripts and palaeography' in Ó Cróinín, *A new history of Ireland*, 1, pp 511–48.

Orschel, V., '*Mag nÉo na Sacsan*: an English colony in Ireland in the seventh and eighth centuries', *Peritia*, 15 (2001), 81–107.

Palmer, J., 'Calculating time and the end of time in the Carolingian world, *c.*740–820', *English Historical Review*, 126 (2011), 1307–31.

Papahagi, A., 'The transmission of Boethius' *De consolatione philosophiae* in the Carolingian Age', *Medium Aevum*, 78:1 (2009), 1–15.

Penrose, R.S., and L.R. Gander, *British Isles & Republic of Ireland turtle strandings & sightings – Annual Report 2011* (September 2012).

Picard, J.-M., 'Bede and Irish scholarship: scientific treatises and grammars', *Ériu*, 54 (2004), 139–47.

Rankin, S., '*Naturalis concordia vocum cum planetis*: conceptualizing the harmony of the spheres in the early Middle Ages' in Clark and Leach, *Citation and authority*, pp 3–19.

Rankin, S., '*Si [...] memoraliter decantare consuescas*: melodies for computistical verse', *Anglo-Saxon*, 2 (forthcoming).

Reynolds, L., *Texts and transmission: a survey of Latin classics* (Oxford, 1983; corr. repr. 1986).

Richter, M., 'St Gallen and the Irish in the early Middle Ages' in Richter and Picard, *Ogma*, pp 65–75.

Richter, M., 'The young Willibrord' in Kiesel and Schroeder, *Willibrord*, pp 25–30.

Richter, M., and J.-M. Picard (eds), *Ogma: essays in Celtic Studies in honour of Próinséas Ní Chatháin* (Dublin, 2002).

Ripper, S., and Lynden P. Cooper, *The Hemington bridges: the excavation of three medieval bridges at Hemington Quarry, near Castle Donnington, Leicestershire* (Leicester, 2009).

Rynne, C., 'Waterpower in medieval Ireland' in Paolo Squatriti (ed.), *Working with water in medieval Europe* (Leiden, 2000), pp 1–50.

Sarton, G., *Introduction to the history of science*, 2:1 (Baltimore, 1931).

Schmid, J., *Die Osterfestberechnung auf den britischen Inseln vom Anfang des vierten bis zum Ende des achten Jahrhunderts* (Regensburg, 1904).

Schola Gregoriana Maynooth in *Saints & scholars, chant from Maynooth College Chapel*, track 7.

Sharpe, C., *Bibliotheca Hibernica: Catalogue of the library of the late Edward O'Reilly... to be sold by auction ... at ... no. 33 Anglesea-Street, Dublin on ... November 30, 1830 ... by Charles Sharpe* (Dublin, 1830).

Simonetti, M., 'De mirabilibus sacrae scripturae. Un trattato irlandese sui miracoli della Sacra Scrittura', *Romanobarbarica*, 4 (1979), 225–51.

Smyth, M., 'Isidore of Seville and early Irish cosmography', *CMCS*, 14 (1987), 69–102.

Smyth, M., 'Once in four: the leap year in early medieval thought' in *Proceedings of the Third International Conference on the Science of Computus in Ireland and Europe, Galway, 16–18 July 2010* (Turnhout, forthcoming).

Smyth, M., 'Selections from Augustinus Hibernicus, *On the miracles of Holy Scripture*' in Carey, *King of mysteries*, pp 51–74.

Smyth, M., 'Seventh-century Irish zoologists?' in A. Handy and B. Ó Conchubhair (eds), *The language of gender, power and agency in Celtic Studies* (Dublin, 2014), pp 59–74.

Smyth, M., 'The date and origin of *Liber de ordine creaturarum*', *Peritia*, 17–18 (2003–4), 1–39.

Smyth, M., 'The dove and the star: enduring ideas from seventh-century Ireland', *Peritia*, 20 (2008), 98–134.

Smyth, M., 'The seventh-century Hiberno-Latin treatise, *Lider de ordine creaturarum* a translation', *Journal of Medieval Latin*, 21 (2011), 107–22.

Smyth, M., *Understanding the universe in seventh-century Ireland* (Woodbridge, 1996).

Stalley, R., 'Sex, symbol and myth: some observations on the Irish round towers' in C. Hourihane (ed.), *From Ireland coming: Irish art from the early Christian to the late Gothic period and its European context* (Princeton, NJ, 2001), pp 27–47.

Stalley, R., *Irish round towers* (Dublin, 2000).

Stevens, W.M., *Bede's scientific achievement* (Jarrow, 1986).

Stevick, R.D., 'The Echternach Gospels' evangelist-symbol pages: forms from "the two true measures of geometry"', *Peritia*, 5 (1986), 284–308.

Stevick, R.D., *The earliest Irish and English bookarts: visual and poetic forms before AD1000* (Philadelphia, 1994).

Tannery, P., 'Notes sur la Pseudo-Géométrie de Boèce', *Bibliotheca Mathematica*, 1 (1900), 39–50.

van de Vyver, A., 'Cassiodore et son oeuvre', *Speculum*, 6 (1931), 244–92.

Verbeke, W., D. Verhelst and A. Welkenhuysen (eds), *The use and abuse of eschatology in the Middle Ages* (Leuven, 1988).

Verbist, P., *Duelling with the past: medieval authors and the problem of the Christian era (c.990–1135)* (Turnhout, 2010).

Warntjes, I., 'A newly discovered Irish computus: *Computus Einsidlensis*', *Peritia*, 19 (2005), 61–4.

Warntjes, I., 'A newly discovered prologue of AD699 to the Easter table of Victorius of Aquitaine', *Peritia*, 21 (2010), 255–84.

Warntjes, I., 'Irische Komputistik zwischen Isidor von Sevilla und Beda Venerabilis: Ursprung, karolingische Rezeption und generelle Forschungsperspektiven', *Viator*, 42 (2011), 1–31.

Warntjes, I., 'Köln als naturwissenschaftliches Zentrum in der Karolingerzeit: Die frühmittelalterliche Kölner Schule und der Beginn der fränkischen Komputistik' in *Mittelalterliche Handschriften der Kölner Dombibliothek. Viertes Symposion, November 2010* (2012), pp 41–96.

Warntjes, I., 'Review: Calvin Kendall and Faith Wallis (trans.), *Bede: 'On the nature of things' and 'On Times'* (Liverpool, 2010)', *The Medieval Review*. https://scholarworks.iu.edu/dspace/bitstream/handle/2022/14623/12.08.01.html?sequence=1: accessed 12 Aug. 2001.

Warntjes, I., 'The *argumenta* of Dionysius Exiguus and their early recensions' in Warntjes and Ó Cróinín, *Computus and its cultural context*, pp 40–111.

Warntjes, I., 'The *Computus Cottonianus* of AD689: a computistical formulary written for Willibrord's Frisian mission' in Warntjes and Ó Cróinín, *The Easter controversy*, pp 173–212.

Warntjes, I., 'The Munich Computus and the 84 (14)-year Easter reckoning', *PRIA*, 107C (2007), 31–85.

Warntjes, I., and D. Ó Cróinín (eds), *Computus and its cultural context in the Latin West,*

AD300–1200: Proceedings of the First International Conference on the Science of Computus in Ireland and Europe, Galway, 14–16 July, 2006 (Turnhout, 2010).

Warntjes, I., and D. Ó Cróinín (eds), *The Easter controversy of late antiquity and the early Middle Ages: its manuscripts, texts, and tables: Proceedings of the Second International Conference on the Science of Computus in Ireland and Europe, Galway, 14–16 July, 2008* (Turnhout, 2011).

Warntjes, I., and J. Bisagni, 'The early Old Irish material in the newly discovered *Computus Einsidlensis* (*ca.* AD700)', *Ériu*, 58 (2008), 77–105.

Warntjes, I., *The Munich Computus: Irish computistics between Isidore of Seville and the Venerable Bede and its reception in Carolingian times* (Stuttgart, 2010).

Wickersheimer, E., 'Figures médico-astrologiques des IXe, X^e^ et XIe siècles', *Ianus* (1914), 157–77.

Winterbottom, M., 'Columbanus and Gildas', *Vigiliae Christianae*, 30 (1976), 310–17.

Wittig, J., 'The "Remigian" glosses on Boethius's Consolation of philosophy in context' in Wright et al., *Sources of wisdom*, pp 168–200.

Wittig, J., 'The Old English Boethius, the Latin commentaries and Bede' in Brown and Voigts, *The study of medieval manuscripts*, pp 225–52.

Wright, C., F. Biggs and T. Hall (eds), *Sources of wisdom: studies in Old English and Medieval Latin in honor of Thomas D. Hill* (Toronto, 2007).

Wyse Jackson, P.N., 'Fluctuations in fortune: three hundred years of Irish geology' in Foster and Chesney, *Nature in Ireland*, pp 91–114.

Internet sources

http://natmus.dk/en/historisk-viden/danmark/moeder-med-danmarks-oldtid/the-viking-age/kongeligt-byggeri/broen-over-ravning-enge: accessed 6 June 2013.

https://scholarworks.iu.edu/dspace/bitstream/handle/2022/14623/12.08.01.html?sequence=1: accessed 12 Aug. 2001.

www.emap.ie/documents/EMAP_Report_5_Archaeology_of_Livestock_and_Cereal_Production_WEB.pdf: accessed 10 June 2013.

www.nra.ie/Archaeology/BestofNRADiscoveries: accessed 5 June 2013.

www.science.ulster.ac.uk/crg/nua/research/shannon/shannon.htm: accessed 5 June 2013.

www.vikingeskibsmuseet.dk/en/research/ship-archaeology/the-skuldelev-ships-i: accessed 10 June 2013.

Index

Index of manuscripts